# RESHAPING THE LANDSCAPE OF SCHOOL LEADERSHIP DEVELOPMENT

CONTEXTS OF LEARNING
*Classrooms, Schools and Society*

# 1

# The Emergence of School Leadership Development in an Era of Globalization: 1980–2002

Dr. Philip Hallinger

*Professor and Executive Director, College of Management, Mahidol University, Bangkok, Thailand*

The past two decades have witnessed an unprecedented period of change in the field of school leadership preparation and development. During this recent era of globalization, school leadership and its development have emerged as major issues in educational reform. Policymakers seemed, all of a sudden, to *discover* the need for more effective school leadership. This realization has led governments in Europe, North America, and Austral-Asia steps to establish programs designed to support school leadership and its development throughout the world.

Looking back just 20 years ago – to 1980 – it would be accurate to assert that no nation in the world had in place a clear system of national requirements, agreed upon frameworks of knowledge, and standards of preparation for school leaders. Indeed, in 1980, the United States was one of the few nations in which school administrators were required to have *any type of pre-service preparation or certification*. And the American system that could be best characterized as hit-or-miss.

Indeed this only described the state of *pre-service* training for school leaders. Once obtaining a position as a school principal, neither the United States nor any other nation could boast of either formal requirements or informal norms that supported continued professional development for school leaders. In the United States, for example, it was documented that there were few if any professional norms supporting lifelong learning among school leaders (Barth, 1986b; Hallinger & Greenblatt, 1990, 1991; Thomson, 1993). The typical pattern of professional development for most American school leaders in 1980 consisted of attendance at the annual principal's association convention. There the principal might attend a workshop on retirement planning; this was an interesting statement about lifelong learning in an educational profession (Hallinger & Wimpelberg, 1992; Murphy & Hallinger, 1987).

In Europe, Asia, and Australia of 1980 a tradition of promoting teachers into administrative positions without pre-service preparation was the norm. In some countries there were administrative training institutes. In others, there were university-based diploma or Master degree programs in educational administration. However, like the United States, these programs maintained weak linkages to the workplace. Overall, it is fair to say that in 1980 both pre-service and in-service training for principals and other school leaders were non-systemic, optional, and sparsely provided globally.

Moreover, very often when training was provided by local or national systems of education, it focused primarily upon conveying government regulations to school administrators. This generally reflected the role conceived for school administrators globally. The principal was viewed as one-part manager, one-part teacher, and one-part government representative at the school site (Cuban, 1988). There was little or no emphasis on a *leadership role* for principals. By leadership role, I refer to the principal's active role in fostering development and improvement of the school as an educational institution.

This picture changed dramatically over the past two decades. Today school leadership development has become a global enterprise. Indeed, it is fair to say that in a short space of time, the preparation and development has emerged as a global industry.

The chapter will serve as an introduction to this book by providing a common basis for readers who may not be familiar with the global context of school leadership preparation and development. In this first chapter, there are three goals:

- To provide an overview of the *leadership* role on which most school leader preparation programs are focused in this new era;
- To highlight and describe the change forces fueling this global interest in school leader preparation and development;
- To identify critiques and patterns of practice that emerged over the past two decades with respect to recruitment and selection, training content, and delivery systems for school leader preparation and development.

## The Rationale for School *Leadership* Preparation and Development

This book focuses upon school *leadership* preparation and development rather than school *administrator* preparation per se. The book employs the term *leadership* because this reflects the emergent trend in this field over the past twenty years. Yet the terms educational *leader*, *manager*, and *administrator* are used quite differently from nation to nation. Therefore, it seems worthwhile to define what is meant by a *leadership perspective* and to describe the rationale for its use.

Although the norms vary from nation to nation, traditionally the School Head (UK), Director (Asia), or Principal (North America) has been viewed as the figurehead or representative of the body sponsoring the school. In most nations this sponsor is most frequently some level of government. Consequently, a fundamental role of school principals entails representing, articulating, and implementing government educational policies at the school level.

Beyond this figurehead role, the principal is also expected to serve as a resource allocator, a representative to the community, and a manager of the school facility, students, personnel and programs (Cuban, 1988; Hall & Southworth, 1997; Lam 2002). Overarching these roles, principals have traditionally been responsible for maintaining the stability of the school as an institution of cultural transmission (Cuban, 1988; Hallinger, 2002).

More recently, however, this role has undergone change. Starting in the early 1980's, forces described later in this chapter created pressure on principals to expand their responsibility for *leading* schools. The terminology of *leadership* makes several assumptions about the emerging role of the principal:

- That principals will work with staff to set a clear direction for improvement of the school and its programs (Caldwell, 1998, 2002; Hallinger & Heck, 1996, 2002; Leithwood, 1994);
- That principals should be actively involved in supporting and developing the curricular and instructional programs of the school (Hall & Southworth, 1997; Hallinger & Heck, 1996; Hallinger & Murphy, 1985);
- That responsibility for student performance rests with the school's leaders (Caldwell, 1998, 2002; Fullan, 1993; Teacher Training Agency, 1998; Thomson, 1993; Tomlinson, 1999);
- That principals play a key role in developing the capacities of staff both to lead and to teach (Barth, 1990, 1997; Hall & Southworth, 1997; Hallinger, 1998; Hargeaves & Fullan, 1998; Leithwood, 1994);
- That the principal must not only be the school's Head Teacher or Principal Teacher, but that the principal should also be the school's 'Head Learner' (Barth, 1990, 1997; Hallinger, 1998; Leithwood, 1994).

Consequently, a perusal of the global literature on school leadership of the past two decades finds a predominance of discussion focused on the principal's roles as an instructional leader, change leader, transformational leader, curriculum leader, and leader of a learning community (Hallinger & Heck, 1996). The focus on *leadership* inherent in each of these roles reflects the assumption that this individual is responsible for moving the school forward, not simply maintaining the status quo (Kotter, 1990).

Note that a leadership perspective on the role of the principal does not diminish the principal's managerial roles (Barth, 1986b; Cuban, 1988; Hall & Southworth, 1997; Lam, 2002). Yet expansion of the principal's role into the realm of leading school improvement has created a demand for new skills, new attitudes, and perspectives (Caldwell, 1998, 2002; Hallinger, 2002). Across the globe, school principals are working in a wholly new context. Leithwood cogently articulates why developing the capacity for school leadership development takes on importance at this particular juncture:

> Some of the reasons for this shift in [educational] emphasis are to be found in the quite recent school restructuring movement's preoccupation with the redistribution of power and responsibility … to both central governments and the local schools … At the school level, this has fostered greater interest in the empowerment of teachers and community members including more shared leadership. …
>
> From this [recent] redistribution of power and responsibility has emerged a decidedly different image of the ideal educational organization … This is an organization less in need of control and more in need of both support and capacity development. Organizational needs such as these seem more likely to

> be served by practices commonly associated with the concept of leadership … than administration. (Leithwood, 1996, p. xii)

In late 1980's, some scholars believed that the trend towards *teacher empowerment* would reduce the leadership role of the principal, this does not appear to have been the case. Indeed, it is interesting to note that principal leadership, though of a different type, appears especially critical in schools that seek to empower their stakeholders (Barth, 1990, 1997; Leithwood, 1994; Leithwood, Jantzi, & Steinbach, 1999). Learning about the structural and legal facets of administration sufficed during eras in which principals (and teachers) were primarily order-takers in hierarchical systems of education. Yet, as Leithwood notes, changes in the context in which school leaders operate have created a demand for a different set of knowledge, skills and attitudes (Hall & Southworth, 1997; Lam, 2002; Murphy, 1990, 1993; Tomlinson, 2002).

But, aside from a focus on improving the school, what does this new *leadership role* of the principal entail? Moreover, is it realistic to believe that principals and other school-level leaders can *make a difference in the schooling of children?* And if so, what are the methods and practices by which they can achieve positive results?

In brief, research conducted over the past 20 years finds that school-level leadership does make a difference in school and classroom climate as well as in the outcomes of schooling (Hallinger & Heck, 1996). Not only does research find that school leaders influence the capacity of schools to change, but they also have a *positive and measurable, though indirect*, effect on student learning. This finding has shaped a central goal of leadership preparation and development programs that have emerged over the past two decades. Simply stated, most of these programs seek to develop the capacity of prospective and practicing principals to bring about positive changes in the learning climate and outcomes of their schools (Caldwell, 2002; Chong, Stott, & Low, 2002; Grier, 1987; Hallinger, 1992, 2002; Leithwood, Jantzi, & Steinbach, 1999; Marsh, 1992; Murphy, 1993, 1999).

Unfortunately, while it may be comforting to know that empirical research supports conventional wisdom that *principals make a difference*, this is of limited utility without elaboration of *how* leadership contributes to school effectiveness (Cuban, 1984). Though still incomplete, researchers have also begun to describe the avenues through which principals and other school-level leaders enhance school effectiveness (Hallinger & Heck, 1996). Principals achieve these positive effects through their efforts to:

1. Create a shared vision and mission for the school,
2. Restructure the formal organization of the school (e.g., class schedules, teacher's time, grade/unit organization) in order to support instructional effectiveness and enhance staff collaboration, decision-making and communication around teaching and learning,
3. Provide stimulation and individualized support for development of the teaching *and* learning capacities of staff,
4. Reshape the school culture in order to emphasize norms of continuous learning and collaborative work. (Hallinger & Heck, 1996)

These general leadership functions of the principal remain to be filled out in greater detail. Indeed, some scholars would prefer to wait until the field can provide more detailed direction as to *how* school leaders contribute to school improvement (e.g., Cuban, 1984). Policymakers – globally – have generally viewed that to be an unproductive course of inaction. The knowledge base will never be complete, but new demands are being made of schools globally – from technology implementation to multi-culturalism and shared decision-making. These demands make developing the *leadership* capacity of our school's administrators and senior teachers increasingly urgent. Thus, a leadership perspective represents the dominant foundation on which preparation and development programs for school principals are being built globally.

## Forces for Change in School Leader Preparation and Development

As noted at the outset of this chapter, opportunities for school leader preparation and development were few and far between outside of the United States as recently as 1980. That has changed dramatically during the ensuing decades. In this section, I examine the global change forces that led to the expansion of programs for school leader preparation and development.

### Change forces in the United States

In the USA, several change forces combined to create a new context for the principalship during the 1980's. The initial impact of globalization led to recurring waves of educational reform in the USA during the 1980's and 1990's (e.g., Carnegie Forum on Education and the Economy, 1986). Concurrently, growth of the educational research industry, centered in the USA but also evident elsewhere in the world, had begun to generate findings about the principal's role in school improvement and effectiveness (e.g., Fullan, 1991, 1993). New demands for accountability at the school resulted in intensification of the work of the principal. These forces increased the focus of policymakers on principals and other school leaders.

#### *Policy reforms as a change force*

In 1982, the Secretary of Education in the USA issued a national report, *A Nation at Risk*. This report detailed the failures of American educational structures and approaches. The report concluded with a warning that unless serious efforts were undertaken to reform the nation's educational system, American economic competitiveness was at risk.

Issuance of this report by the US Dept. of Education led to a period of intense activity in several educational policy arenas. Both federal and state governments undertook massive reforms of the American educational system. This was achieved through reforms in the areas of educational standards, graduation requirements, teacher training, educational management, and curriculum (e.g., Carnegie Forum, 1986; Council of Chief State School Officers, 1996; Murphy, 1990; National Policy Board for Educational Administration, 1989; Thomson, 1993).

Not surprisingly, government policymakers viewed principals as important *conduits* of information concerning the new reforms. Beyond this, however, policymakers also recognized that skillful implementation of reforms would require a higher level of expertise than was typically demonstrated by the existing cohort of school leaders (Thomson, 1993). The policy response was to arm principals with new knowledge to assist in raising the standards of their schools (e.g., Council of Chief State School Officers, 1996).

In the United States, this resulted in the establishment of numerous state-sponsored *leadership academies* designed for upgrading the knowledge and skills of practicing school leaders (e.g., Grier, 1987). The explicit purpose of the state leadership academies was to increase the capacity of local school leaders to implement and sustain education reforms (Hallinger, 1992; Hallinger & Wimpelberg, 1992). Organizational goals, rather than individual needs represented the primary basis for programmatic decision-making in these State-sponsored School Leadership Academies.

*Educational research as a change force*

Concurrent with the formulation of educational reforms in the 1980's, a new body of educational research was being published internationally on school improvement and school effectiveness. During the 1960's and 1970's, researchers conducted wide-scale studies of the process of policy implementation, curriculum change and school improvement. Although not designed to focus on school principals per se, findings from this body of research consistently highlighted the principal's role in implementing successful change (e.g., Bossert, Dwyer, Rowan, & Lee, 1982; Fullan, 1991, 1993; Hall & Hord, 1987; Hall & Loucks, 1979). Schools were identified as the key level for reform implementation and principals as key agents in the change process. This research formulated two important complementary conclusions.

- Principals cannot, by themselves, *make change happen* in schools;
- Change in schools seldom happens without the skillful, active support of the principal. (Hall & Hord, 1987; Hall & Loucks, 1979; Fullan, 1991)

At the same time, other research was generating equally compelling conclusions concerning the role of the principal in instructionally effective schools (e.g., Edmonds, 1979, Rutter, Maugham, Mortimore, Ouston, & Smith, 1979). This body of research found that schools serving poor children in urban areas were often led by principals who assumed a non-traditional orientation towards their roles. Teachers identified these principals as *instructional leaders* rather than as facility managers or system administrators (Bossert et al., 1982; Hallinger & Heck, 1996; Hallinger & Murphy, 1985). They had strong personal visions of education and sought to focus their schools on clear academic missions. They took a more active role in directing and coordinating curriculum and instruction in their schools. Finally, they sought to create positive learning climates that would support effective teaching and learning (Hallinger & Murphy, 1985).

Taken together, these complementary bodies of research on school improvement and effectiveness combined with external pressures for policy reform to create a new context for the school principal. The normative role of the principal began to shift, none too subtly, away from the traditional role as a passive implementer of system regulations and towards a *leadership* perspective. Principals began to be viewed as key agents of change, as instructional leaders, and as individuals who would need to think for themselves and lead other staff towards the improvement of schools (Barth, 1990; Caldwell, 2002; Hall & Southworth, 1997; Hallinger & Heck, 1996; Lam, 2002; Leithwood, 1994).

*Intensification of the principal's role as a change force*

In some sense, both emergent policy reforms and research findings represented *external* sources of pressure for change on the principalship. They prompted calls from external constituencies (e.g., policymakers, lawmakers, system administrators, parents, business groups, academics, community stakeholders) for principals to assume greater responsibility as school leaders and to display greater skill in their work. It is not, therefore, surprising that expectations for higher levels of training were forthcoming.

This picture is, however, still incomplete. It is important to recognize an additional source of pressure for change, one that emerged from *inside* the profession of school principals. Roland Barth, a former school principal, founded the *Harvard Principal's Center* in 1982. Barth was an eloquent spokesperson for the need of principals to gain greater access to support and development. His rationale was not based on making principals more efficient implementers of government reforms. Nor was it based on the desire to make principals more effective instructional leaders (Barth, 1986a, 1986b, 1990, 1997).

Barth's analysis of the principalship derived from examining the principal as an *individual* as well as a *role* in the educational system. He observed first that the role of the principal was becoming increasingly complex. The nature of the principal's work was intensifying through the assumption of greater responsibilities within a context of ever-rising expectations of performance. To succeed in this environment would require continuous learning and a network of support. He asked a basic question: how could a principal lead a learning community if s/he was not learning as well (Barth, 1986b, 1990, 1997; Levine, Barth & Haskins, 1987)?

This led to the emergence of what came to be known as the *Principal's Center Movement*. The *Principal's Center Movement* represented a complementary, though at times parallel, source of normative pressure for increasing opportunities for professional preparation and training among school principals. The *Principal's Center Movement* was a *bottom-up* response to the intensification of the principal's role during this era (Hallinger & Wimpelberg, 1992).

The positive response of principals who participated in the Harvard Principal Center's programs resonated throughout the administrative culture of American schools. Subsequently, variants of the Harvard model were adopted by leadership development centers in the United States and abroad (Barth, 1986a, Endo, 1987; Hallinger & Greenblatt, 1990). These *principal-led* centers were characterized by high levels of client involvement in the identification of needs and goals, in formal governance and in the design and delivery of training.

*Top-down/bottom-up and inside-out/outside-in forces for change*

As a result of these change forces, policymakers as well as school principals began to call for greater attention to principal preparation and development in the United States (Barth, 1986b, Council of Chief State School Officers, 1996; Elmore, 1990; Murphy, 1990, 1992, 1993; National Policy Board for Educational Administration, 1989). Though based in different rationales, forces from inside and outside the profession as well as from the top-down and the bottom-up combined to create new opportunities for training and development for school leaders.

In 1980, leadership development centers and academies were scarce commodities in the United States. School leader preparation and development was almost an exclusive domain of universities (Cooper & Boyd, 1987, 1988). Yet, this too began to change. By 1990, the number of organizations that identified themselves as principals' centers, school leadership development units, or state leadership academies exceeded 150 in the United States (Hallinger & Leithwood, 1994).

Moreover, this phenomenal growth was national in scope. As a result of the federally funded *Leadership in Educational Administrative Development* (LEAD) project, at least one major leadership development center was started in every American state between 1982 and 1988. Although founded with Federal seed monies, over time these Leadership Academies generally came to be supported through state government funds.

It should be emphasized, however, that these federal and state-supported efforts only represented a portion of those organizations that came to focus on school leader preparation and development. Indeed, much of the program expansion during the 1980's and 1990's occurred among service providers other than state government agencies. New programs were sponsored by local level school districts, intermediate agencies, research and development, professional associations, and private non-profit agencies (e.g., Endo, 1987; Grier, 1987; Levine, Barth, & Haskins, 1987; Shainker, 1987).

As compared with most other nations, the United States was an *early adopter* of school leader preparation and development. The reasons for this are two-fold. First, as suggested above, the USA had been one of the few nations with a well-established system of pre-service administrator preparation. Indeed, even by 1980 a Master degree (accompanied by state certification as a school administrator) from an accredited university-based program of preparation had become the norm among school leaders in most American states. Although this tertiary system of pre-service administrative preparation came under scathing criticism during the 1980's (Bridges & Hallinger, 1995; Cooper & Boyd, 1987, 1988; Crowson & MacPherson, 1987; Griffiths, 1988; Hallinger & Murphy, 1991; Murphy, 1990, 1992, 1993; Murphy & Hallinger, 1987; Thomson, 1993), it nonetheless represented an *institutional* and *normative* system unmatched in most other parts of the world.

Second, American policymakers evinced an earlier response perhaps than many other nations to the gathering forces that later came to be referred to as globalization. The *Nation at Risk* report published in 1982 was one of the earlier educational documents linking national economic competitiveness directly to national educational attainment. America's subsequent focus on school leadership preparation and development anticipated what was to become a global trend during the mid-1990's.

Twenty years after publication of the *Nation at Risk* report, school leader training and development is an institutionalized facet of life among school leaders in the United States. Although the system still needs improvement, it is notable that life-long learning is now largely accepted as a part of the profession of school leadership. This change reflects both *normative* changes in the profession and *institutional* changes in the educational structures of American education.

**Change forces outside the United States**

The same forces that prompted educational reform in the United States during the 1980's soon became evident in the rest of the world as well. As noted, in 1980 few nations outside of the US gave attention or resources to school leader preparation and development. By the 1990's, however, pressure arising from the onset of globalization led nations throughout the world to converge upon a common set of educational goals: literacy, communication skills, life-long learning skills, IT literacy, numeracy, critical and creative thinking and problem-solving.

With this common educational direction, educational systems globally increasingly began to adopt a common set of *globally-sanctioned* educational policies and practices (e.g., in England see Dfee, 1998; in Hong Kong see Education Commission, 1996; in Taiwan see MOE-ROC, 1998; in Thailand see MOE-Thailand, 1997a, 1997b). For example, school-based management, originating in New Zealand, soon spread to Australia, parts of Canada, the United States and elsewhere (Burke, 1992; Caldwell, 1998). Learning technologies, student-centered learning, cooperative learning, integrated curriculum, effective school, and parental involvement followed a similar process of global dissemination (Caldwell, 1998, 2002; Hallinger, 1998, 2002). The implementation of these global reforms created a new context for school leaders and, not surprisingly, led to the emergence of a common *global* focus on the training of school leaders.

In Europe, Great Britain initiated serious reform of its educational system in the early 1990's. At that time there was neither a traditional emphasis on school *leadership* nor on school leader preparation and development (Bolam, 2002; Hall & Southworth, 1997; Tomlinson, 1999). However, with the central government's initiation of massive educational reforms, the role of the School Head in the UK changed dramatically almost overnight (Bolam, Dunning & Karstanje, 2000; Tomlinson, 1999).

Suddenly, the preparation of school leaders became a national priority (Barber, 1999; Dfee, 1998; Teacher Training Agency, 1998). Within the span of only a few short years, the *National College of School Leadership* was established to address the pre-service preparation of school leaders. Concurrently new programs were designed for the in-service education of school leaders (see Bolam, 2002; Reeves, Forde, Casteel, & Lynas, 1999; Tomlinson, 1999).

In the Asian region, there was relatively little activity in this domain during the 1980's and early 1990's. Several Asian nations had established administrative training centers during the 1980's with support from the World Bank (e.g., Thailand, Malaysia, Sri Lanka). These institutions, however, tended to operate as government agencies. This shaped their missions towards that of government training institutes rather than centers of *leadership* preparation and development.

Universities in a number of Asian countries were offering Master Degrees in educational management and administration during the 1980's. In contrast to the United States, however, few of these university-based programs were linked to any type of certificate or qualification in the local educational systems (Fwu & Wang, 2001; Lam, 2002; Low, 1999). Thus, while some opportunities for training in educational management were available, there was a general absence of *systemic* focus on school leader preparation or development in the region (Hallinger, 2002).

Around 1995 a wave of interest in school leadership and its development arrived in Austral-Asia. This gave rise to new efforts in Australia, China, Singapore, Thailand, Taiwan, Malaysia and Hong Kong to establish *systemic* policies and coherent programs aimed at the preparation and development of school leaders. In most of these nations, the focus on school leadership only emerged after the passage of significant new educational reforms (Gopinathan & Kam, 2000; Hallinger, 2002; Lam, 2002).

The change forces outlined in this section have led to new policies and programs designed for preparation, selection, development and evaluation of principals throughout the world. Indeed at the turn of the century, it is no exaggeration to say that a global trend has emerged focused on the issue of school leadership preparation and development. In the following section, I will examine dominant patterns of practice and review the important critiques that emerged during this era.

## Global Trends in the Pre-service and In-service Preparation of School Leaders

This section of the chapter examines global trends in the pre-service and in-service training of school leaders that have emerged over the past 20 years. The section will focus specifically on trends in the methods of recruitment and selection, curriculum content, and delivery systems.

### Pre-service training

As noted above, after some delay, the global educational reform spotlight has recently focused on school leadership, especially the preparation and development of principals. Not only in the United States, but also in other countries the traditional university-based delivery system was found lacking in many respect (Bridges & Hallinger, 1995; Cooper & Boyd, 1987, 1988; Crowson & MacPherson, 1987; Hallinger & Murphy, 1987, 1991; Murphy, 1992, 1993; Murphy & Forsythe, 1999; National Policy Board Educational Administration, 1989). Critics and policymakers found evidence of significant problems in almost every phase of the traditional mode of pre-service preparation – from the recruitment of students to the way they are certified for employment.

*Recruitment and selection*

By and large, prospective principals have self-selected their participation in programs. Prior to 1995, there were few programs designed to recruit talented school leaders and procedures for selection were not systematic (Fwu & Wang, 2001; Murphy, 1992, 1993; Murphy & Forsythe, 1999; Pounder & Young, 1996). In most countries admission standards to university-based programs have been lax, set so that almost everyone who wants to prepare for the principalship is able to do so.

Not surprisingly, the quality of applicants is quite low. In one study conducted in the United States, for example, educational administration students ranked 91st out of 94 intended majors listed on the Graduate Record Examination (GRE) form (Griffiths, 1988; Murphy, 1992; National Policy Board for Educational Administration, 1989). Educational administration students not only score poorly on tests of academic ability, but they also tend to be politically conservative and averse to risk taking. This profile does not bode well for the profession during an era in which educational systems claim to need *leaders* who can successfully implement change and improvements.

Yet, change is in the works. There is a clear trend towards more systematic recruitment and selection of participants for preparation programs as well as for the job internationally (Barth, 1997; Bridges & Hallinger, 1995; Caldwell, 2002; Hallinger, 2002; Huber, 2002; Murphy & Forsythe, 1999). Indeed, several chapters in this volume describe new models of pre-service preparation in which the recruitment of candidates recruitment is more systematic than in the past (e.g., Chong et al., 2002; Copland, 2002; Littky & Schen, 2002).

*Training content*
There is a widespread belief that traditional university-based preparation programs have lacked coherence, rigor, and standards (Cooper & Boyd, 1987, 1988; Griffiths, 1988; Hart & Weindling, 1996; Murphy, 1992; Murphy & Forsythe, 1999). Continua of skills and understandings are difficult to discern in many of these programs. Instead, preparation programs are often packages of unrelated courses that, not surprisingly, fail to reveal any overarching design or consistent purpose (Murphy, 1992). Most troubling of all, even those who deliver these programs question whether their graduates are adequately prepared to effectively assume their duties.

A micro-level analysis of training content often reveals an equally disheartening picture. Program content in America's university-based educational administration programs does not reflect the realities of the principal's workplace and work life (Bridges, 1977; Bridges & Hallinger, 1995). When the entire fabric of preparation is examined, one is hard pressed to see many threads that are either practice-related or problem-based (Barth, 1986b, 1990; Bridges & Hallinger, 1995; Littky & Schen, 2002; Murphy & Forsythe, 1999). In addition, one finds that traditional educational administration curricula devote remarkably little attention to the manner in which principals influence teaching and learning in their schools.

*Delivery system*
Critics of the American system of preparation have concluded that the overarching university-based system was a design in failure. Faced with the option of developing preparation programs based on the needs and interests of practitioners or of aligning themselves with the culture and norms of the university, professors in educational administration traditionally selected the latter course of action (Cooper & Boyd, 1987; Murphy & Hallinger, 1987; Murphy & Forsythe, 1999). As a consequence, preparation programs were traditionally constructed using arts and science blueprints rather than professional school models.

The social science/theory movement in education became the paradigm for preparation programs in the United States and has generally been followed in other nations as well (e.g., UK, Hong Kong, Australia, Malaysia). Consequently, the professors who taught in these programs were only distally connected to their counterparts in the field and often unfamiliar with problems of practice (Cooper & Boyd, 1987, 1988; Crowson & MacPherson, 1987; Murphy, 1992; Murphy & Forsythe, 1999). Thus, both program content and delivery methods became largely de-coupled from the realities that principals confront on the job (Barth, 1986b, 1997; Bridges, 1977; Bridges & Hallinger, 1995; Cooper & Boyd, 1987, 1988; Murphy, 1992; Murphy & Forsythe, 1999).

Compounding these problems, instruction in many principal preparation programs leaves much to be desired (Bridges, 1977; Hart & Weindling, 1996; Murphy, 1992; Murphy & Forsythe, 1999). Lecture and discussion continue to be the instructional modes of preference across the largest spectrum of classes in principal pre-service programs. In addition, in most programs, one is hard pressed to see evidence of any systematic application of principles of adult learning. Clinical experiences are also notoriously weak. With some notable exceptions (e.g., the Danforth Programs), these tend to be poorly organized, largely unsupervised, poorly connected to *meaningful* work of practitioners, and de-coupled from other components of preparation programs (Hart & Weindling, 1996; Pohland, Milstein, Schilling, & Tonigan, 1988).

Problems with low, non-existent, unenforced, and inappropriate standards plague many aspects of pre-service preparation programs for principals. Until recently there was a relative absence of standards for selection into these programs and when evaluated against the standard of coherence, programs are often found wanting. The same conclusion can be drawn about the rigor and appropriateness of program content (Murphy & Shipman, 2002). Performance criteria in these programs are also particularly slippery, functioning more as symbolic rituals than entry gates to more advanced work. Not unexpectedly, few students who enter certification programs fail to finish their programs for academic reasons (Hallinger & Murphy, 1991).

Until recently, the absence of meaningful standards compounded problems observed in the school administration curriculum, methods of classroom instruction, and the clinical experiences intended to prepare future leaders (Murphy & Shipman, 2002). Consequently, trainees adequately prepared to assume roles of educational leadership are clearly the exception rather than the rule. Critics may counter that this is an unfair indictment since no program of management training in any field has demonstrated the ability to *fully* prepare novices for leadership in the field. However, it is clear that the traditional systems in use have been operating at far from satisfactory levels of performance, even given the low expectations that have characterized the field of educational administration (Hart & Weindling, 1996; Murphy, 1992, 1993).

It is also true that the past decade has seen the seeds of innovation in the field with respect to the preparation of school leaders in many countries. Some examples are discussed at length in this volume. New approaches to instruction (e.g., Copland, 2002; Hallinger & Kantamara, 2002) curriculum organization (Chong et al., 2002; Huber, 2002; Littky & Schen, 2002), program delivery (Heck, 2002; Littky & Schen, 2002; Tomlinson, 2002) and accountability are being tried in various locales (Leithwood, 2002; Murphy & Shipman, 2002; Tomlinson, 1999). The next section examines the system of professional development that inducts new administrators and enables them to further develop and refine the skills and understandings needed to succeed as school leaders over the course of their careers.

**In-service training**

Traditionally, in-service opportunities for school administrators were haphazard, under-funded, lacking in a viable knowledge base, and limited in both scope and content. In a perverse expression of the global educational culture, there was no widely held expectation that professional development was in the job description of those responsible for leading our schools. Professional development was viewed as a luxury and choice, not as a necessity, for leaders in most school systems (Hallinger & Greenblatt, 1990, 1991). The absence of entry-level norms supporting professional growth was reinforced by an administrative culture that promotes a *pull yourself up by the bootstraps* attitude. This resulted in the predominance of a deficit model of staff development in which programs are designed to remediate problems, rather than contribute to ongoing growth (Barth, 1986b; Hart & Weindling, 1996; Heck, 2002; Levine, Barth, & Haskins, 1987; Shainker & Roberts, 1987).

As noted above, during the past two decades, the field of school leader development has undergone an unprecedented expansion. Individual principals, regional groups, and professional organizations began demanding greater support for efforts to develop on-the-job. Here we will examine recruitment, training content and delivery systems.

*Recruitment and induction*

As noted earlier in this chapter, opportunities, and to a lesser degree, expectations for school leader development have increased globally. In some locales, beginning administrators now have support systems to assist the transition to positions of leadership (e.g., UK, USA, Australia, Hong Kong, Singapore, Malaysia). Unfortunately, this is not true everywhere; sink or swim remains the norm in too many school systems. As with teachers, concern for the professional induction of school leaders should be high on any agenda for reform (Hart & Weindling, 1996; Heck, 2002; National Staff Development Council, 1986; Parkay & Hall, 1992). Positive induction experiences are critical to the development of attitudes, skills, and professional norms that support both current and future growth (Caldwell, 2002; Heck, 2002; Levine, Barth, & Haskins, 1987; Tomlinson, 1999).

Patterns of recruitment or participation in professional development programs are uneven, related to the idiosyncracies of the geographic locale, and dependent upon the guidelines of sponsoring agencies. Grass roots agencies often emphasize voluntary participation in the belief this promotes more effective learning among adults. Participants in such centers typically exercise considerable control over the program content that is geared to locally perceived needs. Participants attend programs based upon their interest in the specific content and their general motivation to develop professionally. This model appears to work well with highly motivated individuals and in those areas where local norms support professional growth (Barth, 1986b, 1997; Hallinger, 1992; Marsh, 1992; National Staff Development Council, 1986).

Mandatory participation in administrative staff development has become more common over the past decade. In the United States, for example, many states now require practicing administrators to complete a certain number of in-service courses in administration over a period of years (Hallinger & Wimpelberg, 1992; Murphy & Forsythe, 1999; Thomson, 1993). The same type of requirement is being implemented in other countries as well (e.g., United Kingdom, Singapore, Hong Kong). In some cases, the courses are left to the discretion of the individual and may be completed through a variety of means. In other cases, school systems have mandated administrator participation in staff development programs designed specifically by the state local agency to promote administrators competence in selected domains of practice.

Anecdotal reports and opinion surveys suggest that administrators generally feel that these in-service experiences, whether state or locally sponsored, voluntary or mandatory, are worthwhile. Administrators report a reduction of isolation from peers, increased knowledge about the field, and in some cases, gains in skills. Little systematic evaluation has been conducted, however, to determine:

- who attends these programs,
- the nature of the curriculum and instruction provided, or
- the impact of training and development experiences on administrator beliefs, knowledge, skill development, implementation of training content, or school related outcomes. (Hallinger, 1992, 2002)

*Training content*

When compared with the pre-service curriculum, the in-service curriculum in school leadership is more varied in approach and more firmly connected to the needs of clients (Barth, 1997; Bolam, 2002; Grier, 1987; Low, 1999; National Association of Secondary School Principals, 1987; National Staff Development Council, 1986; Shainker, 1987; Tomlinson, 2002). A number of nations derive staff development goals and related curricula for administrators directly from reform legislation (Huber, 2002). The approved curriculum is then disseminated to administrators throughout the state via central and/or regional leadership centers. Such curricula strive for high levels of coherence and tend to be linked explicitly to research on principal and school effects. While on the surface, these programs appear to be improvements over much of the pre-service curricula, little is known about their true effectiveness (Hallinger, 1992, 2002).

The increased diversity of curricular and instructional approaches used in in-service programs is to be applauded, as is the focus on school-based problems of concern to principals (Hallinger & Bridges, 1997; Murphy & Forsythe, 1999; National Association of Secondary School Principals, 1987). Other developments may be less positive. For example, centralized, mandated approaches to staff development are often pragmatic but short-sighted. This 'one best model' of leadership training is based on an optimistic, though simplistic, faith in the existence of a clearly defined, scientifically validated knowledge base for school leadership. Unfortunately, the knowledge about leadership contained in the government sponsored curricula often sacrifices contextual sensitivities in order to achieve standardization for a broad audience.

This is a looming problem that has been observed increasingly in the past several years. For example, the rapid growth of training programs in Asia has sent policymakers in search of pre-packaged training *solutions*. These have generally been imported from the United States, the United Kingdom and Australia without sufficient consideration for adaptation and validation of the training in the local context (Bajunid, 1996; Cheng, 1995; Hallinger & Kantamara, 2002).

A notably positive development that emerged during the 1990's, especially in Asia, was the push towards developing an 'indigenous knowledge' in the region (Bajunid, 1996; Cheng, 1995; Hallinger & Kantamara, 2002). Advocates of this perspective asked:

- What does leadership and school improvement look like in the local cultures of the region?
- What are the values that drive leaders and motivate staff in this culture?
- What are 'best practices' that achieve the ends valued in these cultures?

While these research questions are only beginning to be addressed, they have implications far beyond school leadership development in Asia. These questions suggest that in the era of globalization we face critical challenges with respect to the content of both pre-service and in-service preparation programs. We will discuss these challenges at length in the final chapter of this volume.

*Delivery system*

A wider variety of organizations have entered the in-service training and development arena. Traditionally, administrative training was the domain of universities. Today, school systems, intermediate service agencies, research and development centers, professional associations, and education agencies are the most visible providers of training and development services to school leaders (National Association of Secondary School Principals, 1987; National Staff Development Council, 1986).

A notable trend emerging from the increased diversity of providers is the greater involvement of administrators in program governance and planning. Although governance structures vary widely among centers, administrators are less frequently viewed as passive recipients. Many, though certainly not all, staff development programs now seek the active participation of practicing administrators in defining needs, planning programs, delivering instruction, and providing coaching and support. This feature contrasts sharply with pre-service preparation programs that generally limit the involvement of practitioners in program development and implementation.

The impact of this involvement is profound. Instructional approaches reflect an orientation to the work world of administrators and are often delivered by practitioners. Content connections to practice have been strengthened through the use of a wider variety of instructional methods, particularly ones that actively involve the participants and which draw upon the knowledge base gained through administrative practice. These include problem-based learning, simulations, case studies, peer observation and feedback (Bridges & Hallinger, 1995; Copland, 2000; Hallinger & Kantamara, 2002; Walker, Bridges, & Chan, 1996).

The increased involvement of administrators in governance, program planning, and instructional delivery are positive developments. However, our conversations with principals and in-service providers and a perusal of program documents lead to a troubling observation. Organizational structures and training processes designed to support implementation of new knowledge and skills structured into development programs or district operations (Shainker & Roberts, 1987). Thus, the institutionalization of coaching, mentorships for new and experienced principals, cross-school visitation with feedback, and setting of professional development goals typically remain dependent upon individual initiative (Chong et al., 2002; Tomlinson, 1999).

## Conclusion

This chapter has provided an overview of developments in the field of school leader preparation and development from a global perspective over the past two decades. A rationale for focusing upon *leadership* preparation and development was presented. The global implementation of educational reforms has driven the shift towards a *leadership* perspective on the principalship throughout the world.

The change forces that have created this new context derive from the nature of the new educational reforms being implemented globally as well as from research on effective schools and school improvement. Principals in this new era must be skilled implementers of change. The new emphasis on schools as learning communities means that school principals must become head learners, in addition to their traditional roles as managers.

Moreover, the intensification of the work role in this era of global reform means that principals not only need more training to do the job well, but also more effective support systems. Subsequent chapters will describe approaches being taken to address these issues through different methods of instruction (e.g., simulation, problem-based learning), program design (e.g., around sets of standards or around systemic reforms), and organization in different nations.

## References

Bajunid, I. A. (1996). Preliminary explorations of indigenous perspectives of educational management: the evolving Malaysian experience. *Journal of Educational Administration*, 34(5), 50–73.

Barber, M. (1999). *A world class school system for the 21st Century: The Blair Government's education reform strategy*. No. 90 in a Seminar Series of the Incorporated Association of Registered Teachers of Victoria (IARTV), December [ISBN 1 876323 31 0] [reprint of a paper presented at the Skol Tema Conference in Stockholm in September 1999].

Barth, R. (1990). *Improving schools from within*. Jossey Bass, San Francisco.

Barth, R. (1986a). On sheep and goats and school reform. *Phi Delta Kappan*, 68(4), 293–296.

Barth, R. (1986b). Principal centered professional development. *Theory into Practice*, 25(3), 156–160.

Barth, R. (1997). *The principal learner: A work in progress*. The International Network of Principals' Centers, Harvard Graduate School of Education, Cambridge, MA.

Bolam, R., Dunning, G. & Karstanje, P. (Eds.; 2000). *New headteachers in the new Europe*. Munster/New York: Waxman Verlag.

Bossert, S., Rowan, B., Dwyer, D. & Lee, G. (1982). The instructional management role of the principal. *Educational Administration Quarterly*, 18(3), 34–64.

Bridges, E. (1977). The nature of leadership. In L. Cunningham, W. Hack, and R. Nystrand (Eds.), *Educational administration: The developing decades* (pp. 202–230). Berkeley, CA: McCutchan.

Bridges, E. & Hallinger, P. (1995). *Implementing problem-based leadership development*. Eugene, OR: ERIC Clearinghouse for Educational Management.

Bridges, E. & Hallinger, P. (1997). Using problem-based learning to prepare educational leaders. *Peabody Journal of Education*, 72(2), 131–146.

Burke, C. (1992). Devolution of responsibility to Queensland schools: Clarifying the rhetoric critiquing the reality. *Journal of Educational Administration*, 30(4), 33–52.

Caldwell, B. (2002). A blueprint for successful leadership in an era of globalization in learning. In P. Hallinger (Ed.), *Reshaping the landscape of school leadership development: A global perspective*. Lisse, Netherlands: Swets & Zeitlinger.

Caldwell, B. (1998). Strategic leadership, resource management and effective school reform. *Journal of Educational Administration*, 36(5), 445–461.

Carnegie Forum on Education and the Economy. (1986). *A nation prepared: Teachers for the 21st century*. Carnegie Foundation, Washington, D.C.

Cheng, K. M. (1995). The neglected dimension: Cultural comparison in educational administration. In Wong, K. C. & Cheng, K. M. (Eds.), *Educational leadership and change: An international perspective*. Hong Kong University Press, Hong Kong, 87–104.

Chong K. C., Stott, K. & Low, G. T. (2002). Developing Singapore school leaders for a learning nation. In P. Hallinger (Ed.), *Reshaping the landscape of school leadership development: A global perspective*. Lisse, Netherlands: Swets & Zeitlinger.

Cooper, B. & Boyd, W. (1987). The evolution of training for school administrators. In J. Murphy & P. Hallinger (Eds.), *Approaches to administrative training in education* (pp. 3–27). Albany, NY: SUNY Press.

Cooper, B. & Boyd, W. (1988). The evolution of training for school administrators. In D. Griffiths, R. Stout & P. Forsyth (Eds.), *Leaders for America's schools* (251–272). Berkeley, McCutchan.

Copland, M. A. (2000). Problem-based learning and prospective principals' problem-framing ability. *Educational Administration Quarterly*, 36(4), 584–606.

Council of Chief State School Officers (1996). *Interstate school leaders licensure consortium: standards for school leaders*. Washington, DC: Author.

Crowson, R. & MacPherson, B. (1987). The legacy of the theory movement: Learning from the new tradition. In J. Murphy & P. Hallinger (Eds.), *Approaches to administrative training in education* (pp. 45–66). Albany: State University of New York Press.

Cuban, L. (1988). *The managerial imperative and the practice of leadership in schools*. State University of New York Press, New York.

Cuban, L. (1984). Transforming the frog into a prince: Effective schools research, policy and practice at the district level, *Harvard Educational Review*, 54(2), 129–151.

DfEE. (December, 1998). *Teachers: Meeting the challenge of change*. London: DfEE.

Edmonds, R. (1979). Effective schools for the urban poor. *Educational Leadership*, 37,15–24.

Education Commission. (1996). *Quality school education* (EC report No. 7). Hong Kong: Government Printer.

Endo, T. (1987). The Fairfax County (VA) principals' research group. *NASSP Bulletin*, 71(495), 48–50.

Elmore, R. F. (1990). *Reinventing school leadership* (pp. 62–65). Working memo prepared for the Reinventing School Leadership Conference. Cambridge, MA: National Center for Educational Leadership.

Fullan, M. (1993). *Change forces*. London: Falmer Press.

Fullan, M. (1991). *The new meaning of educational change*. Teachers College Press, New York.

Fwu, Bih-jen, & Wang, Hsiou-huai. (2001). *Principals at the crossroads: Profiles, preparation and role perception of secondary school principals in Taiwan*. Paper presented at the International Conference on School Leader Preparation, Licensure, Certification, Selection, Evaluation and Professional Development, Taipei, ROC.

Gopinathan, S. & Kam, H. W. (2000). Educational change and development in Singapore. In T. Townsend & Y. C. Cheng (Eds.), *Educational change and development in the Asia Pacific: Challenges for the future*, Lisse, Netherlands: Swets & Zeitlinger, 163–184.

Grier, L. (1987). The North Carolina leadership institute for principals. In J. Murphy & P. Hallinger (Eds.), *Approaches to administrative training in education* (pp. 115–130). Albany: State University of New York Press.

Griffiths, D. E. (1988). *Educational administration: Reform PDQ or RIP* (Occasional Paper No. 8312). Tempe, AZ: University Council for Educational Administration.

Hall, G. & Hord, S. (1987). *Change in schools*. Albany, NY: SUNY Press.

Hall, G. & Loucks, S. (1979). *Implementing innovations in schools: A concerns-based approach*. Austin, TX: Research and Development Center for Teacher Education, University of Texas.

Hall, V. & Southworth, G. (1997). Headship. *School Leadership & Management*, 17(2), 151–170.

Hallinger, P. (1998). Educational change in the Asia-Pacific region: The challenge of creating learning systems. *Journal of Educational Administration*, 36(5).

Hallinger, P. (2002). School leadership development in the Asia Pacific region: Trends and directions for future research and development. In Y.C. Cheng (Ed.), *Handbook of educational research in the Asia Pacific region*. New York: Kluwer Academic Press.

Hallinger, P. (1999). School leadership development: State of the art at the turn of the century. *Orbit*, 30(1), 46–48.

Hallinger, P. (1992). School leadership development: Evaluating a decade of reform. *Education and Urban Society*, 24(3), 300–316.

Hallinger, P. & Bridges, E. (1997). Problem-based leadership development: Preparing educational leaders for changing times. *Journal of School Leadership*, 7, 1–15.

Hallinger, P. & Greenblatt, R. (1991). Principals' pursuit of professional growth: The influence of beliefs, experiences and district context. *Journal of Staff Development*, 10(4), 68–74.

Hallinger, P. & Greenblatt, R. (1990). Professional development through principals' centers: Why do principals participate? *NASSP Bulletin*, 74(527), 108–113.

Hallinger, P. & Heck, R. (1996). Reassessing the principal's role in school effectiveness: A review of empirical research, 1980–1995. *Educational Administration Quarterly*, 32(1), 5–44.

Hallinger, P. & Heck, R. (2002). What do you call people with visions? Vision, mission and goals in school leadership and improvement. In K. Leithwood, P. Hallinger and colleagues (Eds.), *The handbook of research in educational leadership and administration: Vol. II*. Dordrecht: Kluwer.

Hallinger, P. & Murphy, J. (1985). Assessing the instructional management behavior of principals. *Elementary School Journal*, 86(2), 217–247.

Hallinger, P. & Murphy, J. (1991). Developing leaders for future schools. *Phi Delta Kappan*, 72(7), 514–520.

Hallinger, P. & Wimpelberg, R. (1992). New settings and changing norms for principal development. *The Urban Review*, 67(4), 1–22.

Hargreaves, A. & Fullan, M. (1998). *What's worth fighting for out there*. New York: Teachers College Press.

Hart, A. & Wending, D. (1996). Developing successful school leaders. In K. Leithwood, J. Chapman, D. Corson, P. Hallinger, & A. Hart (Eds.), *International handbook of educational leadership and administration*. Dordrecht: Kluwer Academic Publishers (309–336).

Heck, R. (2002). Examining the impact of professional preparation on beginning school administrators. In P. Hallinger (Ed.), *Reshaping the landscape of school leadership development: A global perspective*. Lisse, Netherlands: Swets & Zeitlinger.

Huber, S. (2002). School leader development: Current trends from a global perspective. In P. Hallinger (Ed.), *Reshaping the landscape of school leadership development: A global perspective*. Lisse, Netherlands: Swets & Zeitlinger.

Kotter, J. (1990). *A force for change: How leadership differs from management*. New York: The Free Press.

Lam, J. (2002). Balancing stability and change: Implications for professional preparation and development of principals in Hong Kong. In P. Hallinger (Ed.), *Reshaping the landscape of school leadership development: A global perspective*. Lisse, Netherlands: Swets & Zeitlinger.

Leithwood, K. (1996). Introduction. In K. Leithwood, J. Chapman, D. Corson, P. Hallinger, & A. Hart (Eds.), *International handbook of research in educational leadership and administration*. Dordrecht: Kluwer Press.

Leithwood, K. (1994). Leadership for school restructuring. *Educational Administration Quarterly*, 30(4), 498–518.

Levine, S., Barth, R. & Haskins, K. (1987). The Harvard Principals' Center: School leaders as adult learners. In J. Murphy & P. Hallinger (Eds.), *Approaches to administrative training in education* (pp. 150–163). Albany, NY: SUNY Press.

Leithwood, K. Jantzi, D. & Steinbach, R. (1999). *Changing leaders for changing schools*. Buckingham, UK: Open University Press.

Littky, D. & Schen, M. (2002). Developing school leaders: One principal at a time. In P. Hallinger (Ed.), *Reshaping the landscape of school leadership development: A global perspective*. Lisse, Netherlands: Swets & Zeitlinger.

Marsh, D. (1992). School principals as instructional leaders: The impact of the California School Leadership Academy. *Education and Urban Society*, 24(3), 386–410.

Ministry of Education-R.O.C. (1998). *Towards a learning society*. Ministry of Education, Taipei, Republic of China.

Ministry of Education-Thailand. (1997a). *Introducing the Office of the National Primary Education Commission*. Bangkok, Thailand: Ministry of Education.

Ministry of Education-Thailand. (1997b). *The experience from the Basic and Occupational Education and Training Programme*. Bangkok, Thailand: Ministry of Education.

Murphy, J. (1993). *Preparing tomorrow's leaders: Alternative designs*. University Park, PA: University Council for Educational Administration.

Murphy, J. (1990). The reform of school administration: Pressures and calls for change. In J. Murphy (Ed.), *The reform of American public education in the 1980's: Themes and cases*. Berkeley: McCutchan.

Murphy, J. (1992). *The landscape of leadership preparation: Reframing the education of school administrators*. Thousand Oaks, CA: Corwin Press.

Murphy, J. & Forsyth, B. P. (1999). *Educational administration: A decade of reform*. Thousand Oaks, CA: Corwin Press in collaboration with the University Council for Educational Administration.

Murphy, J. & Hallinger, P. (1987). Emerging training programs for school administrators: A synthesis and recommendations. In J. Murphy & P. Hallinger (Eds.), *Approaches to administrative training in education*. Albany, NY: SUNY Press.

Murphy, J. & Shipman, N. (2002). Developing standards for school leadership development: A process and rationale. In P. Hallinger (Ed.), *Reshaping the landscape of school leadership development: A global perspective*. Lisse, Netherlands: Swets & Zeitlinger.

National Association of Secondary School Principals. (1987). Peer learning and sharing: The promise and hope of principals' centers. *NASSP Bulletin*, 71(495), 1–67.

National Policy Board for Educational Administration. (1989). *Improving the preparation of school administrators: An agenda for reform*. Charlottesville: University of Virginia, Curry School of Education.

National Staff Development Council. (1986). Staff development for school leaders. *Journal of Staff Development*, 7(2), 5–111.

Parkay, F. & Hall, G. (1992). *Becoming a principal: The challenges of beginning leadership*. NY: Allyn & Bacon.

Pohland, P., Milstein, M., Schilling, N. & Tonigan, J. (1988). Emergent issues in the curriculum of educational administration: The University of New Mexico case. In C. Wendel & M. Bryant (Eds.), *New directions for administrator preparation*. Tempe, AZ: University Council for Educational Administration.

Pounder, D. & Young, P. (1996). Recruitment and selection of educational administrators: Priorities of today's schools. In K. Leithwood, J. Chapman, D. Corson, P. Hallinger, & A. Hart (Eds.), *International handbook of educational leadership and administration* (279–308). Dordrecht: Kluwer Academic Publishers.

Reeves, J., Forde, C., Casteel, V. & Lynas, R. (1999). Developing a model of practice: Designing a framework for the professional development of school leaders and managers. *School Leadership and Management*, 18(2), 185–196.

Rutter, M., Maugham, B., Mortimore, P., Ouston, J. & Smith, A. (1979). *Fifteen thousand hours: Secondary schools and their effects on children*. Cambridge, MA: Harvard University Press.

Schainker, S. & Roberts, L. (1987). Helping principals overcome on-the-job obstacles to learning. *Educational Leadership*, 45, 30–33.

Teacher Training Agency (1998). *National standards for headteachers*. London: Teacher Training Agency.

Thomson, S. D. (1993). (Ed.). *Principals for our changing schools: The knowledge and skill base*. Fairfax, VA: National Policy Board for Educational Administration.

Tomlinson, H. (1999). *Recent developments in England and Wales: The Professional Qualification for Headship (NPQH) and the Leadership Programme for Serving Headteachers (LPSH)*.

Paper presented at the Conference on Professional Development of School Leaders, Centre for Educational Leadership, Hong Kong University, Hong Kong.
Tomlinson, H. (2002). Supporting school leaders in an era of accountability: The National College for School Leadership in England. In P. Hallinger (Ed.), *Reshaping the landscape of school leadership development: A global perspective*. Lisse, Netherlands: Swets & Zeitlinger.
Walker, A., Bridges, E. & Chan, B. (1996). Wisdom gained, wisdom given: Instituting PBL in a Chinese culture. *Journal of Educational Administration*, 34(5), 98–119.

# 2

# A Blueprint for Successful Leadership in an Era of Globalization in Learning[1]

Dr. Brian J. Caldwell

*Professor and Dean, Department of Educational Administration, Faculty of Education, University of Melbourne, Victoria, Australia*

There is universal recognition that the current historically high expectations for education and training will only be met with outstanding leadership. Such leadership is directly or indirectly connected to learning in very powerful ways. There is a universal search for ways to prepare, select, place, develop, assess and reward leaders who can make the link to learning, and help realize these high expectations.

It is my contention that the role of school leader in an era of such high expectations is so demanding, and calls on such high levels of knowledge and skill, that nothing short of licensure on the scale expected of the medical specialist will suffice. This has powerful implications for the preparation of school leaders, for it calls for the mastery of a body of knowledge and skill, and for evaluation and ongoing development, with the professional requirements of specialist medical practice providing the benchmark. There are implications for selection, for this means that only those who are licensed may be included in the pool of those to be selected.

This presents a challenge to traditional approaches to these matters, especially in systems of education that were highly centralized and there was little professional discretion in the work of those appointed to positions of responsibility at the school level. There may have been high expectations for the teacher with leaders, especially the principal, viewed as master teachers, with any special knowledge and skill gained in incremental fashion through years of experience, often supported by a wise and trusted mentor. These may still apply, but as I shall endeavor to show in this chapter, there are new expectations and the stakes are high, as nations realize their future is dependent on the knowledge and skill of their citizens.

---

[1] This chapter is based on an invited address at the International Conference on School Leader Preparation, Licensure, Selection, Evaluation and Professional Developemnt held at the National Taipei Teachers College, Taiwan, March 1–4, 2001.

## Overview

My purpose is to provide a blueprint for successful leadership in an era of unprecedented high expectations for student achievement and increasing globalization in learning. It comprises one vision, three tracks, four dimensions, ten domains and six values (Table 1). The vision refers to the desired outcome of the global transformation that is under way and the emerging consensus on expectations for schools. The tracks refer to the broad directions of change in schools and school systems. The dimensions refer to major classifications of approaches to leadership that should be evident in practice. The domains refer to areas in which leaders should concentrate their efforts. The values are those that underpin a new sense of what constitutes the public good in education and training.

Table 1. The Blueprint.

| Component | | Description |
|---|---|---|
| Vision | 1 | Emerging global consensus on expectations for schools |
| Tracks | 3 | Broad directions for change in schools and school systems |
| Dimensions | 4 | Major classifications of approaches to the practice of leadership |
| Domains | 10 | Areas in which leaders should concentrate their efforts |
| Values | 6 | Underpinning of a new sense of 'the public good' |

## One Vision

Kenichi Ohmae, who coined the concept of 'the borderless world', has captured the new reality in *The Invisible Continent* (Ohmae, 2000), in contrast to the five continents that have clearly defined boundaries, geography that is visible, governments that hold power, and societies with unique cultures. He contends that there are four characteristics that 'help explain why some immigrants thrive on the new continent and others fail to gain a foothold':

1. It is 'cyber-enabled'. The new continent 'easily moves information across all kinds of borders, both national and corporate'.
2. As 'a continent without land, the new continent is easy to enter, but only for those who are willing to give up their old ways of thinking'.
3. 'No nation holds a monopoly on entrance to it. Any nation, any company, any race, any ethnic group, or any individual may enter'.
4. The new continent draws on 'highly individualistic values. Communities and families, or old-style establishment connections, do not determine worth in this world'. (Ohmae, 2000, pp. 16–20)

Ohmae is in no doubt about the place of education. He states that 'The most fundamental lever for success in the new continent is education' and that 'education is the first and foremost priority for any nation'.

> Preparing youngsters to comprehend the invisible continent and compete in its endeavors and explorations is the best investment that a government (or parents, for that matter) can make. (Ohmae, 2000, p. 227–229)

Moreover, a global consensus is emerging on expectations for schools on this invisible continent, if documents from key international institutions, such as UNESCO and OECD, and the espoused policies of governments, are taken as a guide (Barber, 1999; Chapman, 1997; Chapman & Aspin, 1997; Delors, 1996). That consensus may be summarized in these words:

> All students in every setting should be literate and numerate and should acquire a capacity for life-long learning, leading to successful and satisfying work in a knowledge society and a global economy on an invisible continent.

It is important to stress that this is an emerging consensus. Different nations, schools systems and schools will, of course, have their own special expectations.

## Three Tracks

A blueprint for successful leadership at all levels calls for a vision of education on an invisible continent along the lines suggested above. It also calls for recognition that movement in this direction is occurring at different rates in different settings but in the same broad directions.

There seem to be three such directions or tracks (Caldwell & Spinks, 1998; Table 2). Track 1 is the building of systems of self-managing schools. More authority and responsibility are being decentralized to the local level within a framework of centrally determined goals, priorities, frameworks, standards and accountabilities. Track 2 is an unrelenting focus on learning outcomes for students. There is unprecedented concern for assessment of student achievement, with international benchmarking now gathering momentum. Track 3 is the creation of schools for the knowledge society. A characteristic of such a society is that the largest group in the workforce is comprised of knowledge workers, being those who solve problems, manage information, or create new knowledge, products and services.

Table 2. Tracks for Change in the Transformation of Schools (Caldwell & Spinks, 1998).

| | |
|---|---|
| Tracks for change | Building systems of self-managing schools<br>Unrelenting focus on learning outcomes<br>Creating schools for the knowledge society |

## Four Dimensions

Dealing with a vision as sweeping as that suggested here, with change proceeding on all three tracks more or less simultaneously, calls for four broad approaches or classifications of leadership. It is helpful to consider these as dimensions of leadership. Four major dimensions are included in the blueprint: strategic, educational, reflective and cultural, as listed in Table 3.

Table 3. Dimensions of Leadership in the Transformation of Schools (Caldwell & Spinks, 1992).

| Dimensions of leadership | 1. Strategic<br>2. Educational<br>3. Responsive<br>4. Cultural |
|---|---|

These dimensions have a powerful place in the blueprint for successful leadership, as illustrated below.

**Strategic leadership**

Leadership is strategic when it involves:

- Keeping abreast of trends and issues, threats and opportunities in the educational environment and in society at large, nationally and internationally; discerning the 'megatrends' and anticipating their impact on education generally and on the leader's school, in particular;
- Sharing such knowledge with others in the school's community and encouraging other leaders within the school to do the same in their areas of responsibility;
- Establishing structures and processes which enable the school to set priorities and formulate strategies which take account of likely and / or preferred futures, and being a key source of expertise as these occur;
- Ensuring that the attention of the school's community is focused on matters of strategic importance;
- Monitoring the implementation of strategies as well as emerging strategic issues in the wider environment, and facilitating an ongoing process of review.

A capacity for strategic leadership has special priority at this time. The notion of an 'invisible continent', and expectations for schools that are emerging in a global consensus, present challenges that have no counterpart in the history of education if they are to be brought to realization. It requires every leader at every level to do the things listed above. It is no longer sufficient for a single minister, or principal, or president of a professional association, to be out in front saying these things. It is a 'whole-of-government' or 'whole-of-enterprise' approach, with every minister, every school leader and all officials in every professional associations having a capacity for strategic leadership in the specific meaning of that term: seeing 'the big picture' discerning the 'megatrends', understanding the implications, ensuring that other can do the same, establishing structures and processes to bring vision to realization, and monitoring the outcomes.

**Educational leadership**

Educational leadership refers to a capacity to nurture a learning community, again defined broadly to include a nation, state, school system, but especially a school. This is explained in more detail in one of the domains for leadership, but there is a 'hard edge' to the concept. A 'learning community' or a 'learning organization' sounds a very comfortable place in which to work, but the stakes are high if the consensus on expectations for schools is to be realized. 'All students in every setting' means that there will be a learning plan for each student.

With the extraordinary range of learning needs, it calls for teachers to have state-of-the-art knowledge about what works for each and every student. It calls for leaders who themselves will have much of this knowledge, but will certainly them to be able to manage the learning and teaching so that the knowledge is acquired and successfully brought to bear. Once again, this extends to all levels, including government, as well as for leaders of learning teams in the local school setting.

**Responsive leadership**
There is an implication here that leaders will respond to the expectations for schools and will be comfortable in collecting, analyzing and acting on data that lets them know how well things are going. Responsive leaders accept there are many stakeholders who have a 'right to know'. As for other dimensions, this acceptance extends to all levels of leadership.

**Cultural leadership**
All of the above indicates that there will be dramatic change to 'the way we do things around here', at the national, state, local and school levels. Successful leaders at each level will have a capacity to change the culture. This is no easy task, given that the scale of the change and the seriousness of the endeavor are still not broadly understood, let alone accepted, in many settings.

## Ten Domains

Ten domains for action are proposed for those who seek to lead schools that meet high expectations. These are listed in Table 4. Eight lie squarely in the field of education. The ninth and tenth span the fields of education, health and community, and they cross national boundaries on 'the invisible continent', reflecting the view that we cannot close the gap between current impressive achievement, and even higher expectations, unless we can, quite literally, 'span the boundaries'.

Table 4. Domains of Practice in the Exercise of Leadership.

| Domains of practice | |
|---|---|
| | 1. Curriculum |
| | 2. Pedagogy |
| | 3. Design |
| | 4. Professionalism |
| | 5. Leader development |
| | 6. Resources |
| | 7. Knowledge management |
| | 8. Governance |
| | 9. Boundary spanning |
| | 10. International protocols |

The integrating concept is change. Drucker (1999, p. 73) contends that the only ones who will survive in a period when change is the norm will be the change leaders, for 'to be a successful change leader an enterprise has to have a policy of systematic innovation' (Drucker, 1999, p. 84). For this reason, each domain for leadership is considered to be a field of innovation.

**Curriculum**

The emerging consensus on expectations for schools is commendable, but many would argue that the range of outcomes and their measures are much too narrow. The idea of 'multiple intelligences', based on Gardner's *Frames of Mind* (Gardner, 1983) is a helpful starting point.

Handy argues that three intelligences – factual intelligence, analytical intelligence and numerate intelligence – 'will get you through most tests and entitle you to be called clever' (Handy, 1997, p. 211). He suggests eight more: linguistic intelligence, spatial intelligence, athletic intelligence, intuitive intelligence, emotional intelligence, practical intelligence, interpersonal intelligence and musical intelligence' (Handy, 1997, pp. 212–213).

Leadbeater suggests that 'the curriculum needs to encourage creativity, problem solving, team building, as well as literacy and numeracy' (Leadbeater, 1999, p. vi). Beck sets a similar curriculum in the context of globalization:

> One of the main political responses to globalization is ... to build and develop the education and knowledge society; to make training longer rather than shorter; to loosen or do away with its link to a particular job or occupation. This should not only be a matter of 'flexibility' or 'lifelong learning', but of such things as social competence, the ability to work in a team, conflict resolution, understanding of other cultures, integrated thinking and a capacity to handle the uncertainties and paradoxes of the second modernity. Here and there, people are beginning to realize that something like a transnationalism of university education and curricula will be necessary. (Beck, 1999, p. 27)

To do all of this will require the abandonment of much of the existing curriculum. Writing for the UK setting, Seltzer contends that 'we can't just keep piling new expectations and structures on to old ones. Something has to give. We should aim to have reduced the national curriculum by half by 2010, in order to make room for new approaches' (Seltzer, 1999, p. xxi).

**Pedagogy**

The revolution in information and communications technology and the advent of exciting, pedagogically sound approaches to interactive multi-media learning mean that it is possible to learn anytime, anywhere. A revolution is clearly under way.

More fundamentally, and linked to the domain of innovation in curriculum, is how learning occurs. To what extent does the following view of how students learn reflect the norm in our schools?

> They are extremely good at manipulating symbols and working on computers, they are verbally fluent and extremely good at asking questions, but they don't really know anything in depth and they haven't really read anything. The high school curriculum is so chopped up into tiny bits and pieces that the integrating power of a liberal education is somewhat lost. (Sheridan, 1999, p. 274)

Henry Kissinger believes that 'the present generation has the power to tap into astonishing amounts of knowledge on any subject but no ability to integrate it into a knowledge of the past and no ability therefore to project it meaningfully into the future' (cited by Sheridan, 1999, p. 274).

**Design**

Curriculum and pedagogy cannot be constrained as single domains for they influence, and are influenced by, what occurs in other domains. It is here that the concept of 'design' comes into play.

Dimmock (2000) provides a rich inter-cultural perspective from the East and West. He distinguishes 'design' from 're-structure' and 'reform'. The design must have intentionality, connectivity, reinforcement, synergy and consistency. Design elements include societal culture, organizational culture, leadership and management, performance evaluation, personnel and financial resources, organizational structures – all centered on informed learning, informed teaching and an outcomes-oriented curriculum, energized by information and communications technology (Dimmock, 2000, p. 4).

Hill and Crévola adopt the same approach and propose a general design for improving learning outcomes that includes standards and targets; monitoring and assessment; classroom teaching programs; professional learning teams; school and class organization; intervention and special assistance; and home, school and community partnerships – all underpinned and centered on beliefs and understandings (Hill & Crévola, 1999, p. 123).

A design for the third track for change ('creating schools for the knowledge society') may be illustrated in a *gestalt* – a perceived organized whole that is more than the sum of its parts – as in Figure 1.

- Dramatic change in approaches to learning and teaching is in store as electronic networking allows 'cutting across and so challenging the very idea of subject

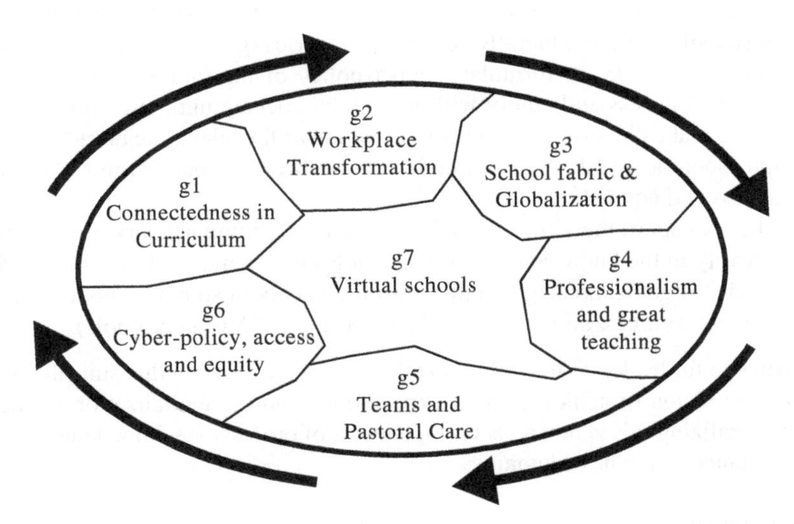


Figure 1. A *Gestalt* design for creating schools for the knowledge society (Caldwell & Spinks, 1998).

boundaries' and 'changing the emphasis from impersonal curriculum to excited live exploration' (Papert, 1993). At risk is the balkanized curriculum that has done much to alienate children from schooling, especially in the middle years of the transition from primary to secondary (g1 Connectedness in curriculum).

- Schools as workplaces are transformed in every dimension, including the scheduling of time for learning and approaches to human resource management, rendering obsolete most approaches that derive from an industrial age, including the concept of 'industrial relations' (g2 Workplace transformation).
- The fabric of schooling is similarly rendered obsolete by electronic networking. Everything from building design to the size, shape, alignment, and furnishing of space for the 'knowledge worker' in the school is transformed. In one sense, of course, the school has no walls, for there are global learning networks, and much of the learning that calls for the student to be located at school occurs in many places, at home and, at the upper years of secondary schooling and, for life-long learning, in the work place. (g3 School fabric and globalization)
- A wide range of professionals and paraprofessionals support learning in an educational parallel to the diversity of support that may be found in modern health care. The role of teacher is elevated, for it demands wisdom, judgments, and a facility to manage learning in modes more complex and varied than ever. While the matter of intellectual capital must be resolved, the teacher is freed from the impossible task of designing from their own resources learning experiences to challenge every student: the resources of the world's great teachers will be at hand (g4 Professionalism and great teaching).
- A capacity to work in teams is more evident in approaches to learning, given the primacy of the work team in every formulation of the workplace in the knowledge society. This, of course, will confound those who see electronic networking in an outdated stereotype of the loner with the laptop. The concept of 'pastoral care' of students is as important as ever for learning in this mode, and in schools that quite literally have no boundaries (g5 Teams and pastoral care).
- Spender's (1995) to formulate 'cyber-policy of the future' is a priority. The issues of access and equity will drive public debate until such time as prices fall to make electronic networks as common as the telephone or radio, and that may soon be a reality, given trends in networked computers (g6 Cyber-policy, access and equity).
- The concept of the virtual organization or the learning network organization is a reality in the knowledge society. Schools take on many of the characteristics of such organizations, given that learning occurs in so many modes and from so many sources, all networked electronically (g7 Virtual schools).

The challenge for leaders is to work with others to create a design that suits the setting, pursuing the vision of achieving high expectations and the globalization of lifelong learning. Realizing this vision calls for high levels of professional knowledge, and that is the substance of the next domain.

**Professionalism**

The unrelenting focus on learning outcomes in the emerging consensus on expectations for schools suggests 'innovation in professionalism', to the extent that teachers' work will be research-based, outcomes-oriented, data-driven, and team-focused, with lifelong professional learning the norm as it is for medical specialists. There is a danger, however, that this may be too narrow a view, just as the concept of world class schools may be too narrow.

A wonderfully rich professionalism is evident in the 'intelligent school' proposed by MacGilchrist, Myers and Reed (1997). This is the organizational counterpart of an individual with 'multiple intelligences'. Professionals in the 'intelligent school' will have contextual intelligence, strategic intelligence, academic intelligence, reflective intelligence, pedagogical intelligence, collegial intelligence, emotional intelligence, spiritual intelligence and ethical intelligence.

**Resources**

The commitment to free and compulsory education in many nations was made when schools consisted of large classes, few professional staff other than teachers, blackboards and slates, and little equipment apart from a few maps and globes. There was considerable community commitment to and 'in kind' support of the local school. Public expectations could be met to the full without a financial contribution from parents, voluntary or otherwise. A similar situation applied to hospitals. In the early 21st century, expectations are rapidly outstripping the capacity or willingness of the community to meet through taxation the full cost of education and health.

The key is to establish and then cost a series of school designs. Interest is growing in several nations and there is now a sturdy methodology for costing the various elements. The International Institute for Educational Planning of UNESCO recently published a report on needs-based resource allocation in education (Ross & Levacic, 1999; see also Goertz & Odden, 1999) that contains examples from several nations of different approaches to allocating resources to schools under conditions of decentralization.

More immediately, however, in the context of decentralization in education, especially where there is school-based management, local management or self-management, school leaders require a capacity to manage the school budget, matching priorities to resources, in an ongoing cycle of goal setting, policy formulation, planning budgeting, implementation and evaluation (see Caldwell & Spinks, 1988, 1992, 1998).

There must be a powerful link between resources and learning in this domain and leaders will need a deep knowledge about how to achieve it. A recent study (Woessmann, 2000) provides evidence on an international comparative scale of the efficacy of approaches such as school-based management, local management or self-management that are set in a centrally determined framework. Along with case study research in particular schools, there is now a relatively robust 'theory of learning' that underpins such developments (Caldwell, 2000; Caldwell & Spinks, 1998; Hillier, 1999; Wee, 1999).

**Leader development**

Attracting, preparing, placing, developing and rewarding school leaders is as much an issue as the nurturing of the profession at large. Around the world there is a crisis in accomplishing these things. The initiative of the Blair Government in creating the National College for School Leadership is remarkable by international standards.

Don't be seduced by the view that leadership is unimportant, or even unnecessary. In the gripping first chapter of Ian McEwan's remarkable novel *Enduring Love* (McEwan,

1998), a hot air balloon containing a small and frightened boy rolls uncontrolled across an open field. A sudden gust bears it aloft while its desperate owner tries to keep it on the ground. He is joined by several passers-by, each of whom take a rope in an endeavor to bring things under control. There is no effort to work together. One after another they release their hold, until one remains. The consequences are devastating. One who let go feels deep guilt. He reflects on the need for leadership, believing that 'No human society, from the hunter-gatherer to the post-industrial, has come to the attention of anthropologists that did not have its leaders and its led …' (McEwan, 1998, p. 11).

Leadership was required in a time of crisis but so was a capacity for all to be committed to a common purpose and to work together. Are these not the requirements for leadership in a time of high expectations for schools? Is it not a crisis for many students, who do not succeed, especially in the early years? Lives are at stake!

**Knowledge management**

The seventh domain calls for innovation in management. Traditional approaches to management are still required in areas such as planning, budgeting and staffing. Consistent with the emergence of the knowledge society, innovation in management calls for 'knowledge management'. This is not just a fad that will pass or a piece of jargon to describe what has always been a requirement in the organization.

According to Bukowitz and Williams (1999, p. 2), 'knowledge management is the process by which the organization generates wealth from its intellectual or knowledge-based assets'. In the case of school education, this may be re-worded, as 'Knowledge management is the process by which a school achieves the highest levels of student learning that are possible from its intellectual or knowledge-based assets'. Success-ful knowledge management is consistent with the image of 'the intelligent school' (MacGilchrist, Myers and Reed, 1997) and the concept of 'intellectual capital' (Stewart, 1997).

Knowledge management involves a school developing a deep capacity among its entire staff to be at the forefront of knowledge and skill in learning and teaching and the support of learning and teaching. This is more than occasional in-service training or professional development. This is a systematic, continuous and purposeful approach that starts with knowing what people know, don't know and ought to know. It assumes an innovative professionalism, as already described, and includes a range of functions such as selection, placement, development, appraisal, reward, succession planning, contracting of services and ensuring that every aspect of the workplace is conducive to efficient, effective and satisfying work for all concerned.

Interestingly, Bukowitz and Williams see the recent loss of middle management as a loss of capacity in respect to these matters. 'Once middle management was "out" it was not surprising that knowledge management was "in". Knowledge management represents an effort to repair past damage and an insurance policy against loss of organizational memory in the future' (Bukowitz & Williams, 1999, p. 7). They conclude that middle managers may be a good idea after all:

> As organizations begin to restore some of these positions, they will do so with the revitalized view of the role of middle managers as orchestrates of knowledge flows … middle managers will increasingly be asked to look across the organization. Their success will hinge on the ability to facilitate communication …

leverage resources, transfer best practices, identify synergies and encourage knowledge re-use. (Bukowitz and Williams, 1999, p. 355)

**Governance**

There have been many innovations in school governance in recent years. The most notable recent development is the 'schools for profit' movement in the United States. It is now a growth industry, with leading players including Edison Schools Inc. which runs 26 independent charter schools and 53 traditional public schools, and Nobel Learning Communities, Inc. that operates 137 private schools (Schnaiberg, 1999, p. 13). The former had an initial public offering (IPO) in November 1999, raising about US$140 million. Edison plans partnerships with universities in teacher education programs (see cover story in Business Week, April 7, 2000 for a comprehensive account of these developments).

Another major development in school governance in the United States has been the emergence of charter schools. These are publicly funded schools that are owned or operated by non-public bodies and generally free from constraining regulations that are applied to schools of a public authority. The number of charter schools has increased from two in one state in 1992–93 to 1484 in 32 states in September 1999. In a recent survey, the most important reasons for applying for charter status were reported to be realizing an alternative vision (59 per cent), serving a special population of students (23 per cent), gaining greater autonomy and flexibility (9 per cent) attracting more students (4 per cent), obtaining increased funding (3 per cent), and securing greater involvement of parents (3 per cent) (Nelson et al., 2000).

**Boundary spanning**

The ninth domain lies in organizational arrangements to design, deliver or in other ways drive the effort. We have had a century or more of largely successful effort in the public sector with responsibility in the hands of discrete government departments each reporting to a particular minister. What occurs in a government or public school is largely a matter for the department of education and a responsible minister for education. Yet the problems to be addressed in closing the gap are complex and demanding of attention of those who work in different departments, or elsewhere outside government and in the private sector.

While inter-department cooperation and freewheeling boundary spanning have been evident, it is only in recent times that signs of a major shift in culture that fosters even higher levels have been seen. That shift has resulted from a backward-mapping approach, starting from a focus on people and a problem, then selecting a strategy to address the problem, then designing and delivering a constellation of services and resources, without consideration of organizational boundaries except where the public good test is not satisfied. This linear process is made more complex because there is rarely a single problem to address and rarely a single solution. Governments that have taken this approach now speak of 'joined up solutions to joined up problems' and advocate breaking down organizational boundaries. They use the metaphor of a silo to describe the isolation of a government department. I should hasten to add that the same metaphor has been adopted to describe different faculties in universities.

The 'full-service' school is making an appearance in many nations, with most serving a high school student population, but concern for linkage is broader, now considered a key element of comprehensive reform.

> The ideal full-service school has administrators, teachers, school counselors and psychologists, physicians or nurse practitioners, school social workers, and other professionals working together to provide one-stop service delivery. (Swerdlik, Reeder & Bucy, 1999)

Lawson (1999) considers full-service schools to be an example of a second mental model of schooling. The first, or dominant, model is 'informed by the metaphor of the factory ... well-established and patrolled professional boundaries and jurisdictions have been established and maintained ... stand-alone schools are standardized as "one best system" thinking, with large bureaucracies as reinforcements'. While recognizing the benefits of full-service schools, he raises questions about sustainability, scale-up, level of responsibility a school may be reasonably asked to bear, and impact on learning. He proposes a third mental model that 'draws selectively and strategically on the benefits of full-service schools and also on the lessons learned'. This model is based on interdependence, collaboration and enlightened self-interest. He concludes that:

> Principals have a clear choice regarding whether to become a full-service community school, or to borrow some elements of it to fit local needs and contexts. Their schools do not have to become social and health service agencies, even though they must be linked strategically to them. This different planning frame fosters educational reform, incorporating plans for anytime – anywhere learning. It reflects enlightened self-interest because it focuses on the special, important accountabilities of schools and the unique responsibilities of educators. (Lawson, 1999)

**International protocols**

The tenth domain calls for leadership in spanning boundaries of a quite different kind. There is increasing awareness that many people in most nations are not reaping the benefits of globalization, and that the riches of many are matched by the deep poverty of many more. How can we look after the interests of all in a borderless world? Who makes the laws of protection? We had these same concerns more than a century ago with the industrial revolution, and it was at this time that a regime of national and state laws, and new systems of education were created. We have reached the same point again, but how much more of a challenge in the invisible continent?

The role of organizations such as UNESCO may become increasingly important in this regard, but so too will be the work of schools and universities that are joining together in global alliances of one kind or another.

## Abandonment

If the integrating theme among the ten domains of leadership is innovation, and if there is not to be an accretion of new tasks on old, it follows that a capacity for systematic abandonment is as important as a capacity for systematic innovation. Drucker (1999) calls for 'organized abandonment' for products, services, markets or processes:

- Which were designed in the past and which were highly successful, even to the present, but which would not be designed in the same way if we were starting afresh today, knowing the terrain ahead;
- Which are currently successful, and likely to remain so, but only up to, say, five years – in other words, they have a limited 'shelf life'; or
- Which may continue to succeed, but which through budget commitments, are inhibiting more promising approaches that will ensure success well into the future.

Virtually every domain of leadership calls for abandonment of one kind or another, regardless of the scenario.

The eight domains for leadership in the field of education call for abandonment of a range of approaches. Innovation in curriculum requires abandonment of some learning areas that have been painstakingly constructed over the last decade. Pedagogy is a domain fraught with dilemmas, but ripe for abandonment of approaches that do not yield outcomes consistent with expectations for schools.

Innovation in design will certainly require abandonment of standard class sizes for all students at every level in facilities built like a collection of boxes, lined end to end or stacked one upon the other. Innovation in professionalism challenges the modest levels of knowledge and skill that sufficed in the past with a vision for values-centered, research-based, outcomes-oriented, data-driven and team-focused approaches that matches, nay, exceeds that of the best of medical practice. Innovation in funding similarly challenges the constrained view of the recent past to call on all of the resources of the community in support of its schools. Innovation in leadership affirms the need for leadership but approaches that do not lead to commitment to a common purpose should be abandoned.

Innovation in management must find a place for knowledge management, suggesting that some tasks should be abandoned, curtailed or shifted to others if the role of middle management, in particular, is to be rewarding for incumbent and organization. Innovation in governance, like innovation in funding, calls for abandonment of a constrained approach, admitting possibilities that have hitherto been unthinkable.

Innovation in the ninth domain of boundary spanning calls for abandonment of the silo metaphor and abandonment of the view that problems occur in simple clusters and can be addressed by networking of services in a single sector. Innovation in the establishment of international protocols calls for abandonment of the notion that a nation or part of a nation can and should keep to itself the writing of a regulatory regime that protects the interests of students to the satisfaction of citizens in separate jurisdictions.

## Six Values

The first five words in the emerging consensus on expectations for schools ('all students in every setting') suggest a place for values in the blueprint for successful leadership. Six are proposed, as summarized in Table 5.

It is suggested that these should provide the basis for a test of 'the public good' in the design and implementation of policy in education.

Table 5. Values That Underpin a New Sense of the Public Good.

| | |
|---|---|
| Values and the public good | 1. Access<br>2. Equity<br>3. Choice<br>4. Growth<br>5. Efficiency<br>6. Harmony |

- *Access*. The policy should ensure all students have the opportunity to gain an education that meets high expectations.
- *Equity*. The policy should provide assurance that students with similar needs will be treated in the same manner in the course of their education.
- *Choice*. The policy should reflect the right of parents and students to choose a school that meets their needs and aspirations.
- *Growth*. Strategies should be in place to ensure that resources are adequate to the task.
- *Efficiency*. Scarce resources should be allocated wisely to optimize outcomes.
- *Harmony*. There should be no fragmentation of commitment and effort in support of policies that reflect these values.

The first five are drawn from a classification proposed by Swanson and King (1997). Three are based on the classic trio of liberty (choice), equality (equity) and fraternity (access).

Our association of the word 'public' with 'government', which has prevailed for more than a century, should give way to a declaration of 'public values' that should underpin the enterprise. Jerome T. Murphy, Dean of the Harvard Graduate School of Education, has this view, believing that 'what will determine whether we call them public schools is not so much the vehicle that's providing the education, but really whether they ascribe to a certain set of public values. Values like equal educational opportunity, values like non-discrimination, and so on. We'll have multiple delivery systems to achieve public values' (Murphy, 1999).

## The Blueprint

This completes the blueprint for successful leadership in an era of unprecedented high expectations for achievement for all students and the globalization of learning. The different components are brought together in Table 6.

Table 6. Blueprint for Successful Leadership in an Era of Globalization of Learning.

| Component | Element |
|---|---|
| Vision for education on an invisible continent | 1. Emerging global consensus on expectations for schools |
| Tracks for change | 2. Building systems of self-managing schools<br>3. Unrelenting focus on learning outcomes<br>4. Creating schools for the knowledge society |

*(continued)*

Table 6. *continued*

| Component | Element |
|---|---|
| Dimensions of leadership | 1. Strategic<br>2. Educational<br>3. Responsive<br>4. Cultural |
| Domains of practice | 1. Curriculum<br>2. Pedagogy<br>3. Design<br>4. Professionalism<br>5. Leader development<br>6. Resources<br>7. Knowledge management<br>8. Governance<br>9. Boundary spanning<br>10. International protocols |
| Values and the public good | 1. Access<br>2. Equity<br>3. Choice<br>4. Growth<br>5. Efficiency<br>6. Harmony |

It cannot, of course, tell us all about the requirements for successful leadership, but a capacity to grasp the vision and work along the tracks in each of the ten domains, exercising leadership in several dimensions, underpinned by a commitment to values that define a new sense of 'the public good', will surely go a long way to bringing the vision to realization.

## Conclusion

Such a complex, demanding and constantly changing role demands a coherent and intensely professional approach to preparation, licensure, selection, evaluation and professional development. Time-honored approaches to simple selection by an employing agency based on the incremental acquisition of skills over many years will no longer suffice. Expectations for the best in specialist medical practice must be matched in educational leadership. In both cases, lives and the well being of the nation are at stake. This is a call to replicate in education, for teachers and, in this instance, their leaders, for legislation and regulation in education to match that in medicine.

There is now a growing literature on domains for leadership and one of the best is by Hedley Beare (2001), one of Australia's most experienced leaders in public education and a distinguished scholar with an exceptional capacity to see the pathways from past to present to future.

He concludes an uplifting chapter about teachers for the school of the future with these words:

> This terrain is *not* for the immature, the shallow, the unworthy, the unformed, or the uninformed, and society needs to be very careful about what people it commissions for this task. (Beare, 2001, p. 185)

Blueprints along the lines proposed here will help ensure that this need can be met.

## References


Barber, M. (1999). *A world class school system for the 21st Century: The Blair Government's education reform strategy.* No. 90 in a Seminar Series of the Incorporated Association of Registered Teachers of Victoria (IARTV), December [ISBN 1 876323 31 0] [reprint of a paper presented at the Skol Tema Conference in Stockholm in September 1999].

Beare, H. (2001). *Creating the future school.* London: Falmer Press.

Beck, U. (1999). Beyond the nation state. *New Statesman,* 6 December, 25–27.

Bukowitz, W. R. & Williams, R. L. (1999). *The knowledge management fieldbook.* London: Financial Times Prentice Hall.

Business Week (2000). *For-profit schools.* Cover story, February 7. [http://www.businessweek.com]

Caldwell, B. J. (2000). *A theory of learning in the self-managing school.* Keynote presentation on the theme 'School-based Management: Theory and Practice at an international conference hosted by National Hualien Teachers College, Hualien, Taiwan, October 19 (available: http://www.edfac.unimelb.edu.au/EPM/StaffProfile/BCaldwell.shtml).

Caldwell, B. J. & Spinks, J. M. (1998). *The self-managing school.* London: Falmer Press.

Caldwell, B. J. & Spinks, J. M. (1992). *Leading the self-managing school.* London: Falmer Press.

Caldwell, B. J. & Spinks, J. M. (1998). *Beyond the self-managing school.* London: Falmer Press.

Chapman, J. (1997). Leading the learning community. *Leading & Managing,* 3(3), 151–170.

Chapman, J. & Aspin, D. (1997). *The school, the community and lifelong learning.* London: Cassell.

Delors, J. (Ed). (1996). *Learning: The treasure within.* Paris: UNESCO.

Dimmock, C. (2000). *Designing the learning-centred school: A cross-cultural perspective.* London: Falmer Press.

Drucker, P. F. (1995). *Managing in a time of great change.* Oxford: Butterworth Heinemann.

Drucker, P. F. (1999). *Leadership challenges for the 21st century.* Oxford: Butterworth Heinemann.

Gardner, H. (1983). *Frames of mind.* London: Heinemann.

Goertz, M. E. & Odden, A. (Eds). (1999). *School-based financing.* Thousand Oaks, CA: Corwin Press, Inc.

Handy, C. (1997). *The hungry spirit.* London: Hutchinson.

Hill, P. & Crévola, C. (1999). The role of standards in educational reform for the 21st century. In Marsh, D. (Ed). *Preparing our schools for the 21st century.* ASCD Yearbook 1999, Alexandria, Virginia: ASCD (Chapter 6).

Hillier, N. (1999). *Educational reform and school improvement in Victorian primary schools 1993–1999.* Unpublished thesis for the degree of Doctor of Education, University of Melbourne.

Kotter, J. (1990). *A force for change: How leadership differs from management.* New York: The Free Press.

Lawson, H. A. (1999). Two new mental models for schools and their implications for principals' roles, responsibilities, and preparation. *NASSP Bulletin*, 83, 611, 8–27. [The entire issue of this publication of the National Association of Secondary School Principals (USA) is devoted to full-service and similar schools]

Leadbeater, C. (1999). It's not the economy, stupid. *New Statesman*. Special Supplement on the theme Knowledge is Power! iv–vi.

McEwan, I. (1998). *Enduring love*. London: Vintage.

MacGilchrist, B., Myers, K., & Reed, J. (1997). *The intelligent school*. London: Paul Chapman.

Murphy, J. T. (1999). Remarks on the theme 'The next millennium: Now What?' CNN special program (transcript).

Nelson, B., Berman, P., Ericson, J., Kamprath, N., Perry, R., Silverman, D. & Solomon, D. (2000). *The state of charter schools 2000*. Fourth-Year Report, Washington, DC: Office of Educational Research and Improvement, US Department of Education.

Ohmae, K. (2000). *The invisible continent: Four strategic imperatives of the new economy*. London: Nicholas Brealey Publishing.

Papert, S. (1993). *The children's machine: Rethinking school in the age of the computer*. New York: Basic Books.

Ross, K. N. & Levacic, R. (Eds.; 1999) *Needs-based resource allocation in education via formula funding of schools*. Paris: UNESCO.

Seltzer, K. (1999). A whole new way of learning. *New Statesman*. Special Supplement on the theme Knowledge is Power! 27 September, xvii–xix.

Schnaiberg, L. (1999). Seeking a competitive advantage. *Education Week*. 1, 12–14.

Sheridan, G. (1999). *Asian values Western dreams*. St. Leonards: Allen & Unwin.

Spender, D. (1995). *Nattering on the net: Women, power and cyberspace*. North Melbourne: Spinifex.

Stewart, T. A. (1997). *Intellectual capital: The new wealth of organizations*. London: Nicholas Brealey.

Swerdlik, M. E., Reeder, G. D. & Bucy, J. E. (1999). A partnership between educators and professionals in medicine, mental health, and social services, *NASSP Bulletin*. 83, 611, 72–79. [The entire issue of this publication of the National Association of Secondary School Principals (USA) is devoted to full-service and similar schools]

Swanson, A. D. & King, R. A. (1997). *School finance: Its economics and politics*. Second Edition, New York: Longman.

Walsh, M. (1999). Report card on for-profit industry still incomplete. *Education Week*, 1, 14–16.

Wee, J. (1999). *Improved student learning and leadership in self-managed schools*. Unpublished thesis for the degree of Doctor of Education, University of Melbourne.

Woessmann, L. (2000). *School resources, educational institutions, and student performance: The international evidence*. Kiel Institute of World Economics, University of Kiel (available: http://www.unikiel.de/ifw/pub/kap/2000/kap983.htm). This paper was presented at the annual conference of the Royal Economic Society, Durham, April 9–11, 2001.

# 3

# The Changing Roles and Training of Headteachers: Recent Experience in England and Wales

Dr. Raymond Bolam

*Professor, Department of Educational Administration, Cardiff University, Cardiff, Wales, UK*

Systems of school leader preparation, licensure/certification, selection, evaluation and professional development are necessarily rooted in the particular context of a single country. They are the product of a unique set of circumstances – political, economic, social, cultural, historical, professional and technical – in that country and may usefully be conceptualised as complex organisational innovations. Accordingly, those attempting to promote international exchange and learning about such matters must take seriously what we know from research and experience about the adaptation and use of complex innovations across cultures.

It is in this spirit that this chapter has been written. It presents a, largely descriptive, account of selected aspects of *School Leader Preparation, Licensure and Certification, Selection, Evaluation and Professional Development* in England and Wales and should be read alongside the complementary chapter by Professor Harry Tomlinson (cf).

Although the terms leadership and management were, until recently, used interchangeably in England and Wales, in this chapter:

> … I take educational leadership to have at its core the responsibility for policy formulation and, where appropriate, organisational transformation; I take educational management to refer to an executive function for carrying out agreed policy; finally, I assume that leaders normally also have some management responsibilities. … (Bolam, 1999)

### School Leaders in Context

Schools in the United Kingdom operate mainly in four, increasingly distinctive, systems: England, Wales, Northern Ireland and Scotland. So, to simplify what would otherwise be a complex and lengthy account and analysis, this chapter focuses on schools in the state-maintained sector in England, with some reference to Wales. In England and Wales, there are about 24,000 primary schools with 4.7 million pupils and 223,000 teachers, 82% of whom are female; 4,700 secondary schools with 3.5 million pupils and 233,000 teachers, 48% of whom are female; 1,800 special schools with 110,000 pupils; and 2,500 independent schools with 595,000 pupils. Approximately 130,000 teachers have formal leadership and management responsibilities in their primary, secondary or special school. These include about 30,000 headteachers, 40,000 deputies and 70,000 middle managers, with a departmental, section or subject specialist role.

Central responsibility for state-maintained schools currently rests with the Department for Education and Skills (DfES) and the Welsh Assembly Education Department (WAED). Local Education Authorities (LEAs) continue to have some responsibilities for education although these are changing. Most schools are financed *via* local government from three sources: from central government *via* the Revenue Support Grant (43%) and specific grants (12%); from a local Council Tax (20%); from a local Business Rate (25%). However, the actual money generated varies considerably: for example, in 1998/99, although the average funding per primary pupil in England was £2248, the range was from £2006 in Leicestershire to £3377 in Tower Hamlets. Under the system of local management of schools (LMS), local authorities are now expected to delegate 85% of this money to schools using a pupil number-based formula.

Primary (i.e., infant and junior) schools have pupils aged 5–11 years and range in size from approximately 40–500+. A typical staffing structure for such schools is: headteacher, deputy head, and approximately 1–10 teachers. The average class size is 25.8 for seven year-old pupils and 28.3 for ten year-olds. Secondary schools have pupils aged 11–18 years and range from approximately 300–1000+. The average class size is 21.8 for fourteen year-old pupils. A typical staff structure is: headteacher, deputy head, and approximately 10 subject leaders, 5 middle managers and up to 50 teachers. The typical school day is from 9.00 am to 4.00 pm, but this varies between schools.

The salary scales of teachers and headteachers in early 2001 were as follows:

| | |
|---|---|
| Teachers<br>(via nine annual increments) | £15k–£24k |
| Threshold<br>(depending on performance assessment) | £26k–£30k |
| Subject leaders/middle managers<br>(increment depending on extra duties) | £1.5k–£9.5k |
| Deputy heads<br>(depending on school size) | £28k–£46k |
| Headteachers<br>(depending on school size) | £32k–£76k |

In making comparisons across countries, the following benchmark criteria may be useful: the price of a new, small car is about £11,000; and, compared to other professionals, most classroom teachers find it difficult to buy their own homes, especially in London and other large, prosperous cities.

The selection process for headteachers and other school leaders is as follows. When a headteacher vacancy occurs, the job is advertised in the national press and, increasingly, on the Internet (e.g., www.tesjobs.co.uk). Applications are open to any suitably qualified candidate and are normally processed by the LEA. Potential applicants are sent an information pack, which includes details about the school and a specific job description. Short-listed candidates are then interviewed by a panel of school governors, on which the LEAs representative has an advisory role. This stage of the interview and selection process normally takes one day but can last up to two or even three days, especially in the case of secondary schools, and can include a variety of tasks and observations. Consultants and head-hunters are occasionally involved for larger schools.

The main qualifications required for the position relate to previous, successful experience as a deputy head or as a headteacher in another school. Candidates will be expected to have a good first degree, a teaching qualification and successful teaching experience. Most will have a Masters level degree, usually in school management, and will also have attended many professional training courses although neither of these is a requirement. In future, the NPQH (see below) will be a required qualification for all candidates. For other school leadership positions (e.g., deputy head or subject leader) in both primary and secondary schools, candidates apply directly to the school itself. The interviews are conducted by a panel that includes school governors, but the most influential role in making an appointment is that of the headteacher. According to a recent study of new headteachers in Wales (Dunning, 2000, p. 136), most had been appointed in their early forties and had nineteen years teaching experience. 98% of secondary and 88% of primary heads had been deputy heads. Two thirds of the secondary heads were male but the position in primary schools was more balanced (52% men; 48% women). All heads continued to do some classroom teaching. In small schools this could be substantial (i.e., 50% or more).

Wallace and Huckman (1999) concluded that several factors encouraged the growth of shared leadership and management in schools. The creation of comprehensive secondary schools from 1960s to 1980s led to larger institutions and more extensive management tasks. Senior management teams (SMTs) thus became almost universal in secondary schools and the long tradition of headteacher as go it alone top leader and manager was tempered by more recent concern among other staff to participate in decisions affecting them. Moreover, the post-1986 education reforms (see below) also increased the range of management tasks, providing additional incentive for headteachers to share leadership and management, even prompting large primary schools to create an SMT.

So, SMTs are now common in most schools, except the very small two or three teacher schools in rural areas. They consist of the headteacher, deputy or deputies and one or more of the senior teachers. Their typical working practice is to meet together regularly, usually weekly in the headteachers office, to brief each other on what is happening in school and to discuss new policies (e.g., literacy hour), the ongoing implementation of existing policies and day-to-day administration (e.g., procedure for wet playtimes). Their typical role is to lead and manage the school on internal matters, under the overall leadership of the headteacher and within policy framework agreed by the Governing Body.

In a study of primary schools, Wallace & Huckman (1999) found that SMT responsibilities were distributed in three main ways – school wide management tasks (e.g., managing staff development), department management (e.g., infants/key stage one) and curriculum coordination (i.e. advising all teachers on one or more subject areas, e.g., religious education). All SMT members, except headteachers and a few deputies, also taught a class and most had multiple management responsibilities. In an earlier study of secondary schools, Wallace & Hall (1994) found that the level of funding allowed a higher proportion of promoted posts than in primary schools and that management responsibilities were based on specialist subject departments and the separate pastoral/administrative support system. Typically, there was a deputy headteacher (curriculum) responsible for liaising with subject departments, often grouped as faculties, and deputy headteacher (pastoral) responsible for year group teams. This system gave the SMT shared oversight of the work of all other staff, and gave headteachers oversight of other SMT members work. In addition, each SMT member had a subject teaching responsibility, usually 20–60% of a normal workload.

## The Evolving National Context: Post-1986 Reform and Re-structuring

The jobs of classroom teachers and, especially, school leaders have changed dramatically since 1986. Successive British Governments, like the governments of other European and OECD countries, have pursued an agenda for achieving improved quality of schooling, sometimes referred to as restructuring. The approach adopted in England and Wales concentrated on the outputs of schooling, raising standards, education as a contributor to economic performance, vocational training, increasing the influence of the community in the management of schools, improving institutional accountability, economy, efficiency and effectiveness and achieving national targets.

The approach can best be understood by relating it to three inter-related sets of underpinning, and controversial, political ideas and beliefs. First, the adoption of market (or quasi-market) mechanisms and the extension of consumer choice are the best way of ensuring that education is provided as cost effectively as possible and of promoting improvements in overall quality. Second, business management techniques should be adopted in education for reasons of efficiency and effectiveness. Third, although some decisions and powers are best de-centralised to schools, greater strategic control over education must be centralised to the level of national government.

From the mid-eighties, the first two of these ideas increasingly influenced education reforms internationally. However, compared to most other OECD countries, the approach in England and Wales was noteworthy both for the scale and scope of the post-1986 reform programme, which covered all schools in the country, and for the shift towards centralisation, when many other European countries were introducing de-centralising measures.

In summary, the main features of the 1986–96 reforms, under successive Conservative governments, were as follows. First, came the introduction of local management of schools (LMS) with site-based control over delegated budgets, pupil recruitment, strategic policy and planning, the hiring and firing of staff, staff development and buildings. This was closely linked to increased management powers for school governing bodies with the headteacher cast in the role of chief executive and the chair of governors in the role of chair of the board of directors. The introduction of a quasi- or regulated market involved parents (as customers/consumers) exercising choice and schools (as providers) competing for their custom (in the form of pupil numbers). It also involved mechanisms designed to extend parental choice (e.g., open enrolment) and to inform their choice, for example the annual publication of raw test and examination results, subsequently transformed into school league tables by the local and national media, and published inspection reports. Significant reductions in the powers of LEAs were accompanied by substantial increases in those of central Government. The latter included the national curriculum and national testing related to four Key Stages (i.e., for pupils aged 7, 11, 14 and 16), regular external inspections by a privatised inspectorate, and strengthened accountability and quality assurance roles for a range of national agencies. Existing national salary negotiating machinery was abandoned in favour of Government-imposed, national salary scales, conditions of service and career-ladders, together with compulsory appraisal, for all teachers and headteachers. The Teacher Training Agency (TTA) was made responsible for initial teacher training and continuing professional development, including headteacher training. It promoted an extended role for schools in initial teacher training by requiring all training institutions to pay for training places and mentor support in their partner schools, and by funding a minority of selected schools to run school-based initial teacher training schemes themselves, with the power to buy in help from universities, as necessary.

Since May 1997, successive Labour Governments have continued with much of the reform thrust of its predecessors but the policy emphasis has shifted away from marketisation to focus on the control of inputs, processes and outputs. The main initiatives, each with implications for school leadership, include the requirement that

> All LEAs produce an Education Development Plan; the introduction of literacy and numeracy schemes in primary schools in which time, content and pedagogy are specified; a requirement that schools produce their own tough targets at Key Stages 2 and 3; the modernisation of the teaching profession through the introduction of performance management, performance-related pay, improved opportunities for continuing professional development and General Teaching Councils in England and Wales; the overhaul of school leadership training, including the creation of a National Leadership College.

## The Impact of the Reforms on School Leadership and Management

There is general consensus that the reforms resulted in extensive and radical changes in the roles and responsibilities of headteachers and other senior staff involved in school management and that what amounts to a shift in the professional culture occurred. Since 1986, headteachers have been required to exercise strategic leadership, planning, marketing, evaluation and development skills; to focus much more on student learning and assessment; to operate as a quasi chief executive in relation to school governors; to deal with and respond to external inspections; and to cooperate, as well as compete, with neighbouring schools.

These issues have been the focus of considerable research. For instance, a unique, ten-year, longitudinal study (Weindling, 1999) offered insights into the cumulative impact of the reform process on a cohort of British secondary headteachers. In 1987, 80% of the sample said their role was very different from when they had started the job in 1982 and, in 1993, 90% said their role had continued to change significantly over the previous five years. The main areas of difference concerned the introduction of LMS, which had pushed finance-related issues up their list of concerns, together with the other mandated changes. A European study found that Welsh heads were much more likely than their counterparts in The Netherlands, Norway and Spain to see government reforms as causing them substantial problems (Bolam et al., 2000).

Campbell and Neill (1994) reported on research, using diaries and questionnaires, into the work of a volunteer sample of 384 secondary school teachers, drawn from 91 LEAs in England and Wales. On average, their respondents worked 54.5 hours each week during term time, seven and a half hours per week more than for the 1978 sample, an increase that they concluded was largely due to the impact of the reforms. Comparing their findings with those for other countries and occupations, they concluded that their sample of teachers, '*actually work significantly longer hours than most other non-manual and manual workers in Great Britain and Europe*' (Campbell & Neill, 1994, p. 68). Managers, especially deputy heads, spent much more time on relatively low-level clerical administration and less on teaching than classroom teachers.

The salary scales presented above demonstrate that there are considerable financial incentives for teachers to apply for middle and senior management roles. After all, a headteacher can earn from twice to five times the salary of a beginning teacher, depending on the size of the school. However, there is also contrary evidence that teachers are less motivated to apply for promotion to deputy headships and headships; that the number of such senior management vacancies is increasing; that this increase is frequently due to early-retirements caused by stress or ill-health; and that workload pressure is having a negative impact on teachers motivation and morale (School Teachers Review Body, 1996, p. 15; Travers & Cooper, 1996).

The reforms resulted in the creation of new managerial posts, often associated with the national curriculum and assessment. Thus, in primary schools specialist posts for coordinating Key Stages 1 and 2 assessment became common as did posts related to the literacy and numeracy hours and analogous new posts were also created in secondary schools. In addition, the roles of existing heads of department were extended. For example, a head of mathematics in a secondary school or a head of the infants department in a primary school may now be responsible for managing a devolved budget, including a component for staff development. They play a major role in the inspection process, are often responsible for staff appraisal and development and may also draft the relevant section of the schools annual development plan. Their responsibility for the performance of their department is also much more explicit, largely because the evidence from inspections and research demonstrates that there can be significant differences in the performance of, say, mathematics departments in schools with similar intakes. Thus, they are now held accountable for the standard achievement tests (SATs) targets for students and examination results at each key stage. These new and enhanced responsibilities are included in the National Standards (TTA, 1998).

One important theoretical perspective on these reforms is that of managerialism. The policies summarised above were, according to many commentators, pursued by both Conservative and Labour Governments as part of broader strategies across the public sector as a whole. Reforms in health, social services and housing, as well as education, were said to have a common technical/ideological core, often referred to as managerialism, rational management or new public management. The main features, in education, were said to include: the centralisation of national decision-making; an increased emphasis on the line management of teachers and on managerial control of teachers work in the interests of efficiency; the weakening of teacher autonomy; the development of new divisions of labour by delegating routine tasks to lower qualified/ paid; the need for new skills and responsibilities; the emergence of a more distinct managerial layer in schools; a marked shift in the nature of the professional organisation from professional to managed or corporate bureaucracy; a consequential emphasis on target setting, rational management and accountability; proceduralism in the form of targets, development plans, performance management; and, arising from all of these changes, work intensification.

We should be cautious about such claims and generalisations and not exaggerate the impact of these developments on teachers and schools. In reality, the content of managerialism belies agreed, precise operational definition because the concept refers to an assemblage of beliefs and values, only some of which may be emphasised by particular policy makers, groups and writers. Hence, in spite of much, sometimes polemical, writing to the contrary, how far a managerialist-style ideology has actually underpinned policy making in government and, more importantly, how far a managerialist agenda has actually been implemented, is best treated as problematic (Hood, 1995).

## The Background to Improvements in School Leadership Training

Improvements in school leadership training since the early 1980s are best understood in the light of developments in related areas of training policy and practice. First, there were significant developments in the organisation, funding and provision of in-service training for teachers in general that also applied to management training for headteachers. In summary, the main features were: the introduction, in 1987, of a regulated-market for in-service training in which schools receive annual funding to provide and purchase training and consultancy; a substantial increase in the number of professional associations and unions, private trainers and consultants and other agencies specialising in the provision of training; more flexible and market-driven university structures, including modularisation, credit accumulation and transfer, accreditation of prior learning and experience, professional development profiles, distance and open learning programmes; a substantial increase in the number of specialist education management degree courses available at universities throughout the country, including new taught doctorates; most recently, the extensive development and use of the Internet for a wide range of professionally relevant information exchange (see selected Web sites listed below). In addition, in 1995–96 Teacher Training Agency initiated a framework for continuing professional development based on national standards at five career stages: newly qualified teachers; expert teachers, including special needs; subject leaders/heads of department; headteachers.

There were also many innovations in the content and methods of school management training, including the encouragement of dissertations focused on school-management problems and school improvement projects; the increasing application of theories of experiential learning and reflective practice, often related to the use of Professional Development Profiles, action research, action learning as well as of coaching, mentoring and other forms of peer-assisted learning, especially for new headteachers; the adoption by many schools of the *Investors in People* scheme, based on a Total Quality Management approach to human resource and organisation development.

Underlying these developments were criticisms of the contribution of universities to school leadership training, largely on the grounds that it was too theoretical. One useful perspective on this issue is to distinguish between two types of theory: *theory for understanding and theory for action*. Universities have traditionally been seen as good at the former but poor at the latter. The best Masters degree courses on school management adopt an approach based on the use of informed professional judgement, suggesting practical guidelines for action (and training) based on research and practitioners experience together with available theoretical insights, particularly those from contingency theory. Contingency theorists argue, broadly, that there is no single correct way either to structure or lead an organisation since the leaders, the followers and the organisation itself have distinctive, even unique, characteristics and invariably operate in a turbulent environment that compounds the already unpredictable nature of their work. Hence, in effective organisations, the structures and leadership are adapted to meet their particular circumstances. One practical implication for training is that school leaders should be equipped to learn and use a repertoire of techniques and styles, an approach that is now widely adopted.

## Developments in School Management Training and Development

It is important to recognise that the National Professional Qualification for Headteachers (NPQH), described below, only emerged after a period of about fifteen years during which there was experimentation with different forms of organisation, funding and provision for school management training in England and Wales. In 1981, a national survey concluded that the provision, organisation and funding of school management training across the country was patchy and needed to be rationalised. This led, from 1983–88, to the Government-funded National Development Centre (NDC) for School Management Training, based in a university. The NDC coordinated over 40 university-based regional centres, which offered 90 twenty-day basic courses (aimed at new headteachers) and one-term training the trainer courses (aimed at experienced headteachers). The NDC validated and evaluated these courses that were attended by over 6,000 headteachers and deputy headteachers. Since 1987, school management training has been a national priority area for national funding and, from 1989–92, the Government funded the School Management Task Force, centrally based in the Department for Education, to support the implementation of its national reform programme. It did so by promoting more practical forms of management training, controlled by schools. From 1991, a compulsory system of biennial appraisal for headteachers and deputy headteachers, designed in part to identify their individual training needs, was in operation and, in 1992, the Government funded national pilot schemes to trial mentoring for new headteachers and forms of competency-based training.

An important question for policy makers is: what kind of national organisational structure is appropriate? In the mid-1980s, the idea of a single national centre for training all headteachers in England and Wales was considered and rejected as being impractical. Several other types of organisational structure were also tried. Initially, the Department for Education, controlled directly by a junior government minister, funding all universities to run their own courses. Next, a university based national centre coordinating courses run by selected regionally-based colleges and universities and this was followed by a national Task Force based in the Department for Education. The system outlined below was, therefore, introduced after a long period of somewhat unsystematic experiment and development. The advent of the Internet was undoubtedly a significant new factor in making possible the particular form of the National College for School Leadership (see below).

In addition, it is also important to consider the nature and scope of the professional infrastructure in England and Wales that made possible and supported these developments. Thus, two headteachers unions (the National Association for Headteachers and the Secondary Heads Association) have, for many years, been active in promoting and delivering training and development programmes for headteachers. In the higher education sector, the Universities Council for the Education and Training of Teachers (UCET) has also been active in this field for many years while the British Educational Research Association (BERA) has promoted and disseminated relevant research *via* its annual conference and its journals. Similarly, the British Educational Management and Administration Society (BEMAS), an open professional society for policy-makers, headteachers, academics, trainers, consultants and researchers, has actively sought to influence government policy, in part by promoting networking and the dissemination of information and research via national and international conferences and published articles and books.

## Developments in Preparation, Certification and Induction

The National Headship Development Programme now consists of three stages or components: preparation, induction and in-service training. All three are now co-ordinated by the recently established National College for School Leadership (NCSL). The original model for the preparatory programme, the NPQH, is described below together with a new induction scheme for Welsh headteachers, The Professional Headship Induction Programme (PHIP), and introduced in 2001. In England, the induction phase was initially covered by the HEADLAMP Programme, currently under review. The latest version of the NPQH and the in-service component, the Leadership Programme for Serving Headteachers (LPSH) are described in Professor Tomlinson's chapter, which also includes a brief account of the NCSL.

### Preparation: The National Professional Qualification for Headteachers (NPQH)

The NPQH was introduced by the Teacher Training Agency (TTA) in 1995–96 as a national qualification for aspiring headteachers in England and Wales (TTA, 1998).

It was designed in the light of extensive consultation with the profession and employers and involved two components – assessment and training – each of which was conducted by independent regional centres. These centres were designated by the TTA, following a process of competitive national bidding by over eighty consortia of universities, local authorities and private companies. Most of the NPQH candidates were deputy headteachers who wished to become headteachers; candidates were funded nationally and applied *via* their LEA. To date it has been a voluntary scheme and, although the revised NPQH is planned to be mandatory by 2003, the difficulties in recruiting headteachers (see above) make this problematic.

The key principles underpinning the NPQH were said to be that it was rooted in school improvement, drew on best practice in education and industry and was based on agreed national standards. It was intended to signal the successful candidates readiness for headship but not to replace the selection process. It was also intended to be rigorous and flexible, to help aspiring headteachers to prepare for the role and to provide a baseline for their future development as leaders. Its core purpose was to enable headteachers to provide professional leadership for a school in order to secure its success, to ensure high quality education for all its pupils and to improve standards of learning and achievement. The National Standards for the NPQH were based on satisfactory performance on modules in five key task areas: strategic direction and development of the school; learning and teaching; people and relationships; the development and deployment of human and material resources; accountability for the efficiency and effectiveness of the school.

The programme itself consisted of six components or stages:

- an initial, diagnostic needs assessment at a Regional Assessment Centre;
- an individual action plan agreed with the Regional Assessment Centre;
- a compulsory training and development module run by a Regional Training and Development Centre;
- if necessary, in the light of the initial needs assessment, one or more of the three further modules run by recognised training agencies and/or via industrial attachments;
- assessment against all the national standards for each module;
- a final overall assessment at a Regional Assessment Centre.

The initial and final assessments were based on self-evaluation tests; psychometric tests; personal interviews; observed group discussion exercises; presentations; simulations; practical exercises in schools and/or industrial and commercial settings; a file of evidence including independent reports from colleagues and headteachers. All assessors and trainers had to be recognised by the TTA and were drawn from schools, local education authorities, universities, industry and commerce and independent management consultancies.

### The Professional Headship Induction Programme (PHIP) in Wales

The underlying rationale for the PHIP is the belief that newly appointed headteachers can have a substantial impact on raising standards and school improvement. It is offered as an entitlement, supporting newly appointed headteachers during their first two years of headship. It is designed to be flexible, allowing newly appointed headteachers to access a national programme but also providing each one with a framework to build their own development programme. It aims to develop the newly appointed headteachers as a professional group by establishing and nurturing a network of mutual support and professional links and to make a clear connection between their professional development and school development planning and improvement.

It has four components:

a) Professional Headship Profile to help new heads to reflect on their practice, identify needs and inform their professional development;
b) Peer Networking and Support system to provide structured opportunities to discuss issues and share ideas about successful practice and to reduce professional isolation. This involves 6 days over two years with each head hosting a meeting in their own school;
c) Mentoring Programme to provide newly appointed headteachers with support from experienced colleague heads, trained as mentors. This involves up to 6 days over two years;
d) National Training and Development Directory to inform newly appointed headteachers about training opportunities offered by registered providers.

In addition, it is planned that newly appointed headteachers will have access to the services of the National College, particularly those offered on-line via the Internet (www.ncslonline.gov.uk).

## Conclusion

This chapter, together with Professor Tomlinson's chapter, presents an outline account of the current state of *School Leader Preparation, Licensure/Certification, Selection, Evaluation and Professional Development* in England and Wales. This account may usefully be conceptualised as a description of a complex, evolving innovation consisting of several component parts. Anyone seeking to learn from this experience needs to bear in mind the context in which it was developed. Hence, the earlier parts of this chapter offer a brief historical analysis of the main features of the context that influenced the changing nature and scope of this complex innovation.

An account of this kind can only touch upon the many policy, research and theoretical issues inherent in the development of these programmes. For example, disappointingly, few evaluations of this programme have been carried out, and still fewer published. Moreover, comparisons with parallel schemes in other countries would obviously be instructive. These and other related issues are considered in Bush et al. (1999) and Brundrett (2000, 2001).

## References

Bolam, R. (1999). Educational administration, leadership and management: Towards a research agenda. In R. Bolam, G. Dunning and P. Karstanje (Eds). *New headteachers in the new Europe*. Munster/New York: Waxman Verlag.

Bolam, R., & van Wieringen, F. (Eds., 1999). *Research on educational management in Europe*. Munster/New York: Waxman.
Bolam, R., Dunning G., & Karstanje, P. (Eds., 2000). *New headteachers in the new Europe*. Munster/New York: Waxman Verlag.
Brundrett, M. (2000). *Beyond competence: The challenge for educational management*. Dereham, Norfolk: Peter Francis Publishers.
Brundrett, M. (2001). The development of school leadership preparation programmes in England and the USA: A comparative analysis. *Educational Management and Administration*, 29(2), 229–45.
Bush, T., Bell, L., Bolam, R., Glatter, R., & Ribbins, P. (Eds.; 1999). *Re-defining educational management*. London, Paul Chapman Publishing Ltd.
Campbell, R. J., & St. John, N. (1994). *Secondary teachers at work*. London: Routledge.
Dunning, G. (2000). New heads in Wales. In R. Bolam, G. Dunning and P. Karstanje. (Eds). *New headteachers in the new Europe*. Munster/New York: Waxman Verlag.
Hood, C. (1995). Contemporary public management: a new global paradigm? *Public Policy and Administration*, 10(2), 104–117.
School Teachers Review Body. (1996). *Fifth report*. London: HMSO.
Teacher Training Agency. (1998). *National standards for headteachers*. London: TTA.
Travers, C., & Cooper, C. (1996). *Teachers under pressure: Stress in the teaching profession*. London: Routledge.
Wallace, M., & Hall, V. (1994). *Inside the SMT: Teamwork in secondary school management*. London: Paul Chapman.
Wallace, M., & Huckman, L. (1999). *Senior management teams in primary schools: the quest for synergy*. London: Routledge.
Weindling, D. (1999). Stages of headship. In R. Bolam, G. Dunning and P. Karstanje. (Eds). *New Headteachers in the new Europe*. Munster/New York: Waxman Verlag.

## Selected Web Sites

Department of Education in Northern Ireland: www.deni.gov.uk
DfEE: www.dfee.gov.uk
General Teaching Council for England: http://www.gtc.org.uk
National College for School Leadership: http://www.ncslonline.gov.uk
National Assembly of Wales: http://www.assembly.wales.gov.uk
OFSTED: http://www.ofsted.gov.uk
Qualifications and Curriculum Authority: http://www.qca.org.uk
Queens Belfast: http://www.qub.ac.uk/edu/leadership
Scotland SQH: http://www.sqh.co.uk
Standards Issues and Resources for Teachers: http://www.standards.dfee.gov.uk
Teacher Training Agency: http://www.tta.gov.uk

# 4

# Reconceptualizing Administrative Preparation of Principals: Epistemological Issues and Perspectives

Dr. Joseph Meng-chun Chin

*Professor, Department of Educational Administration, Graduate School of Education, National Taiwan University, Taiwan, R.O.C.*

Questions of how to establish a suitable knowledge base and how to transmit the required knowledge to prepare future and incumbent principals have been of long-standing interest in educational administration. Given that principals play a critical role in educational development and innovation, comprehensive preparation programs for them both of pre-service and in-service are central to educational effectiveness and efficiency.

The development of administrative preparation programs for school administrators embodies a commitment to epistemological assumptions that entail with different way of knowing or forms of knowing. One of the ways in which epistemology affects preparation programs is by influencing views of the kinds of knowledge sought and the strategies for seeking it. From an epistemological point of view, this paper deals with the knowledge and content infused into administrative preparation programs of principals. It starts by examining the dominance of logical positivism in educational administration, which began to dominate in the mid-1950's. This is followed by exposition of criticisms of that dominance, mainly from post-positivism and post-modernism. Finally, the emergent trends and future development of the preparation programs for principals are analyzed and recommendations offered.

## Foundationalist Epistemologies and Logical Positivism

Administration of education as a field of study was oriented to quite practical approaches prior to advent of the Theory Movement in the 1950's. The knowledge base employed was comprised mainly of the experiences of former administrators, few of whom had interest or experience in academic research and theory building. Training consisted largely of prescriptions of what to do in specific administrative jobs. Subsequent critics asserted that preparation programs lacked theoretical foundations and 'relied too heavily upon naïve empiricism' (Halpin, 1957, p. 156).

With the support of Daniel Griffiths, a central scholar of the Theory Movement, concepts of Feigl (1953) were adopted as a rationale for creating a science of educational administration. Feigl was a former member of the Vienna Circle and had given the Circle's philosophy a specific label: *logical positivism.* Feigel believed that the quest for scientific knowledge is regulated by the specific ideals that include

- intersubjective testability,
- reliability,
- definiteness and precision,
- coherence,
- comprehensiveness. (Feigl, 1953, pp. 11–12)

Feigl's philosophical work met the urge among a new generation of American emerging educational administration scholars to contribute to the creation of a new science of administration. Griffiths later restated the underlying logic of this approach:

Inquiry in administration must come to be characterized by objectivity, reliability, operational definitions, coherence or systematic structure, and comprehensiveness. The content of administration is capable of being handed in a scientific manner even though it is not now being handed in that manner. (Griffiths, 1959, p. 45)

One can easily recognize the similarity between Griffith's five requirements for conducting scientific inquire in administration and Feigl's five features of science. Feigl's brand of logical empiricism became a critical foundation in the development of the Theory Movement in educational administration. In turn, it is not an exaggeration to say that this movement reshaped the mainstream of educational administration between the 1950's and 1980's. Indeed, the basic assumptions of the Theory Movement have become institutionalized in most of administrative preparation programs throughout the world.

Epistemologically and methodologically, logical empiricism set five assumptions that came to underlie the new science of educational administration. They included:

1. A theory is a hypothetico-deductive structure where less general and singular claims, including expected observations, are deduced from more general claims.
2. The methodology of natural sciences is appropriate for social sciences and the goal of researchers is to develop law-like explanations for human behavior.
3. Theories are justified by meeting certain conditions of empirical testability and true knowledge was obtained through empirical verification and analytic logic.
4. All the theoretical terms of theory must be able to give operational definitions.
5. Scientific theories of educational administration exclude substantive ethical claims since they cannot be verified empirically.

Logical positivism follows the philosophical traditions of foundationalist epistemologies since it insists that theoretical statements can be confirmed only if they logically entail propositions asserted in a theory-neutral language. Foundationalist epistemologies view knowledge as a representation of reality that is grounded. Theoretical arguments are constructed from propositions that serve as premises for conclusions.

Research in educational administration rooted in such views is most likely to emphasize 'law-like generalizations' because they can be empirically justified. Moreover, these generalizations are designed to predict the consequences of action. It is, however, interesting to note that the social science methods of this type of research seldom meet these requirements. Also noteworthy is that ethical and value issues – central to school administration – are mostly excluded from this type of research as they are difficult to study using conventional 'scientific' methods. In sum, from the standpoint of logical positivism, the study of educational administration should be oriented to the social disciplines and should seek empirical verification of laws of administrative and organizational behavior.

Under the dominant influence of the Theory Movement, the main trend of studies in the field came to be characterized by:

- empirical research;
- research that relied heavily on using quantitative approaches;
- the widespread use of survey instruments for data collection;
- statistical analysis to measure related variables. (Bridges, 1982; Everhart 1988)

After examining proposals submitted to the Administration Division of the American Educational Research Association in 1985, Griffiths (1988) found that only 3 of the 230 proposals employed qualitative research methods. He concluded that there was little diversity methodologically in the study of educational administration.

## Criticisms of Foundationalist Epistemologies

The logical positivist foundation of the Theory Movement in educational administration came in for criticism from many sides during the 1980's and 1990's. These included criticisms from alternative philosophical schools: subjectivism (Greenfield, 1975), paradigmatic approach (Kuhn, 1962), critical theory (Bates, 1983; Habermas, 1972), and postmodernism (Foucault, 1972; Lyotard, 1988). Many of these critiques also focused on the assumptions of foundationalist epistemologies more generally.

### Critiques from subjectivism

During the 1960's, when the Theory Movement was at its peak, scholars such as Quine (1960) and Feyerabend (1963) openly rejected the concept of theory-neutral research. These critics sought to demonstrate the complexity of relationships between empirical evidence and the theoretical language used to interpret findings. However, it was Greenfield (1975) who launched the most significant attack on the philosophical and methodological assumptions underlying the Theory Movement. Greenfield came to be identified as the major proponent of subjectivism in educational administration.

As Culbertson (1988) said, Greenfield was the first person to fire a lethal shot at the Theory Movement. He insisted that organizations are not things but human constructs. He rejected the testability assumption that claimed that researchers must test their theoretical view of world by collecting empirical data through operationally-defined procedures. Greenfield believed that truth is mind-dependent and can be perceived only subjectively. He declared:

> The process of truth making in the academic world... does not differ materially from what goes into truth making in the world at large. Truth is what scientists agree on or what the right scientists agree on. It is also what they can get others to believe in. (Greenfield, 1978, p. 8)

Greenfield (1975, 1986) further argued that research in educational administration should not exclude values, emotions and personal travails because these behaviors of administration were social phenomena. He strongly asserted that human action and organizational behavior are *subjectively created* rather than the reflection of natural laws. Since all observation is theory-laden and all theory underdetermined by available evidence, extra-empirical criteria are needed for justification. Simply stated, in the social sciences data that are inadequate to the task of answering the proposed research questions tends to force scholars to fall back on their own assumptions (and biases) when interpreting the results of their empirical research.

In addition, Greenfield contended that quantitative research is largely irrelevant to educational administration because it does not have the capability to address major themes in the field. For example, this type of research cannot explore themes such as values, will, and ethics. He treated the subject matter of educational administration as fundamentally *normative*, instead of *descriptive*. He, therefore, viewed qualitative approaches as more appropriate to understanding the complicated processes of social interactions in organizations.

**Critiques from values theory**

Like Greenfield, Hodgkinson (1978, 1983) also challenged the application of behavioral science assumptions in educational administration. In particular, he identified values as one of the fundamental domains of administrative operation. He claimed that, 'a value can exist only in the mind of the value-holder and it refers to some notion of the desirable, or preferred state of affairs, or to a condition which ought to be' (Hodgkinson, 1978, p. 105). In opposing positivism and its application to educational administration, Hodgkinson asserted that administration is essentially humanistic, not scientific.

Hodgkinson (1978) has for over 20 years worked on developing a hierarchical conception of values in administration. Three types of values are distinguished:

1. transrational values that are based on the will rather than the reasoning faculty;
2. rational values that are justified either by quantitative-based consensus in a collectivity or through a future set of consequences held to be desirable;
3. subrational values that are self-justifying in that they reflect an individual pattern of preferences.

In Hodkingson's theory, transrational values (based on will) are superior to rational values (based on reason) that are superior to subrational values (based on emotion). Ultimately, administrative behavior mainly involves the expression of transrational values. Moreover, he views these transrational values as unsuited to justification or verification by logical argument and empirical investigation.

### Critiques from paradigm theory

Another major objection against foundationalist epistemologies came from the analysis of theoretical paradigms. This drew largely on the work of Thomas Kuhn who employed what he called *paradigms* to describe the process of conceptual change in the physical sciences. According to Kuhn (1970), scientific revolutions are arbitrary. He likened paradigm shifts to gestalt shifts: 'the switch of gestalt… is a useful elementary prototype for what occurs in a full-scale paradigm shift' (p. 85).

This critique is significant because it calls into the question the validity of a given body of knowledge. That is, the body of knowledge built by any community of scientific scholars may achieve dominance, only to be replaced by a completely different body of knowledge resting on different assumptions. This critique highlighted the hubris of 'scientific scholarship' and the reoccurring tendency of scholars to assert their particular brand of knowledge.

Each scientific paradigm contains a separate but integrated set of beliefs and a distinctive vocabulary through which its claims are justified. Therefore, when paradigms conflict, there is no way to reach universal agreements on standards of interpretation since they are specified by the languages and logics of different paradigms. Moreover, they cannot be tested by confirming the truth of an empirical assertion affirmed by one and denied by the other.

Kuhn challenged the notion that scientific knowledge is uncontaminated by human subjectivity. He argued that paradigms, as different ways of knowing, are incommensurable since, 'when paradigms change, there are usually significant shifts in the criteria determining the legitimacy both of problems and of proposed solutions' (Kuhn, 1962, p. 109). As indicated, each paradigm constructs its own rival standards, but there are no omnipresent epistemic points from which different paradigms can be assessed.

### Critiques from critical theory

While Kuhn was trying to conceptualize explanations of theory change in the physical world, critical theorists like Bates (1983) were focusing on power relationships in the political nature of scientific knowledge. Influenced by the early works of Habermas, Bates and others asserted that a science of administration is essentially manipulative and represents a technology of control.

The critical theorists have questioned the appropriateness of the positivist perspective by arguing that regularities observed in the world are human artifacts. They deny their existence as a reflection of natural laws. Furthermore, in view of critical theorists, intellectual findings generated by the hypothetico-deductive model of positivism merely, 'provided the illusions necessary for the continued employment of techniques of hierarchical administrative control that perpetuate the injustices of an unequal society' (Bates, 1983, p. 30).

### Critiques from post-modernism

Other anti-foundationalist epistemologies including postmodernism are, in their more radical forms, characterized by anti-rational and anti-systems attitudes. They are best known for the destruction of language and meaning (Lyotard, 1984). They claim that different ways of knowledge are constantly developed and that researchers have no privileged position from which to judge them (Rorty, 1979).

Rather than focusing on a specific methodological paradigm for understanding phenomena, postmodernists suggest that there is no clear window into the inner life of subjects. This has opened up a debate of how practitioners will apply the results of the educational research. Eloquent but controversial, this strand of thought has not emerged to exert a major influence in educational administration (Willower, 1994).

**Summary Assessment of the Critiques of the Theory Movement**

These preceding critiques attack the assumptions of foundationalist epistemologies, including logical positivism. While inconsistent in the nature of their critiques, they often share the following beliefs:

1. There is no theory-neutral language whereby propositions require no further justification and no theory-free measurement procedures. Rationality is local and heavily dependent on human purposes and on substantive assumptions. There are no foundations, only ongoing arguments.
2. Justification is construed as epistemologically progressive learning, not just from its contiguity to a particular observation procedure. Inquiry into human behavior and organizations is a highly *social* activity. There is no such thing as evidence per se, only evidence relative to current concepts.
3. Rationality involves a dialectical relationship between theory and experience and cannot be separated from the language used for interpretation. Each language has its own rules and is adapted to its own form of meaning. Different languages may be incommensurable.

Methodologically, foundationalist epistemologies have provided the virtue of clarity and developed powerful means by which to discern between reason and unreason. The Theory Movement associated with logical positivism has exerted a puissant influence on the justification, development, and content of administrative theories in education. Although criticized as relativistic and tending towards unbounded skepticism, the non-foundationalist epistemologies have to some degree liberated scholarship from the control of 'scientific method' of logical positivism.

At the same time, while their influence on research and discourse in educational administration cannot be disregarded, the impact of these approaches on practice has been less dramatic. Arguably subjectivism and values theory have drawn the most attention from scholars involved in developing new approaches to program preparation. However, it is fair to say that, in practical terms, the summary result of the past 25 years of critiques of the Theory Movement and its influence on administrative preparation has been a mixed bag. No single philosophical approach has emerged as a dominant force in the field.

## The Evolution of Administrative Preparation Programs

Research findings have indicated that school principals play a critical role in the efficiency and effectiveness of schools. This has become the rationale for the importance given to the development of effective administrative preparation and in-service training. During the twentieth century, the scope and content of preparation programs in the United States varied considerably, reflecting the different developmental phases of educational administration as a field of practice and a field of study. Most of the Western world and much of Asia has followed the American approach to administrative preparation. Therefore, I will focus on trends in the development of the field in the USA. Murphy (1992) identified three major phases of administrative preparation in the USA: the prescriptive era (1900–1946), the behavioral science era (1947–85), and the dialectic era (1986–present).

**The prescriptive era**

The prescriptive era was highly influenced by the spread of scientific management and the human relations approach across the corporate landscapes (Cooper & Boyd, 1987). During the prescriptive period, the scope and content of preparation was highly practical. Theories, techniques, and operations from the business world were adopted in preparation programs. The knowledge base employed was prescriptive and mainly drawn from the personal experiences of executives, rules of thumb, and testimonials of estimable administrators (Murphy, 1992). Both professors and students in the preparation programs concentrated primarily on the technical and mechanical aspects of administration. They generally ignored theoretical and conceptual frameworks in their development of curricula (Campbell et al., 1987).

**The behavioral science era**

As noted earlier, the ascendance of the Theory Movement was marked by harsh critiques of the unabashed practical bent that characterized previous approaches to administrator preparation. Between the 1960's and 1980's, the Theory Movement held the upper hand. There was a resultant emphasis in preparation programs on teaching students to understand theories of administration, to interpret findings from empirical study of the field, and to generate empirical research.

The behavioral science era was characterized by the pursuit of theory-based knowledge, drawn mainly from the behavioral sciences into preparation programs. Strong dissatisfaction with the prescriptive knowledge base led scholars to attempt to build a knowledge base grounded in hypothetico-deductive theory. The perspective of logical positivism led scholars to focus on development of *value-free knowledge* for use in preparation programs for school administrators.

Consequently, the knowledge base underlying preparation programs changed substantially during this era. Three specific trends were evident. First, knowledge drawn from the behavior sciences received increasing accentuation in preparation programs. Similarly, the inquiry tools employed by students and instructors were borrowed frequently from social sciences disciplines (Miklos, 1983).

Second, interdisciplinary approaches became a la mode in university-based preparation programs. Students were required to take courses drawn from psychology, political science, law, sociology, economics, business, and related areas (Culbertson, 1964).

Third, new strategies and instructional approaches in preparation programs were introduced. This furnished a richer pedagogical environment. In particular, field training and case study strategies, not widely available during the prescriptive era, provided new modes of learning for the students.

Also, during the behavioral science era, the number of institutions offering preparation programs increased significantly. In the USA, for example, a master degree became required by most states for the position of principal. In the 1980's, most preparation programs were state-controlled, credit-driven, and certification-bound (Cooper & Boyd, 1988).

Faculty members in educational administration became more academically-oriented than in the past. They came to view themselves as generators of new knowledge to underlie the knowledge in the field. Consequently, they offered more theoretical and research-based courses aimed at analyzing and understanding organizational behavior.

**The Dialectic era in preparation**

The dialectic era, the third period of administrative preparation program, emerged in response to critiques of preparation programs based on the behavioral science model. In the mid-1980s, almost all aspects of these programs were under serious attack and scrutiny by reformists in educational administration. They began to ask questions about how leaders create meaning within organizations and insisted that the characteristics of leadership are difficult to uncover through quantitative inquiry.

Other critiques focused on program content that was often perceived as irrelevant and poorly connected to the tasks of school leadership. Faculty, largely trained as social science researchers, were criticized as being similarly disconnected from the life of schools and the practice of school administration and leadership.

Highly regarded foundations such as Carnegie, Danforth, Ford, and Kellogg subsequently supported a variety of initiatives in the field of leadership preparation. For example, the Danforth principalship program was offered by 22 selected universities from the year of 1987. Participants in the program were local teachers evaluated by their school districts as having the potential to be successful educational leaders. Different from traditional preparation programs, the Danforth program featured the use of student cohort groups, an emphasis on clinical experiences, collaboration with school districts, and a coordinated curriculum across courses (Milstein et al., 1993).

Other trends have emerged during this era. These included problem-based learning, simulations, and apprenticeship models (Bridges & Hallinger, 1995; Hallinger, 1999; Hallinger & McCary, 1991). While these models all differ in important respects they are similar in their effort to connect preparation more successfully to the practice of school leadership.

It is difficult to predict the future direction of preparation programs. Nonetheless, based on literature reviews and case studies, Murphy (1990, 1992, 1993) has asserted that preparation programs in the dialectic era may be moving toward a grounding of programs in more facilitative and democratic conceptions of a leadership, toward a greater emphasis on reality-based learning formats and materials, and toward a reconnection of the practice and academic arms of the profession.

## The Debate on Content and Methods

Depending on the needs and circumstances within and between countries, there is noteworthy diversity in the contents and methods of preparation programs. Nevertheless, a general core of knowledge and skills has been addressed by different formats and methods. It usually includes topics as the following:

1. The strategies of communication and decision-making in the school environment including team building and operation.
2. The legal aspects of school management including collective bargaining.
3. Critical management tasks like strategic planning, total quality management, and policy analysis.
4. School leadership both in instruction and administration.
5. Curriculum management and teaching methods.
6. Business management of financial and material resources.
7. The external relationship with parents, education authority, and special interest groups.
8. The strategies of school evaluation of effectiveness and efficiency.
9. The strategies of school innovation and development.

Although new approaches have been developed and implemented in preparation programs since the early 1980's, the content and delivery of successful preparation programs remain problematic. Dissatisfaction with conventional approaches, consisting of formal curriculum from university and external short courses, led to the criticism of being too theoretical and mechanistic (Barth, 1997). The recipe-style knowledge that was extracted from textbooks and professor's handouts does not reflect constructively on the reality of the school administration (Bridges, 1977; Bridges & Hallinger, 1995, 1997; Schon, 1987). There is a widespread skepticism that these types of preparation programs can properly equip to meet the challenges in school settings.

These uncertainties are rooted in the problems inherent in a knowledge base generated largely through behavioral science methods based in logical positivism. Positivism remains a potent, albeit declining, force in the shaping of administrative theory (Foster, 1994). Over the past decade, however, there has been increasing disillusionment with the usefulness of the positivist approach for creating knowledge perceived as useful for preparation programs. In the words of Murphy, the lack of relevance of the positivist knowledge base reflects 'the concomitant failure to stress inductive approaches and to use qualitative lenses to examine organizational phenomena' (1990, p. 284).

The courses taught in most preparation programs are provided by discipline-based scholars brought up under the tenets of the Theory Movement. They are frequently criticized by school practitioners for not making knowledge relevant to the reality of principal leadership (Barth, 1997; Bridges, 1977; Bridges & Hallinger, 1995). As Cambron-McCabe (1993) has indicated, the positivist paradigm diverted attention away from moral issues related to the purposes and values of education. Accordingly, reconstructed preparation programs are needed to provide more attention to cultural diversity and social activism (Maniloff & Clark, 1993).

An emphasis on shifting the focus from school manager to educational leader is particularly noteworthy across preparation programs that have radically redesigned the contents of curriculum. Grounded in efficiency, scientific predictability and certainty, the managerial approach has trained the school administrators 'not to challenge the status quo, but to maintain it, not to reconceptualize schools but to reproduce them' (Cambron-McCabe et al., 1991, p. 4).

For the past two decades, school-based management and the decentralization of authority to school sites have provided an impetus for moving the role of administrator from manger to leader. In contrast with management training, which is still prevalent in many institutions, the main purpose of reconstructed leadership preparation is to develop school administrators as group motivators who can initiate reforms and improvements in the school sites. According to the Crews and Weakley (1995), it emphasizes 'decision making, problem solving, team building, goal setting, encouraging innovation, self-assessment, delegating, and conflict resolution' (p. 7).

Consistent with the new curricular reforms of preparation programs, identification of the core knowledge base is considered a prerequisite to transforming the leadership curriculum. In the 1990's, leading American institutions have moved the focus of the administrative preparation curriculum from foundationalist to non-foundationalist perspectives (Achilles, 1994; Milstein et al., 1993). The adult learners are urged to actively construct knowledge based on individual school needs. The teacher is no longer viewed as the person to dispense knowledge to students (Wilson, 1993). Instead, faculty members in most of the redesigned preparation programs have worked jointly with students to develop and identify coherent curriculum (Daresh, 1994).

In the learning process, problem-based instruction strategies and collaborative learning activities such as cohorts groups provide opportunities for students to assess the complexities of the school reality (Bridges & Hallinger, 1995, 1997; Hallinger & McCary, 1991). Norris and Barnett (1994) found that cohort groups help students to understand the benefits of dynamic instruction in a group of learners while fostering a sense of community among students and faculty.

Bridges and Hallinger (1995) noted that problem-based approaches to leadership development start with the problems of schools as the focus of the curriculum. Scholars then seek to connect theory, empirically-derived knowledge, and the practical wisdom of schools to the understanding and solution of these problems. The theoretical assumptions behind these approaches are grounded in cognitive psychology. The notion is that when learners seek knowledge to address real problems that they perceive as meaningful, they will be more likely to understand, retain, and spontaneously access the knowledge when confronted by similar problems on the job (Bridges & Hallinger, 1993, 1995, 1997). Research on the impact of such programs is in its early stages. However, results from outside of education (Bridges & Hallinger, 1993) and from educational administration (Bridges & Hallinger, 1995, 1997; Copland, 2000) suggest promise for this line of inquiry and preparation.

Cambron-McCabe (1993) indicated that these redesigned programs have put more attention on social and cultural context of schools. There is an increasing emphasis on embedding multiple perspectives towards understanding educational issues. Traditionally excluded perspectives such as ethical issues are being included in preparation programs reporting curricular redesign.

The National Council of Professors of Educational Administration (NCPEA, 1989) has recommended that the curriculum of preparation programs include topics:

- societal and cultural influences on schooling,
- teaching and learning processes and school improvement,
- organizational theory,
- methodologies of organizational studies and policy analysis,

- leadership and management processes and functions,
- policy studies and politics of education, and
- the moral and ethical dimensions of schooling.

A survey conducted in 1994 reported that the expansion of ethics instruction was a major change in preparation programs while receiving strong support from faculty and school heads (McCarthy & Kuh, 1997).

Although there have been considerable calls for redesigning preparation programs, institutions involved in substantial programmatic changes are still viewed as outliers (McCarthy, 1999). Over the last decade, more rhetoric has been generated than actual changes in practice. The continuity of traditional curriculum offerings can be partly explained by the nature of the faculty members.

Boyan (1981) has indicated that the faculty members in the preparation programs during the 1970's and early 1980's became increasing specialized and identified with subfields, such as leadership, law, organizational theory, principalship, economics and finance, and supervision of instruction. Basically, faculty only can offer what they know; thus, faculty may resist redesigned curriculum based on new paradigms simply because they do not know how to teach.

On the whole, conventional preparation programs are too theory-oriented to reflect the school reality and provide explicit practical help to people preparing for administrative roles in changing schools. Administrative theories have their place in shaping preparation programs; however, they should not be viewed as a knowledge base that is likely to guarantee the high quality of the school leaders and administrators.

Multifaceted and alternative models of preparation curriculum are needed to emulate the social context of schools. An integrated curriculum that reflects workplace problems and promotes the development of problem-solving skills is the prerequisite of a successful preparation program. Additionally, it is argued that external forces, predominantly government-led educational reforms, have long determined the aims and contents of preparation courses. It is necessary to make a shift to self-development approach of learning that put its stress on amalgamation of theoretical knowledge and practical experiences (Barth, 1997).

## Emerging Trends: Non-foundationalist Perspectives

This chapter has noted that foundationalist knowledge, generated largely as a result of the Theory Movement, continues to influence the mainstream curriculum in programs of administrative preparation in the USA and elsewhere around the world. Yet, non-foundationalist perspectives suggest a number of implications for the preparation programs of school administrators as well as educational leadership.

First, non-foundationalist theorists argue that rationality is heavily dependent on human purposes and on substantive assumption. Hence, there is no 'true' value-free knowledge to manage an organization. Positivistic approaches are sentenced to failure not only because of the lack of predictability in human behaviors, but also because they neglect the value and ethical dimensions of educational administration. To understand an organization, then, requires comprehension of the social history and context that individuals bring with them and how they interpret their worlds.

It should be kept in mind that organizations are not stagnant bureaucracies, but dynamic and interactive constructs of individuals and social settings. Everyone has values and beliefs that influence his or her behaviors and performances profoundly. What are the beliefs and values that are held? How do they impact upon the everyday practice of administrators? How are school leaders shaped by their background variables such as gender, ethnicity, culture, and educational experiences? In what way do they interact with the society?

Instead of trying to find value-neutral laws and general maxims of administrative behavior, preparation programs should focus on the less predictable but more realistic value conflicts that attend the real world of schooling. There needs to be more emphasis on helping prospective school leaders understand the social context of schooling and how school leadership and administration can adapt to meet local circumstances.

Second, there is a movement to redefine the very concept of educational leadership (Leithwood, 1994). Traditional preparation programs tend to assume a managerial role for school administrators. They are often viewed as order-takers responsible for coordination and control of bureaucratic operations of school institutions. The non-foundationalist approach emphasizes the participatory involvement of all members in the organization. Proponents suggest that leadership must involve all parties of the institution, particularly those conventionally disfranchised by the system. Burns (1979), for example, indicated that leadership requires voluntary cooperation of members of the organization in a mission recognized mutually by the leaders and followers. There will be no real change in the institution unless the consciousness of followers is elevated with the intent of consummating genuine innovations (Leithwood, 1994).

The perspective addressed above suggests that it becomes less important to promote strategies and techniques for controlling behaviors of subordinates in the content of the preparation programs. School administrators, especially principals, should prepare to focus on the processes of the participatory involvement of all parties in the schools rather than desperately searching for order in the value-laden organization. Leadership is not just power from an administrative position but the promotion of the voluntary cooperation of all the members. Successful preparation programs require an understanding of how to elevate the motivation and consciousness of all parties as well as to achieve the organizational goals identified mutually by followers and leaders.

Third, acceptance of the belief that rationality involves a dialectical relationship between theory and experience leads to emphasis on connection between theory and practice (Bridges & Hallinger, 1995; Hallinger, 1992, 1999). There is a need to continue conversations about philosophical foundations, the meaning of leadership, and empirical methods in academic and practitioner forums. Regarding this connection, notable new approaches to preparation programs such as coaching, peer-assisted learning, and problem-based training deserve more attentions (Hallinger 1992, 1999).

Generally speaking, although administrative theories have provided basic understanding, they contribute little practical help either for school leaders or their trainers (Bridges, 1977). Effective educational leaders are those who are able to resonate constructively on their practical experience. Therefore, preparation programs ought to help them to engage in reflection on their practice (Barth, 1997).

While theory-based training should not be ignored completely, skill-based and values-based approaches are worthy of implementation as well. Significant new training strategies such as portfolios and cohort grouping learning were implemented successfully during the 1990's. These approaches have begun to reconceptualize the delivery of preparation programs in their quest to maintain a continuous and meaningful focus on the connection of theory and practice.

The fourth and final implication is that field-based and self-development approaches to learning should be adopted in the preparation programs. These give emphasis not simply to the development of individual leaders but also to the overall management of the school (Barth, 1997). The intent is to balance the needs of individual administrators with those of the organization.

A professional administrator should be prepared and educated within a self-development environment, rather than simply trained to follow centrally determined polices (Barth, 1997). The effective performance of schools fundamentally relies on the quality of its administrators and other members. A congruous self-development strategy for staff and administrative development is a prerequisite for ensuring that these professional staffs are of high quality.

The preparation programs should, therefore, be designed and implemented in order to fulfill the needs of both the leaders and followers according to their demographic status, school type, job stage, and the contents of the job. Instead of being taught mainly through conventional university-based lectures and externally controlled workshops, training courses can be designed and provided by the school itself. The approach of the workplace as the main setting for learning allows the practitioners to engage in systematic inquiry into their own works and helps each other to tackle their problems by peer-assisted learning and cooperation in the school setting.

## References


Achilles, C. M. (1984). Searching for the Golden Fleece: The epic struggle continues. *Educational Administration Quarterly*, 30, 6–26.

Barth, R. (1997). *The principal learner: A work in progress*. The International Network of Principals' Centers, Harvard Graduate School of Education, Cambridge, MA.

Bates, R. J. (1983). *Educational administration and the management of knowledge*. Geelong: Deakin University Press.

Boyan, N. (1981). Follow the leader: Commentary on research in educational administration. *Educational Researcher*, 10(2), 6–13.

Bridges, E. (1982). Research on the school administrator: The state-of-the-art, 1967–1980. *Educational Administration Quarterly*, 18(3), 12–33.

Bridges, E. (1977). The nature of leadership. In L. Cunningham, W. Hack, & R. Nystrand (Eds.), *Educational administration: The developing decades* (202–230). Berkeley, CA: McCutchan.

Bridges, E. & Hallinger, P. (1993). Problem-based learning in medical and managerial education. In P. Hallinger, K. Leithwood, & J. Murphy (Eds.), *Cognitive perspectives on educational leadership*. New York: Teachers College Press.

Bridges, E. & Hallinger, P. (1995). *Implementing problem-based learning in leadership development*. Eugene, OR: ERIC Clearinghouse.

Bridges, E. & Hallinger, P. (1997). Using problem-based learning to prepare educational leaders. *Peabody Journal of Education*, 72(2), 131–146.

Burns, J. M. (1979). *Leadership*. New York: Harper and Row.

Cambron-McCabe, N. (1993). Leadership for democratic authority. In J. Murphy (Ed.), *Preparation tomorrow's school leaders: Alternative designs*. University Park, PA: University Council for Educational Administration.

Cambron-McCabe, N., Mulkeen, T. & Wright, G. (1991). *A new platform for preparation school administrators*. St. Louis: The Danforth Foundation.
Campbell, R. F., Fleming, T., Newell, L. & Bennion, J. W. (1987). *A history of thought and practice in educational administration*. New York: Teachers College Press.
Cooper, B. S. & Boyd, W. L. (1987). The evolution of training for school administration. In J. Murphy & P. Hallinger (Eds.), *Approaches to administrative training in education*. Albany, New York: State University of New York Press.
Copland, M. A. (2000). Problem-based learning and prospective principals' problem-framing ability. *Educational Administration Quarterly*, 36(4), 584–606.
Crews, A. C. & Weakley, S. (1995). *Hungry for leadership: Educational leadership programs in the Southern Regional Education Board (SERB) states*. Atlanta: SERB.
Culbertson, J. A. (1988). A century's quest for a knowledge base. In N. J. Boyan (Ed.), *Handbook of research on educational administration*. New York: Longman.
Daresh, J. C. (1994). Restructuring educational leadership preparation: Identifying needed conditions. *Journal of School Leadership*, 4, 28–38.
Everhart, R. B. (1988). Field work methodology in educational administration. In N. J. Boyan (Ed.), *Handbook of research on educational administration: A project of the American Educational Research Association*. New York: Longman.
Feigl, H. (1953). The scientific outlook: Naturalism and humanism, In H. Feigl & M. Brodbeck (Eds.), *Readings in the philosophy of science*. New York: Appleton-Century-Crofts.
Feyerabend, P. K. (1963). How to be a good empiricist. In B. Baumrin (Ed.), *Philosophy of science: The Delaware seminar*. New York: Interscience.
Foster, W. (1986). *Paradigms and promises: New approaches to educational administration*. Buffalo, New York: Prometheus.
Greenfield, T. B. (1975). Theory about organization theory: A new perspective for schools, In M. G. Hughes (Ed.), *Administering Education: International challenge*. London: Athlone Press.
Greenfield, T. B. (1980). The man who comes back through the door in the wall: Discovering truth, discovering self, discovering organization. *Educational Administration Quarterly*, 16(3), 26–59.
Greenfield, T. B. (1986). The decline and fall of science in educational administration. *Interchange*, 17(2), 57–80.
Griffiths, D. (1959). *Administrative theory*. New York: Appleton-Century-Crofts.
Hallinger, P. (1992). School leadership development: Evaluating a decade of reform. *Education and Urban Society*, 24(3), 300–316.
Hallinger, P. (1999). School leadership development: State of the art at the turn of the century. *Orbit*, 30(1), 46–48.
Hallinger, P. & Heck, R. (1997). Exploring the principal's contribution to school effectiveness. *School Effectiveness and School Improvement*, 8(4), 1–35.
Hallinger, P. & McCary, C. E. (1991). Using a problem-based approach for instructional leadership development. *Journal of Staff Development*, 12(2), 6–12.
Halpin, A. W. (1957). A paradigm for research on administrator training. In R. F. Campbell & R. T. Gregg (Eds.), *Administrative behavior in education*. New York: Harper & Row.
Habermas, J. (1972). *Knowledge and human interests*. London: Heinemann.
Hodgkinson, C. (1978). *Toward a philosophy of administration*. Oxford: Blackwell.
Hodgkinson, C. (1983). *The philosophy of leadership*. Oxford: Blackwell.
Kuhn, T. S. (1970). *The structure of scientific resolutions* (2nd Ed.). Chicago, Illinois: University of Chicago Press.

Leithwood, K. (1994). Leadership for school restructuring. *Educational Administration Quarterly*, 30(4), 498–518.

Lyotard, J. F. (1984). *The postmodern condition: A report on knowledge*. Minneapolis, Minnesota: University of Minnesota Press.

McCarthy, M. M., Kuh, G. D., Newell, L. J. & Iscona, C. M. (1988). *Under scrutiny: The educational administration professorate*. Temple, Arizona: University Council for Educational Administration.

McCarthy, M. & Kuh, G. (1997). *Continuity and change: The educational leadership professorate*. Columbia, MO: University Council for Educational Administration.

Maniloff, H. & Clark, D. (1993). Preparing effective leaders for schools and school systems: Graduate study at the University of North Carolina-Chapel Hill. In J. Murphy (Ed.), *Preparing tomorrow's school leaders: Alternative designs*. University Park: University Council for Educational Administration.

Miklos, E. (1983). Evolution in administrator preparation programs. *Educational Administration Quarterly*, 19(3), 153–177.

Milstein, M. & Associates (1993). *Changing the way we prepare educational leaders: The Danforth experience*. Newbury Park, CA: Corwin.

Murphy, J. (1990). The reform of school administration: Pressures and calls for change. In J. Murphy (Ed.), *The reform of American public education in the 1980s: Perspectives and cases*. Berkeley, California: McCutchan.

Murphy, J. (1992). *The landscape of leadership preparation: Reforming the education of school administration*. Beverley Hills, California: Corwin.

National Policy Board for Educational Administration (NPBEA) (1989). *Improving the preparation of school administration: An agenda for reform*. Charlottesville, VA: NPBEA.

Norris, C. J. & Barnett, B. (1994). *Cultivating a new leadership paradigm: From cohorts to communities*. Paper presented at the annual convention of the University Council for Educational Administration, Philadelphia.

Quine, W. V. (1960). *Word and object*. Cambridge, Mass: M.I.T. Press.

Rorty, R. (1979). *Philosophy and the mirror of nature*. Princeton, New Jersey: Princeton University Press.

Schon, D. A. (1987). *Educating the reflective practitioner: Towards a new design for teaching and learning in professions*. San Francisco: Jossey-Bass.

Willower, D. J. (1994). Values, valuation, and explanation in school organizations. *Journal of School Leadership*, 4(5), 466–485.

Wilson, P. (1993). Pushing the edge. In M. Milstein & Associates (Eds.), *Changing the way we prepare educational leaders*. Newbury Park, CA: Corwin.

# 5

# Developing Standards for School Leadership Development: A Process and Rationale

Dr. Joseph Murphy

*Professor, Department of Leadership and Organizations, Peabody College, Vanderbilt University, Nashville, TN, USA*

Dr. Neil J. Shipman

*Director, Inter-state Consortium for School Leadership, USA*

Reform efforts for educational leadership in the United States since the mid-1990's have coalesced around the work of the Interstate School Leaders Licensure Consortium (ISLLC). Since the inception of ISLLC in 1994, new points of view are redefining the structure of school leadership for the twenty-first century. Initiatives from many school leadership interest groups – policy makers, professional associations, universities, and foundations – are giving a new look and purpose to school leadership. There are many who believe that if attention is not given to leadership in the schools, no reform will really occur. All but one of the fifty states in the United States require some type of license to practice educational administration. However, until ISLLC, no common standards were available to use as criteria for licensure. In this chapter the most significant of school leadership reforms undertaken by the ICLLS will be discussed – establishing *Standards for School Leaders*, and then providing tools and momentum for the Standards to become the focal point for school leadership issues in the United States.

In the remaining parts of this chapter, we will describe the central elements of the ISLLC design. We will start with a picture of the infrastructure that supports the ISLLC work. An analysis of both the changing environment in which schooling occurs and the evolving postindustrial concepts of education follows. We then unpack the concepts of leadership that rest on these foundations. We show how these concepts of leadership were then translated into a set of standards for strengthening school leadership throughout the nation's educational systems. Finally, we provide examples of how the Standards are being used as strategies for creating more effective school leaders.

## Foundations Underlying the ISLLC Standards for School Leaders

In 1994, the ISLLC Consortium of 23 states and 12 major professional associations with interest in the area of school administration began discussion of the design issue by building ideological scaffolding for their conception of school leadership. The pillars of this infrastructure took form from two bodies of information: knowledge about appropriate models of schooling and the changing environment in which schooling takes place (see Murphy, 1992).

The ten national educational leadership associations that comprised the National Policy Board for Educational Administration (NPBEA) shared the belief that common standards for principal performance were an essential component of professionalism. However, no such set of standards existed. This resulted in an invitational national forum held in Reston, Virginia in January 1994. Eighty-one representatives from 37 state education departments and licensing boards and 10 national associations participated.

At that meeting, participants generally agreed on the need for common leadership standards. They identified the ideal or key characteristics of licensure standards for principals and developed an initial work plan calling for subsequent meetings to begin drafting standards. Several goals were established:

- improving student performance,
- reforming and restructuring education,
- enhancing the profession,
- enhancing principal performance, and
- providing accountability.

The time was ripe for the educational leadership community to begin looking at standards for school leaders as another tool for school improvement.

Approximately 70 representatives from a variety of state agencies and national associations attended a second meeting sponsored by NPBEA and the Council of Chief State School Officers (CCSSO) in August 1994. The purpose of this meeting was to discuss the need for model standards for school principals. The participants focused on developing a mission and work plan for the new consortium. They also examined the changing context of education, including important changes in the institutional, managerial, and technical core of schooling, and the evolving role of leadership for tomorrow's schools.

### Evolving conceptualization of the educational environment

The forces propelling the reform of schooling and the redefinition of school leadership are primarily located in the educational environment – in the political, economic, and social dynamics in which the educational system in the United States is enmeshed. These dynamics gave rise to the social and technical infrastructure that supports ISLLC's understanding of leadership for schools of the 21st century. Basic changes in these dynamics necessitate new views of the role of school leaders.

The new school leader will have a great deal to say about how schools respond to the changing political, economic, and social dynamics that characterize the nation as we end one century and begin another. Leaders of schools and school systems will also be influential in creating 'communities' where hierarchy is now the norm and to reconstruct the core technology of schooling so that the promise of success for all students is a reality (see Murphy, 1992).

*Political forces*
New ideas in the political environment are shaping understandings of school leadership and education. One key element is the notion of locus of control. Once referred to as *democratic localism*, it is now known as *localization* or, more simply, *decentralization*. Whatever it is called, it represents a backlash against centralized, bureaucratic forms of education organization.

A second political force is the recasting of democracy, replacing representative governance with more populist conceptions. A third political shift favors greater control in governance by lay citizens and less power from the state and local boards of education. The emphasis has turned to the power of parents by recognizing their rights to be directly involved in the schooling of their children. This all leads to a gradual decline of control by professional managers and teacher unions.

*Economic forces*
The focus of the economic environment for schooling is now on having graduates prepared for the global economy in the new millennium. One side of the problem is the belief that systems that simply maintain the status quo in today's world are in reality in decline. Where others see stability, these critics see 'increasing obsolescence of the education provided by most U.S. schools' (Murnane & Levy, 1996, p. 6). The other side of this productivity issue is that because of the changing nature of the economy, the level of outcomes students need to reach need to be increased significantly (Murnane & Levy, 1996): 'Critics find that schools are not meeting this new criteria for productivity: "American schools are not providing students with the learning that they will need to function effectively in the 21st Century"' (Consortium on Productivity in the Schools, 1995, p. 3).

*Social forces*
The Consortium's understanding of schooling in the new century and the characteristics of the leaders needed for those schools was also formed by the need to address the changing social dynamics in America and to repair an ever-widening tear in the social fabric of the nation. The first issue is a concern with the demographic shifts that threaten 'our national standard of living and democratic foundations' (Carnegie Council on Adolescent Development, 1989, p. 27).These changes threaten to overwhelm the capacity of schools – as they are presently designed – to meet societal expectations.

The proportion of less advantaged young people and minority enrollment in schools in the U.S. are both on the rise. The number of students whose primary language is not English is also rising rapidly. There is no longer a majority of families representing the traditional family structure of one parent working and one parent staying home to care for the children. Daily news reports of dysfunctional families and young people are rampant. Teen pregnancies, violent behavior, drug abuse, and unemployment are but a few examples of decreasing social stock and a change in societal values.

## The changing nature of education

*Technical core*
There are indications that a greater understanding of the education production function is beginning to be seen in new ways of thinking about teaching and learning.

Behavioral psychology – the strongest theoretical and disciplinary influence on education – is being pushed off center stage by constructivist psychology and newer sociological perspectives about learning.

Central to this new vision of schooling are radical changes in assumptions about intelligence and knowledge. The new educational design considers 'knowledge not as somehow in the possession of the teacher, waiting to be transmitted to the student or to be used to treat the students' problems, but as mutually constructed by teacher and student' (Petri, 1990, p. 17–18).

New views about what is worth learning are also emerging. An emphasis on *learning to learn* and on the *ability to use knowledge* is replacing the traditional focus on acquisition and retention of information. New perspectives on the context of learning are being developed, directing attention to active participation by the learner. The desire for independent work and competition during the 20th century is taking a backseat to cooperative learning. Reformers in the forefront of redesigning schooling are also taking on ineffective practices such as tracking, an emphasis on covering the content, rote learning of basic skills, and a textbook orientation.

At the same time, in these newly structured schools, traditional teacher-centered classrooms are being replaced with learner-centered teaching. Teachers are no longer the *sage on the stage* where they are seen as content specialists who transmit their knowledge to students through lectures. Instead they become the *guides on the side* and act as facilitators for student learning. In the 21st century school, students will be 'producers of knowledge' and teachers 'managers of learning experiences' (Hawley, 1989, p. 23). The focus will be on learning, not on the delivery system.

*Managerial level*

There is increasing concern that the administrative structure for America's schools is failing. There is a feeling that school systems during the last century produced 'bureaucratic arteriosclerosis, insulation from parents and patrons, and the low productivity of a declining industry' (Tyack, 1993, p. 3). Reformers of this century are concluding that the bureaucratic system of administration is not capable of addressing the problems of the schools in the United States.

The hierarchical bureaucracies that have administered schools for the past century are being replaced by decentralized structures that result in shifts in roles, relationships, and responsibilities. School leadership is connected to competence for specific tasks rather than to formal positions. More focus is being placed on the human element as indicated by isolation being replaced by cooperative teams. Principals are no longer viewed as managers but rather as facilitators; teachers not as workers but as leaders. Thus there is a reorientation from control to empowerment.

*Institutional level*

Researchers argue that the public monopoly approach to public education has resulted in 'the belief in almost complete separation of schools from the community and, in turn, discouragement of local community involvement in decision making' (Burke, 1992, p. 33). Indeed, there is a considerable body of research that supports the notion that a major function of school administrators is to buffer the schools from parents and other community members (Meyer & Rowan, 1975).

There is an emerging vision of school improvement designs being driven by economic and political forces that substantially increase the salience of the market. In this vision the traditional dominant relationship of professional educators in control and parents as cheerleaders, agitators, or simply passive spectators is replaced by new paradigms that favor substantive involvement of the consumers. Four elements of this new picture of transformed management of schools are prevalent:

- choice in selection of a school,
- a strong voice in school governance,
- parents as partners in the education of their children, and
- enhanced membership of parents in the school community.

## The Changing Nature of Leadership

The discussion thus far about the fundamental shifts in the context of education and changing dynamics of schooling necessitates new models of school leadership also. In this section of this chapter we describe school leadership for the 21st century portraying fundamental changes from what leadership is today to what leadership needs to be in tomorrow's schools (see Murphy, 1992).

### Leader as community servant

The greatest challenge for school leaders of tomorrow's schools is to lead the transition from the bureaucratic model discussed above, to a post-industrial adaptive model with the goal of success for all students in the educative process. This means a reorientation towards leadership rather than simply management. Even competent managers will be unlikely to meet the challenges of leading schools in this new age. School leaders need to change from risk-evaders to risk takers, from a focus on educational and managerial process to a concern for outcomes, and from implementers to initiators. Moreover, they will need to do so within acceptable beliefs of the new heterarchical systems they are seeking to create. They must learn to lead not from the apex but from the nexus and with people rather than through them. Their influence must be based on professional expertise and dispositions rather than primarily from position authority. They will need to learn to lead through empowerment rather than through control.

Servant leadership will differ from the traditional view of management in additional ways. Symbolic, spiritual, and cultural leadership are key leadership forces. This style of leading has as much heart as it does head. It is grounded more on teaching than on telling, more on learning than on knowing, and more on modeling and clarifying values than on giving directions to subordinates. At the heart of servant leadership is relationships built on trust. The servant leader is reflective and self-critical.

### Leader as organizational architect

From our discussion above about organization and management for tomorrow's schools, it is clear that the organization and governance of the present school systems must be systemically redesigned. The school leaders of tomorrow will be faced with the difficult challenge of helping to define and give birth to these new forms of organization and governance.

As the architects of these new organizations, school leaders must have a focus on change rather than the traditional focus on stability. They will act as change agents rather than managers. They will need to disavow bureaucratic tenets of controlling, directing, supervising, and evaluating, and implement principles associated with heterarchies such as cooperation, empowerment, community, and participation.

The greatest challenge will be to design blueprints based on the best available knowledge of how children learn. We know that 'the organization of schooling appears to proceed as if we had no relevant knowledge regarding the development of children and youth' (Goodlad, 1984, p. 323). Thus, the 'main challenge facing educational leaders is ... to reconstruct conceptions of authority, status, and school structure to make them more instrumental to our most powerful conception of teaching and learning' (Elmore, 1990, p. 63).

**Leader as social architect**

Future school leaders need to address the changing social context of education that we addressed above. The social fabric is changing. Persons of color and persons from linguistically different cultures are increasingly populating society. Citizens are aging yet becoming more mobile. Income is being distributed less equitably. And the traditional concept of family is no longer the norm thus youngsters may come to school with less support. School leaders will have a significant role in determining whether schools are successfully reshaped to address these new social contexts.

The role of school leader as architect is clearly to design more responsive schools. 'They must invent and implement ways to make schools into living places that fit children rather than continuing to operate schools for "good kids" who adapt to the living structure' (Clark, 1990, p. 26). In addition, because schools are increasingly responsible for social support needed by students, school leaders in their role as social architects must see schooling as one part of the larger picture in dealing with problems faced by students. Besides being busy redesigning the purposes and organization of their own institutions to better serve the changing student population, they will also need to develop integrated networks of services with other agencies.

**Leader as moral educator**

Throughout most of the 20th century the field of educational administration has gravitated toward scientific images of business management. This is a limited concept, and fortunately, there is recognition that the pendulum must swing back to school administrators being called upon, first and foremost, to be educational leaders.

As moral educators, school leaders need to be much more heavily involved in setting purposes than simply in managing what is already there. Moral leadership means that school leaders must engage staff, students, and community in reinterpreting and placing new priorities on guiding values for education and in redesigning the organization to match the desired goals and values of the school community.

School leaders of tomorrow must provide students with a more complex and demanding educational experience. These leaders must reach many students who have not met with success in the present system. The new school leader will need to be much more knowledgeable about the core technology of education. They will need to truly put instruction and curricular leadership at the forefront, and, as leaders, keep the school's focus constantly on teaching and learning. In a dramatic shift from established practice, school leaders will need to exercise intellectual leadership not as head teachers but rather as head learners.

Central to school leadership seen as moral educational leadership is the notion that activities of leaders are deeply intermeshed with ethical issues. This will necessitate changing schooling to be responsive to the needs of historically disenfranchised and undereducated children rather than attempting to mold them to fit already dysfunctional schools. Indeed, it calls for school leaders to ensure success for all students, and to nurture the development of learning, professional, and caring communities through reflective inquiry and democratic participation.

Having provided a description and analysis of the principles and core beliefs that provide the architecture for the ISLLC Standards, let us now move to the development and use of the actual standards.

## The ISLLC Standards

During the initial phases of the work of the Consortium, it was recognized by the membership that standards provided an appropriate and powerful leverage point for systemic reform. Standards could help to drive improvement efforts for licensure, preparation program approval, accreditation, and assessment. At that time there simply was no set of common standards for school leaders that cut across all levels of schools throughout the United States. Let us move now to a discussion of major decisions and procedures from small and large group developmental meetings of ISLLC.

### ISLLC meetings to develop national standards

*August 19–20, 1994 meeting of the ISLLC*

Three major questions were addressed at this meeting.

1. What are the pressures to reform the principalship?
2. What characteristics are desired for principals of tomorrow's schools?
3. What are the leverage points for change, especially those related to licensure?

Three key decisions were made.

1. All participating states would continue to pursue common standards through the Consortium.
2. Several source documents would inform the development of the standards initially.
3. School district leadership roles as would as the principalship would be addressed.

*March 27–28, 1995 meeting of the ISLLC*

Discussions among the consortium membership on these days focused on building a draft of principles on which all the standards work would stand as well as an early draft of generalized standards for principals. The participants were invited to answer two questions individually and then in small groups.

1. What type of leaders do we need for tomorrow's schools?
2. What guiding set of standards should shape licensure?

Following this development of their own visions of standards, the group reviewed standards development work that had been completed by several national organizations and commissions, examined research on effective school leadership, and discussed the changing face of the education industry. They were charged with looking for ways used to describe standards. Thus, the work of ISLLC became part of a long tradition of regularly upgrading the field and became a central pillar in forging a vision of educational leadership for the schools of the future. From these activities were crafted the initial drafts of principles and standards on which the remaining two years' work was based.

*June 4–5, 1995 meeting of the ISLLC*

Twenty states and four associations met in Washington, DC to accomplish two major tasks: (1) to review and affirm a draft of the standards from the March 1995 meeting and (2) to develop a deeper understanding of the draft standards. Recognition by the members of the Consortium that they were producing standards for a broad range of users as well as defining leadership broadly led to a reaffirmation of the decision by the membership that the model standards should apply to all school leaders. The Consortium acknowledged that there are differences in leadership that correspond to roles, but the Consortium members were unanimous in their belief that central aspects of the role are the same for all school leadership positions, from principal through superintendent.

It was also agreed during this meeting that indicators should be written to define each standard; that such indicators would allow for differentiation among school leaders; and that a common set of knowledge statements, performances, and dispositions could be determined. States could add to the indicators, if, for example, they wanted to differentiate between a district leader and a local school leader. A writing team was to be convened comprised of professors, practitioners, Consortium representatives, and the chair and director of the Consortium. The National Association of Elementary School Principals (NAESP), the National Association of Secondary School Principals (NASSP), and the American Association of School Administrators (AASA) were contacted for suggestions for names of practitioners.

*August 9–11, 1995 meeting of the ISLLC*

The team charged with writing indicators was convened for three intensive days of indicator identification. Initial time was spent in helping the group understand the standards. Then a series of small group work times resulted in drafts of indicators for each standard.

*September 23–24, 1995 meeting of the ISLLC*

At this meeting of the full Consortium major tasks to be accomplished included a review of the process used to identify a preliminary draft of indicators from the writing team as discussed above and a plan for getting feedback from stakeholders in the states and associations.

*March 12–13, 1996 meeting of the ISLLC*

Among the tasks the leadership team wanted to accomplish during this meeting were a review and synthesis of feedback generated by focus groups since the September 1995 meeting, revisions of the standards and indicators as deemed to be appropriate by the group, and development of a plan for obtaining feedback from a much broader audience for the purpose of finalizing the standards.

*November 1–2, 1996 meeting of the ISLLC*
The culmination of two years' work occurred at this final meeting of ISLLC devoted to standards development. On November 2, 1996, *the 21 states in attendance voted unanimously to adopt the Interstate School Leaders Licensure Consortium Standards for School Leaders* (CCSSO, 1996). This was a milestone in the reform of educational leadership in the United States. The adopted Standards present a common core of knowledge, performances, and dispositions that help link leadership more forcefully to productive schools.

**Major components of the standards development**
Outlined above was a chronology of key meetings and decisions made at the meetings for the Standards development. In this section a focus on major developmental components will help the reader understand the basic conceptualization for the identification of the standards and indicators, and the principles on which both are based. As conceived by the ISLLC Chairperson, the standards rest on four legs:

1. a thorough analysis of what is known about effective educational leadership at the school and district levels;
2. a comprehensive examination of the best thinking about the types of leadership that will be required for tomorrow's schools;
3. synthesis of the thoughtful work on administrator standards developed by various national organizations; and
4. in-depth discussions of school leadership standards by leaders within each of the states involved in ISLLC.

*Principles*
An early task of the Consortium was to craft principles that would consistently be used to guide actual development of standards. As noted by Murphy (CCSSO, 1996, p. 7), the members saw that the principles in reality could serve two functions. First, throughout development they acted as a touchstone to which the group regularly returned to test the scope and focus of emerging products. Second, they helped to give meaning to the standards and indicators. The guiding principles privileging the standards are:

- Standards should reflect the *centrality of student learning.*
- Standards should acknowledge the *changing role of the school leader.*
- Standards should recognize the *collaborative nature of school leadership.*
- Standards should be high, *upgrading the quality* of the profession.
- Standards should inform performance-based systems of *assessment and evaluation for school leaders.*
- Standards should be *integrated and coherent.*
- Standards should be predicated on the concepts of *access, opportunity, and empowerment for all members of the school community.*

*Standards*
Formal leadership in schools and districts is a complex, multi-faceted task. ISLLC set out to honor that reality, and recognized that standards would acknowledge that effective

leaders often think, believe, and behave differently from normal expectations within their profession. Reflected in the standards should be the concepts that effective school leaders are strong educators; that they are advocates for children and the communities in which they serve; and that these leaders make strong people connections, valuing and caring for others as individuals (CCSSO, 1996, p. 5). These central concepts are embedded throughout the development of the leadership standards.

There was little debate during the standards development meetings about the importance of knowledge and performances in the framework, but the inability to 'measure' dispositions caused a good deal of consternation throughout the process. However, the belief became stronger and stronger throughout the group's deliberations that dispositions often were central to development of the standards, and in fact, often give nourishment and meaning to performance. As time moved ahead, the members of the Consortium grew to understand that the three parts belong together, and that a leader's dispositions form the foundation that under girds one's actions (CCSSO, 1996, p. 8).

*The single, most important factor in the development of these leadership standards was that learning and teaching and the creation of powerful learning environments resonate throughout the standards.* The members concurred that in forging the standards there must be evident connections to student learning. The success of students is paramount, as indicated by the opening stem for each standard, 'A school administrator is an educational leader who promotes the success of all students by…' The standards define school administrators as EDUCATIONAL leaders. There was also overwhelming endorsement of this concept from all sources of feedback throughout the developmental process. The emphasis resonates throughout the standards that the professional practice of school leaders must be firmly grounded in the knowledge and understanding of the practice of learning and teaching.

*Indicators*

While there are only six general standards, there are nearly 200 indicators of knowledge, performances, and dispositions that define the standards and provide specificity. The indicators writing team attacked each standard separately. Small groups determined the key concepts and elements of each standard and then unpacked them into key ideas. Previously developed vignettes demonstrating specific leadership behaviors were given to the groups to read. Then, based on discussions, their review of the vignettes, and each writing team member's personal expertise, individuals developed indicators for each standard using the framework of knowledge, performances, and dispositions.

The small groups then reviewed and aggregated the work of the individuals for each of the six draft indicators. The ISLLC Chairperson took the work of the small groups and 'blended' their efforts into a set of indicators for each standard. The full Consortium then worked with these blended indicators at the September 1995 meeting, making appropriate revisions, additions, and deletions.

A subcommittee was formed to meet in November 1995 to make further revisions of all materials – principles, standards, and indicators. Their charge was to:

1. provide consistency of language throughout all of the draft material,
2. develop a logical order/organization to the indicators,

3. reformulate the indicators to reflect the collaborative expectation for leadership, and
4. reflect the proactive nature of the principles in the indicators.

*Feedback*
In order to develop comprehensive standards and indicators it had been established early in the process that attempts would be made to reach out to a broad audience of stakeholders (practitioners, community members, professors in leadership programs, etc.) and state and national opinion leaders for input and affirmation. Several methods were used to obtain data.

The ISLLC leadership team made presentations at most of the major national conferences/conventions – NAESP, NASSP, AASA, University Council for Educational Administration (UCEA), American Education Research Association (AERA), etc. – and solicited reactions. Attempts were also made to publicize the Consortium's work through major professional publications such as *Education Week*. The executive directors of each organization belonging to NPBEA were asked to gather input from their respective memberships.

Feedback was formally sought from the states and associations in two ways-focus groups and a written survey. The first effort was to meet with small, intensive groups within the member states and associations for face-to-face discussions. Types of groups would include smaller leadership meetings such as executive boards of associations, groups who were focusing on parallel issues, executive directors from professional associations, and/or 'beginners.'

With the focus, at that time, that these were emerging standards and indicators in draft form, some of the questions used with focus groups were:

- What is the purpose of standards?
- Are these draft standards conclusive, comprehensive, and inclusive enough?
- What are their relationships with whatever is happening in your state or association?
- How would this work be of use in individual states? Are the proposed standards clear and concise?'

It was also emphasized that the intent of the standards is to get at the essence of what it means to be a successful school leader, not to generate long lists of every conceivable skill a leader may ever use.

The data received from the focus groups and others was analyzed and synthesized by the ISLLC director for use at the September 1995 ISLLC meeting. Representatives from each of the member states and allied professional groups that obtained feedback shared highlights of the reactions. Five themes emerged – professional development, assessment, entry level, breadth of the job, and relationships to teacher standards. Small groups then assessed whether changes should be made and gave recommendations that they believed would strengthen the document. As an outcome of that meeting, revisions were made based on the discussions of the feedback.

In July of 1996 a survey of the latest draft of the standards and indicators was sent to licensing offices in all 50 states, the five extra-state jurisdictions, the District of Columbia, and The Department of Defense schools. The entire draft was also put on the CCSSO web site. Reactions were compiled and used during the November 1996 meeting to make one final attempt at meaningful revisions prior to formal adoption. The feedback was extremely positive. It is significant and a commentary on the thoroughness by which these standards were crafted, that only minor revisions were believed to be necessary from this broadly distributed survey.

*The ISLLC standards*

The six standards emerging from the above process follow:

A school leader is an educational leader who promotes the success of all students by:

- … facilitating the development, articulation, implementation, and stewardship of a vision of learning that is shared and supported by the school community.
- … advocating, nurturing, and sustaining a school culture and instructional program conducive to student learning and staff professional growth.
- … ensuring management of the organization, operations, and resources for a safe, efficient, and effective learning environment.
- … collaborating with families and community members, responding to diverse community interests and needs, and mobilizing community resources.
- … acting with integrity, with fairness, and in an ethical manner.
- … understanding, responding to, and influencing the larger political, social, economic, legal, and cultural contexts.

## Putting the ISLLC Standards to Work

The heart and soul of the standards project for school leaders is how they are used. Development was an involved process that led to six broad standards defined by 200 indicators. But what good are these standards, unless they are implemented and utilized in major reform efforts? These model, performance-based standards were designed to be used for:

- licensure of prospective school leaders,
- relicensure of practicing school leaders,
- revision and/or validation of existing state standards,
- recruitment and induction activities for new school leaders,
- guidelines in development of assessments for aspirants to school leadership positions,
- reshaping, state approval, and national accreditation of preparation programs for potential administrators,
- certification of school leaders who demonstrate advanced competencies,
- developing a process for school leader evaluation,
- and in guiding professional growth opportunities for existing school leaders.

Uses vary from state to state, and as this chapter is written, usage is still emerging. After all, states may adapt or adopt the standards to their individual needs. However, several trends are already evident. We provide examples of some of these trends below.

*State standards development*

Nearly 40 states have adopted or adapted the ISLLC Standards. For example, Kentucky, Illinois, and Ohio have adopted the ISLLC Standards in their entirety while others such as Mississippi and Louisiana have adapted them to fit their states' unique needs.

*Licensure*

The most significant use to date has been a collaborative effort among several states, ISLLC, and the Educational Testing Service (ETS) to develop assessments for the initial licensing of beginning principals and/or superintendents. As this book goes to press, nine states are administering the School Leaders Licensure Assessment (SLLA) to prospective principals and eleven others are in the process of adopting it. We anticipate that by the end of 2002 at least half of the 50 states in the United States will be administering the SLLA. The School Superintendents Assessment (SSA) was administered for the first time in Missouri in the fall of 2001.

*Relicensure*

ISLLC, ETS, and several states also collaborated to develop a portfolio to be used for licensure renewal and/or professional development. This licensure portfolio is presently being piloted by several hundred practitioners in Missouri, Ohio, Mississippi, Indiana, and North Carolina, and will be available for full implementation by the summer of 2002.

*Recruitment and induction*

A few states, local school districts, and leadership associations are beginning to use the Standards for identifying and nurturing potential school leaders. For example, the Montgomery County, Maryland school system is grounding their professional development of new high school assistant principles on the ISLLC Standards.

*Preparation program improvements*

The Standards are becoming the major drivers in curriculum and preparation program revisions in institutions of higher education. Use of these standards for a framework for developing an educational leadership program will require major and long overdue adjustments, including new approaches to delivery of instruction such as problem and/or performance based pedagogy, internships, cadres of students, and assessment of programs.

A longer time line for preparation and development may also be necessary. There will be a need for reflective practice, feedback to students, less seat time, and more on-the-job experiences in a variety of positions. Guidance for departments of leadership is beginning to appear in the literature (e.g., Van Meter & Murphy, 1997). These preparation program reform efforts run the gamut from individual universities such as Central Arkansas State University to whole-state initiatives as in Maryland, Iowa, and Mississippi.

*Accreditation*

The National Council for Accreditation of Teacher Education (NCATE) is revising their process for national recognition of preparation programs for school leaders around the ISLLC framework.

*Professional development*
States (e.g., Ohio and Missouri) and professional associations (e.g., NAESP) are also using the Standards to frame continuing development for school leaders. The ISLLC Consortium developed the Collaborative Professional Development Process for School Leaders (CPDP), an exceptionally powerful professional development tool for self-growth.

*Evaluation*
Districts and states (e.g., Delaware) are using the Standards to create new evaluation systems for school leaders.

## Conclusion

There are, indeed, a few speed bumps to be gone over slowly as the road is traveled to implementation of the leadership standards. Any change to long established practices is slow and difficult to accept. There is always a degree of cynicism about innovation – the fear of the unknown. After all, the current system is familiar and comfortable. Will these new standards improve school leadership or will they simply gather dust on shelves as so many other reform efforts have done? These national Standards will, of necessity, require building support for a new paradigm of school leadership. This will be a static process – the final version of the Standards will never really be final. As needs of the profession change, so too should the Standards change.

Alignments among content standards, teacher performance-based standards, and administrator performance-based standards are still very unclear. Roles must be clarified. The importance of leadership in the schools must be emphasized. Without careful attention to the reform of educational leadership there will be no lasting reform of K-12 education. Members of the profession need to constantly remind the learned societies, the professional associations, the state and local boards, the public, and the unions about the research that consistently points to the importance of the principal in effective schools.

ISLLC, through the use of a thorough process approach to development by personnel from 24 states and several national associations, has crafted Standards for School Leaders. Literally thousands of practitioners and other stakeholders provided input into their development. The ISLLC Standards are contemporary, based on research about effective school leaders, visionary in an approach to the preparation of school administrators, and refocus the role of school administrators on learning and teaching. They are firmly anchored in the belief that all students can learn.

## References

Burke, C. (1992). Devolution of responsibility to Queensland schools: Clarifying the rhetoric critiquing the reality. *Journal of Educational Administration*, 30(4), 33–52.
Carnegie Council on Adolescent Development. (1989). *Turning points*. Washington, D.C.: Author.

Clark, D. L. (1990). *Reinventing school leadership* (pp. 25–29). Working memo prepared for the Reinventing School Leadership Conference. Cambridge, MA: National Center for Educational Leadership.

Consortium on Productivity in the Schools. (1995). *Using what we have to get the schools we need*. New York: Columbia University, Teachers College, The Institute on Education and the Economy.

Council of Chief State School Officers (1996). *Interstate school leaders licensure consortium: standards for school leaders*. Washington, DC: Author.

Elmore, R. F. (1990). *Reinventing school leadership* (pp. 62–65). Working memo prepared for the Reinventing School Leadership Conference. Cambridge, MA: National Center for Educational Leadership.

Goodlad, J. I. (1984). *A place called school: Prospects for the future*. New York: McGraw-Hill.

Hawley, W. D. (1989). Looking backward at education reform. *Education Week*, 9(9), 23, 35.

Meyer, J. W. & Rowan, B. (1975). *Notes on the structure of educational organizations: Revised version*. Paper presented at the annual meeting of the American Sociological Association, San Francisco.

Murnane, R. J. & Levy, F. (1996). *Teaching the new basic skills: Principles for education children to thrive in a changing economy*. New York: Free Press.

Murphy, J. (1992). *The landscape of leadership preparation: Reframing the education of school administrators*. Thousand Oaks, CA: Corwin Press.

Petrie, H. G. (1990). Reflecting on the second wave of reform: Restructuring the teaching profession. In S. L. Jacobson & J. A. Conway(eds.), *Educational leadership in an age of reform* (pp. 14–29). New York: Longman.

Shipman, N. & Murphy, J. (Spring 1999). *ISLLC update*. UCEA Review. XL. 13, 18.

Tyack, D. (1993). School governance in the United States: Historical puzzles and anomalies. In J. Hannaway & M. Carnoy (eds.), *Decentralization and school improvement* (pp. 1–32), San Francisco: Jossey-Bass.

Van Meter, E. & Murphy, J. (1997). *Using ISLLC standards to strengthen preparation programs in school administration*. Washington, DC: Council of Chief State School Officers.

# Section II

## Global Approaches to School Leader Preparation and Development

# 6

# Developing School Leaders: One Principal at a Time

Dr. Dennis Littky

*National Director, Principal Residency Network, The Big Picture Company, Providence, RI, USA*

Dr. Molly Schen

*Director of Program Development, Principal Residency Network, The Big Picture Company, Providence, RI, USA*

The program's design focuses on leadership learning for moral courage, moving the vision, and relationship building and is individualized to meet the needs of each aspiring principal. It is not enough to get the right people into the right program around core leadership challenges. The Principal Residency Network, based at the Big Picture Company, is also dedicated to changing the conditions of work by designing and partnering with small, personalized schools where the rewards of leadership can be realized.

This chapter is a principal's personal account of how an apprenticeship-based preparation program evolved out of his experience. Having a mentor, knowing oneself well enough to take important stands, cultivating leadership in others – these all figured prominently in Dennis Littky's work experience and were confirmed by research in leadership development. The creation of the Principal Residency Network, its mission, philosophy, design and program components, are detailed.

The program relies on a careful selection process of both aspiring principal and mentor principal, with attention to encouraging people of color into school leadership. Recognizing that people will only want to become school principals if the job is doable and rewarding, the Principal Residency Network acknowledges the need for small, personalized schools and, with its umbrella organization, the Big Picture Company, is working to do just that.

### Raising Bulldogs

I walked into his office at 6 a.m. on February 4, 1969 after my first day of work as a school principal, to analyze a minor crisis. 'Help,' I asked. 'What went wrong?' From that day, Rhody became my mentor. He was a man fighting for African American students in New York City. He believed these kids had the right to be taught and cared about in the same way that white students were taught and cared about in the suburbs.

Rhody was committed to saving these urban children and worked tirelessly and boldly to carry out his dream.

Most importantly, Rhody would not compromise when it came to saving students' lives. He acted like a bulldog. (He also raised bulldogs, which always made me wonder: Who was the mentor in this situation?) Rhody was not afraid to die for his cause. He was a black man ahead of his time. He threatened people by his mere presence. He knew he could die for his cause and he was still willing to fight. In Rhody I observed moral courage in its purest form. Nothing is a risk if you are willing to die for your actions. Because Rhody was willing to die for his cause, he was free to stand up for what he believed. He did not have to compromise his views and actions. Rhody is still living today.

What a first mentor! It not only allowed me to observe moral courage in its purest form, but also influenced every decision I made as a twenty-five year old. I did not realize at the time the power that mentorship would have in the rest of my life. It was the beginning of my learning the real meaning of moral courage. I was not and am not today willing to die for my cause … but nor am I afraid to be fired. So in my own way, I have freed myself up to fight for the same urban children, thirty years later in a different city. In fact, in each of my jobs, there were major attempts to fire me. These situations gave me strength and freed me up to fight.

As a small school principal for twenty-five years, I have had the opportunity to witness and encourage many teachers in their pursuit of the principalship. I was a good role model for these teachers. I loved my job, worked hard and wouldn't compromise when it came to providing learning opportunities for my students and support for my staff. My focus is and has always been building relationships with kids. It is both as simple and as complicated as that.

It makes sense to involve others in decisions affecting the school, and so in the schools I have led, teachers work through hiring choices and budget priorities and the direction of curriculum with the kids in mind. I realized early on that this student-focused, collaborative style of leadership appealed to many teachers and at the same time gave them the opportunity to hone their own decision-making and leadership skills. All of the teachers became leaders in their own way. Tom Peters, management guru, in his PBS video *Leadership Alliance*, looked at what we were doing and said that he found that all the teachers in this school had become outstanding leaders. He concluded that good leaders help develop good leaders.

For years, I watched teachers in my schools – at least one dozen teachers – make the decision to pursue the principalship and then embark on many semesters of evening coursework at a nearby college. Typically, these courses had little connection to teachers' work in the school. There may have been a project here or there, or an interview or occasional job shadow. But there was a gulf between theory and practice, between their course-based principal preparation and the long-term, many-layered complexities of taking the lead for something consequential in the life of school. I couldn't help but note the chasm between the preparation of principals and the work of principals. As an avid reader of leadership literature in many fields, I have no bone to pick with reading lists or with theory. I just think that, on balance, people need a lot more practice in walking like a leader.

For years, I garnered occasional commentary from these same teachers in my schools, many of whom are now principals with schools of their own. They told stories about how they had truly learned the art of the principalship. It was not from their courses, although they gleaned important tidbits of legal knowledge, historical perspective and analytic skills from their coursework. (In truth, many could not even remember their courses.)

Their practice, they said, had been influenced by working with me, watching me, and reflecting on why we were doing the things we were, and how things were turning out. They talked about how the most crucial learning – how to be strong, not to back down, to persist in doing what is right for kids – could not be taught through textbooks. They spoke of my vision and how they remembered me pushing the staff to stay on track. They spoke of the tenacity we had as a staff to examine and re-examine our practice. And they talked about the important role that communication played in our school, with three group meetings each week, individual meetings with me once a month, weekly newsletters to which all staff contributed, journal sharing between them and me, regular retreats and summer workshops.

These principal colleagues appreciated the stances I took and structures I created. They understood that I needed to exercise moral courage to sustain the vision for our kids. They realized that the various forums for meeting with one another – through writing as well as conversation – were necessary vehicles for communicating with different people. It was clear that their in-school opportunities to be teacher-leaders and my modeling of leadership qualities, more than formal coursework or short-term internship, had shaped these aspiring principals.

## Creating The Big Picture

Then the time came when I became more than casually interested in their comments. In my own work, I moved from a school reformer to a school creator to a school designer. I developed an eye toward growing many schools with certain features, schools we now dub Big Picture Schools. Each of our new Big Picture Schools would need a principal. Suddenly it was apparent that I had landed squarely in the land of leadership preparation. Even as I put a tremendous amount of stock in leadership, I put very little stock in leadership preparation as it was traditionally carried out.

A rapid scan of the field was rewarding. Fellow principals of democratic, student-centered schools reported that they also had an unusually high number of teachers go on to become principals. They, too, found that their aspiring principals relied heavily on mentoring and modeling to learn the craft of the principalship. Research confirmed these anecdotes. In a study by Brent, Holler and McNamara, principals rank formal coursework last in relation to impact on their practice. Our aspiring principals were the lucky few who had an experienced mentor help them learn to lead through the cycle of action and reflection, augmenting the knowledge and skills they were acquiring in their courses.

### Time to formalize

My colleagues and I were heartened by the possibility of preparing principals differently, and we decided to get together to dream up the ideal way for principals to be prepared. Our group included Roland Barth, founder of the Harvard Principals' Center; Elliot Washor, Co-director of the Met School; and exemplary principals from across the USA.

Together, we dreamed up a new model of school leadership training. We met a creative, entrepreneurial dean from Lewis and Clark College, Jay Casbon, who was willing to grant certification to promising aspiring principals who took part in our intensive, school-based principal residency. (Since then, Northeastern University, Johnson & Wales University, Rhode Island College, Providence College, and Keene State College have also put their college seals on the program.) We have been assisted considerably in our work by the support of the Wallace-Readers' Digest Funds, and by Rhode Island's Business Education Roundtable and Human Resource Investment Council.

A group of exemplary principals from around the country who believed in mentoring were brought together in Rhode Island to design the specifics of a new model of school leadership training. It was a pretty simple idea, as I told Bess Keller of *Education Week*: 'You get the best people in the field, and the students get trained by them' (Keller, 2000).

Four years and fifty-three aspiring principals later, the design that was once radical has already become a state-approved program. The article continues by saying that our program is 'more radical than other programs,' but the good news is that it 'is far from the only recent attempt to move the training of principals in new directions (Keller, 2000).

Quietly, under the cover of educational lingo and university requirements, reformers around the country have established beachheads of clinical education for principals. The programs view schools – not university lecture halls, as the proper training ground for future leaders' (Keller, 2000). In truth, we do hope to help reshape principal preparation around the country.

In a *Commonwealth* article, Ann Duffy, associate commissioner of education in Massachusetts, says we already have. According to Duffy, the Principal Residency Network has 'influenced Massachusetts Department of Education's rules governing administrator certification' and adds, 'They are the benchmark. We are really looking hard to see what we can learn from the success they've had' (Gerwin, 2001, p. 34).

## Rigor in the Residency

What does the program look like? In this section, we describe the heart of the program – what we are training people to be able to do as school leaders – as well as the design principles and program components. Aspiring principals in the Principal Residency Network learn the craft of the principalship by working as full-time interns under the guidance and with the support of a strong mentor principal for at least a year.

At the core of the Residency are consequential school-based projects that contribute to the school while fostering the individual's leadership learning. Through this project work, aspiring principals 'walk the high wire' of school leadership, taking risks in their project work. They know full well that they may take a fall, but trust that the safety net beneath will prevent lasting damage to them and to the community. The mentors are close by, coaching, pushing, encouraging, and sometimes pulling the plug when the risk is too great. In this model of real-world training, performance assessment is inevitable and meaningful. Aspiring principals recognize that their work impacts and is evaluated by the entire community. The goal of project work is not to earn a grade or credit, but to contribute to the school community, to earn the trust of one's staff and the respect of one's community.

The Principal Residency Network is based on the belief that people learn best through an ongoing cycle of action and reflection. Experience alone is not sufficient for growth and development as a leader, but it is essential. Similarly, theory without practice is ineffectual in the action-oriented world of school leadership. The Residency is designed to allow aspiring principals to engage in consequential action while ensuring that they have ample opportunities to think, talk, read, and write about their work.

We formulated a mission, philosophy and design that flow from our experience and from research on adult development. The Principal Residency Network's mission is to 'develop a cadre of principals who champion educational change through leadership of innovative, personalized schools.' Our philosophy is grounded in teaching 'one student at a time' at all educational levels. We believe that the best learning takes place in small communities that integrate academic and applied learning, promote collaborative work, and encourage a culture of lifelong learning. That goes for the schools that we help to create at the Big Picture Company, and it goes for the design of the Principal Residency Network for aspiring principals as well.

## Moral Courage and Other Leadership Qualities

In its three-year history, the Principal Residency Network has identified a set of leadership skills that its founders and mentor principals hold dear. We understand well that school principals need skills that extend well beyond management of the three 'B's': buses, boilers and budgets. Our six leadership areas are:

- moral courage,
- moving the vision,
- instructional leadership,
- relationships and communication,
- management through flexibility and efficiency,
- public support.

These six areas 'cross-walk' onto the Interstate School Leaders Licensure Consortium (ISLLC) standards and various state competencies. That is, even as the program emphasizes these desired qualities, aspiring principals have been able to organize their portfolios around state requirements. Each of these qualities takes much-treasured traditional leadership qualities (such as having a vision, decision making, teambuilding, supervision and evaluation, community involvement) and builds on them. The program pushes them to the place where the leader is both thoughtful and dynamic, and where planning and reflection are spurs to immediate action. Below we describe each of the leadership qualities in more detail.

*Moving the vision*

An effective school leader must develop and maintain a consistent vision and inspire others to work towards it. S/he is able to say no to ideas that do not support the vision, for he understands the direction in which the school is moving and is able to predict the desired outcomes.

*Moral courage*
An influential principal has the courage to stand alone. S/he has a commitment, above all else, to doing what is best for children despite the dictates of others. S/he challenges assumptions and traditions and helps others do so as well.

*Instructional leadership*
A successful principal creates joy around learning. S/he deals well with adversity, loves leadership, and thrives on her work. S/he is committed to continual learning and growth for himself.

*Relationships and communication*
An exemplary school leader pays attention to the personal. S/he is thoughtful, understanding, and just. S/he shows respect for and trust in his staff and promotes the spirit of democratic collaboration. S/he knows how to support, delegate, and offer input. S/he listens carefully to the thoughts, feelings, and concerns of others. S/he knows the questions to ask and is able to collect and share information as needed. S/he is tactful yet direct and has a talent for both one-on-one and group communication.

*Management through flexibility and efficiency*
An effective principal is able to juggle many tasks and thoughts at once. She is patient but willing to move, organized, and good at following through on tasks.

*Public support*
A strong principal has a gift for public relations. She is able to present and articulate school results and is an effective fundraiser.

## Design Principles of the Principal Residency Network

We have already given some clues about the central importance of the schoolhouse. Indeed, the Residency is where the aspiring principal learns most of the craft of leadership. The other design principles underscore the importance of reflection, of assessing competence and skill as close to the action as possible, and our commitment to recruiting people of color and women into the principalship.

### The residency

The schoolhouse is the best place for developing the next generation of school leaders. Aspiring principals, selected for their leadership capacity and commitment to school change, learn the craft of the principalship through full-time, site-based residencies, under the guidance of a mentor principal. Aspiring principals serve their school communities for twelve to twenty months and are awarded principal certification upon completion of the residency. Rigorous, individualized learning plans guide their experiential study and ensure that project work is supported by substantive reading, writing, and reflection. The program is committed to building an authentic curriculum around in-school stewardship that is relevant and responsive to the needs of the school community and the aspiring principal.

**Learning from experience is not inevitable**

Learning from the experiences of the schoolhouse requires deliberation, self-awareness, and constant feedback. Aspiring principals are trained to be reflective practitioners who derive insight from their experiences and know how to modify their practice accordingly. Journal writing and regular, in-depth discussions with the mentor principal are critical to this process. Retreats, readings, and school visits add perspective and depth to reflective practice.

**Real-world assessment**

Aspiring principals learn through the real-world consequences of their projects. They document their efforts and results and create extensive portfolios that illustrate project work, writing, research, and reading. At least twice each year, they give formal exhibitions through which they publicly present their work and reflect upon their learning goals and growth. Aspiring principals receive ongoing written and verbal feedback from colleagues in the program cohort and from a school-based feedback circle.

**The greater the diversity, the greater the learning**

The Principal Residency Network recognizes that respect for equal opportunity enhances learning and promotes the common good. In order to develop a cadre of leaders that reflects the diversity of our student populations, the program works directly with schools to find and recruit talented educators who have not traditionally been represented in school leadership positions.

## Program Components and Requirements

For those interested in what aspiring principals are responsible for doing, in a more concrete way, the description that follows may be helpful. There is individual work, group work, and a considerable responsibility for showing one's work in different ways for assessment purposes.

### Individual work

*Learning plan*

Aspiring principals craft individualized learning plans with their mentor principals, incorporating state standards and school needs with professional and personal learning goals.

*Project-based learning*

Each aspiring principal initiates a consequential project to address a challenge or need in the school. Through this project, aspiring principals must analyze school-based data, develop and implement strategies for change, evaluate program outcomes, and make mid-course corrections as necessary. Aspiring principals review critical literature associated with their projects and visit other schools to inform their work.

*Writing*
Aspiring principals continually reflect on their work and leadership development through formal and informal writing. They share journal entries with their mentor and the program cohort and are expected to produce a publishable piece of writing during the course of the program.

**Group work**

*Team meetings*
Each aspiring and mentor principal team meets daily for at least half an hour and weekly for an extended meeting to allow for ongoing planning and deep reflection. These meetings provide aspiring principals with the opportunity to connect scholarship to practice, revise learning goals, and continually assess progress.

*The network*
Aspiring and mentor principals come together for monthly network seminars and quarterly institutes to share best practices, provide support and critical feedback, and discuss theory and research related to educational leadership. Seminars regularly involve book and article discussions, journal sharing, or subject-specific workshops on topics such as law or finance. At institutes, aspiring principals present their work and receive in-depth assessment of their learning plans, project-work, and portfolios.

*Cross-school visits*
As part of the program, aspiring principals must visit a number of schools both within and outside the program network. These visits expose participants to a range of school practices, designs, and cultures and help build critical friends' groups within the network.

**Performance assessment**

*Portfolios*
Participants develop extensive portfolios that illustrate project work, writing, research, and reading. Graduates use the portfolio to document and demonstrate their leadership experiences and work readiness.

*Exhibitions*
Twice each year, aspiring principals present their project work to a panel of fellow participants, mentors, university faculty, and school community members. Exhibitions allow participants to publicly reflect on their growth, demonstrate mastery of learning goals, and identify competency areas that demand further work.

*Mentor narratives*
Twice each year, mentor principals write detailed narratives that assess the aspiring principal's service to the school, scholarship, growth, and leadership potential.

*Feedback circle*
Aspiring principals enlist community members from the school to provide on-going in-house evaluation. Members of this feedback circle check in regularly with the

participant and meet formally at least three times in the course of the year. Members write a detailed evaluation at the end of the program.

**'Who – me?' tapping teacher leaders (especially people of color) for the principalship**

> 'If it weren't for all of the leadership opportunities in our school, I never would have considered becoming a principal.' – Amy Bayer
> 'I was honored to be tapped for this program by my principal.' – Renee Lamontagne

It is not easy to get into the Principal Residency Network. The selection process involves not just the aspiring principal, but an understanding and commitment on the part of a mentor principal, superintendent, and district that this person is heading for a principalship. The demands on entrance into the program, while complex, are quite deliberate. We want the best people to be tapped. We also want to attract more people of color to the principalship. While the challenge of recruiting diverse leadership talent is daunting due to the lack of diversity in the teacher ranks, we have made a commitment to finding and recruiting a diverse group of aspiring principals. Our strategy is to work directly with superintendents, schools, and principals to identify those people who have leadership promise but might not have otherwise considered administration.

People go into the principalship for all kinds of reasons – to make more of a difference, to make more money, or because it is one of the few shifts from the classroom that is possible in education. They might start taking classes in school administration, one or two at a time, perhaps without fully committing to the idea of becoming a school leader, but just to dip toes in the water of leadership. If they are unsure about their goals, it is quite likely that those around them, their colleagues and supervisors, are even less aware of potential aspirations. We have met more than one superintendent who did not know that several teachers in the district had principal certification. (In fact, some researchers say that there is no shortage of certified principals in the country; rather, the problem is that many people received certification without any intention of entering the job.)

In the Principal Residency Network, aspiring principals 'go public' early on. They might step forward with interest in the program, or they might be tapped by a principal or a superintendent. Either way, prior to applying, the principal and the aspiring principal discuss the applicant's leadership potential. The superintendent is involved in these discussions as well. Several superintendents and principals report that they debate their choice of aspiring principal candidate from among several possibilities. Some principals talk about the program and application process with their entire staff.

These conversations distinguish our pre-application process from traditional programs. In a very important way, schools and districts actively participate in growing their own future leaders. One aspiring principal says, 'You have faith in the people you've chosen.' In the Principal Residency Network, the superintendent and principal know from the outset who is aspiring to the principalship. In fact, they have a chance to recruit and select their successors.

The project director still has to screen applicants carefully, both on paper and through site visits to the schoolhouse. It is up to the project director to ensure that applicants have the interest, confidence, and entrepreneurial spirit to be leaders of innovative schools. We believe that our program can take good people and make them very good, and we can take very good people and make them exceptional. However, we are not able to take applicants with weak skills and work magic.

Also, mentor principals have the very highest level of competency, commitment and capacity to train an aspiring principal. Because the mentor principal is the primary teacher, it is crucial that he/she be both an exemplary leader and an exemplary coach, able to observe, support, listen and give feedback to, the aspiring principal. As one researcher cautions, 'Well-intentioned individuals who are genuinely interested in mentoring others may not possess the skills to effectively do so' (Medeiros, 2001, p. 84).

We look for candidates who have the spirit of inquisitiveness, openness, and a mixture of patience (with process) and impatience (to get important things done). The project director looks carefully at the aspiring principal candidate to see if he/she has initiative. If accepted into the program, the aspiring principal will have to be a self-starter in order to be successful. We believe that knowledge and performance can be readily coached, but new dispositions are difficult to coach in a one-year program.

Walking through the hallways and into some classrooms with the aspiring principal and mentor principal gives us an initial assessment of their rapport with staff and students. Because this program holds a very high value on relationship building, this qualitative information is crucial. Without real strength in human relations, school leaders are unlikely to succeed. At the same time, the project director is learning as much as possible about the school context, since 90% of the learning is done through real-world projects at the school. In essence, the aspiring principal candidate and mentor principal and school are all applying to the program, for we need to select the right people to be aspiring principal, the best principals to be the primary teachers in the program, and healthy school conditions for optimal leadership learning.

Contrast this entry process with the more traditional course-based program. The benefit of this highly selective and complicated selection process is that people start to sit up tall when they hear about the program. Applications are flowing in at a far greater rate than available space (and we are determined to keep the programs small). Word is getting out: this is a great way to learn to be a leader!

## The Art of Mentoring

We have described the importance of leaders doing the right thing, of designing the right kind of program, and of selecting the right people for school leadership. Once aspiring principals are selected, the program rests largely on the mentoring within the schoolhouse. How should the mentoring principal coach the aspiring principal?

So much has been written about the isolation of school leaders (and, indeed, of teachers as well) that it is well worth examining how mentor principals and aspiring principals work together. It all starts with the relationship between the mentor principal and the aspiring principal. The relationship is key. I have to open up who I am and what I do, and I have to get to know someone else exceedingly well. We all lead from who we are. When I mentor someone, I commit myself to a relationship characterized by candor, mutual responsibility and trust. To mentor well, I have to give thought and time to the aspiring principal and really establish a working relationship.

A recent study of mentoring in the Principal Residency Network found that mentor-mentee relationships are indeed characterized by strong relationships. The researcher, Medeiros, found that 'open communication with reciprocal feedback' was an important element of the relationships (Medeiros, 2001, 64). Aspiring principals said of their mentor principals, 'There is continuous feedback,' 'We debrief and reflect … every day,' and appreciated that the mentor was 'willing to be vulnerable to share disasters as frequently … [as] successes' (Medeiros, 2001, 61–62). 'Experts in the field of mentoring confirm the importance of building trusting relationships as the basis for learning' (cited in Medeiros, 2001; Allen & Poteet, 1999; Daresh & Playko, 1990).

Thought and time may not be enough: there is also the issue of sharing space. Many mentor principals work in more cramped quarters, or even move into a different room large enough for two desks, so that they can be in close proximity to the aspiring principal. Aspiring principals value this closeness, prize the in-the-moment reflections with the mentor and access to the complete range of the principal's work. Closeness in time and space opens the possibility for sharing oneself. Chris Hempel, who graduated from the Principal Residency Network, notes that the program provided for 'personalizing the mentoring' and that the mentor 'is able to say the right thing at the right time.'

I know, myself, that people learn from powerful role models. Careful observation and thoughtful listening teach one a great deal! As Charlie Plant, a graduate of our program, now a principal, said recently of his mentor, my colleague Elliot Washor, 'I watched how he [Elliot] worked with the kids … and he really modeled how to do that work.'

Another graduate, Jill Homberg, who is also now a principal, learned about 'the balance between when you let a thing play out, versus making a call.' She learned by watching 'how Dennis does it.' Sometimes she finds herself imagining how I would handle something if I were there, as a way of approaching a situation. While she has her own quiet style of warmth, she reflected on one situation and remarked, 'It was funny. I behaved exactly as Dennis would have, if he'd been there.'

It is one thing for an aspiring principal to observe me as I talk with kids and parents and staff; it is quite another for me to explain aloud why I'm doing what I'm doing. It is one thing for the aspiring principal to facilitate a staff meeting; it is quite another for us to analyze the agenda together, and debrief the meeting when it is over. Our conversations have to be candid if we want powerful learning to emerge.

That means that I have to trust the aspiring principal, be confident of his or her abilities even as I think of ways to coach him or her to be bold, to be organized, to be strategic for the long haul, to be more reflective – whatever it is that the aspiring principal is working on. I have to be comfortable in my own skin as a mentor, ready and willing to hear all questions – including those that make me rethink my practice! In short, it means I have to be a teacher.

Jill remembers the modeling – and conversation – about the difference between how one feels and how one behaves as a leader. As Jill said, I really 'let it all out [in private]. Alone with her, I was able to say, "I can't believe so-and-so …" She marveled, then when she saw me address the individual later in the school day: "And then hearing you talk to the person in calm tones … "'

The contrast was noteworthy for her, as she figured out the importance of behaving as a leader. Medeiros found that, 'the skills or qualities of effective mentors ... in this study included the ability to model, the ability to listen and reflect, and the ability to question and teach' (Medeiros, 2001, p. 78). Indeed, mentors need strong communication skills to help the aspiring principal make sense of the modeling.

Powerful as it is to observe and gain insights from a skillful mentor, a residency needs to get the aspiring principal into the action. We want aspiring principals to be doing, to be taking on important school projects, not just observing. What is a mentor principal to do, exactly? A principal intern in another program, Teresa Gray, suggests several tips for mentor principals, such as 'integrate the intern into the school' and 'provide time for continuous evaluation' (Gray, 2001, p. 663).

Over the past three years, we have learned a lot from mentor principals in the Principal Residency Network about the seemingly little moves that turn out to be incredibly significant in the mentoring process. Here are a few actions we have found quite powerful, actions that speak volumes about the mentor's commitment to and trust in the aspiring principal.

- The mentor principal takes home a 'dialogue journal' on Friday afternoons, writing brief reflections and questions to the aspiring principal and returning it on Monday morning.
- The mentor principal and aspiring principal meet for breakfast at 7 a.m. at the local diner each Tuesday morning for 90 minutes of in-depth reflection and analysis of one or two moments (where either the aspiring principal or mentor principal is the key actor) as well as careful planning of an upcoming event.
- The principal invites the aspiring principal to lead a key piece of the August professional development days with returning staff.
- The principal is matter-of-fact about having the aspiring principal as a welcome colleague at every meeting, every walk through the halls, hearing every telephone conversation, for the first weeks or month of school – until the aspiring principal begins to work on his or her own consequential project.
- Recognizing that this is an internship in preparation for the principalship, the principal protects the aspiring principal from responsibilities normally consigned to the 'assistant principal.'
- Together with the aspiring principal, the mentor principal talks with the teachers' union representative about ways that the aspiring principal can gain experience in supervision and evaluation, under the guidance of the mentor.

Probably the most poignant aspect of the relationship – and the most significant – is how so many of the mentors are committed in an enduring way to the aspiring principal. And aspiring principals need this! No matter how terrific the experience of the aspiring principal, no matter how vast the learning during their year of internship, it is far different being a principal than being an aspiring principal.

Now, instead of talking as they walk down the hall together, there are flurries of phone calls and emails. The mentor principal typically asks, 'How are things going?' If the phone call comes from the aspiring principal, there are questions such as 'How would you have handled this?' and 'What else should I consider, before I make this decision?'

The project directors of the Principal Residency Network have a commitment to the graduates, too. We realize the program is not over when a student gets her certification.

The real work and learning in the school has just begun, and the importance of the Network now comes into play in important ways. The Network helps with job referrals, suggestions for those just entering the Network as aspiring principals, newsletters, continued staff development, and support in graduates' new positions.

## Should We Find Charismatic Principals or Change the Job?

We know we need to attract the right people to the principalship and give them the right kind of training so they have the courage and relational skills to do the right thing for kids. We know we need powerful mentoring. But that sidesteps the question of whether the job of principal is doable, whether the conditions of work are even tenable.

The current lament about the principalship is loud. 'Most principals are little more than building managers, overseeing everything from bus schedules to lunch duty. At the very moment when there is huge pressure to improve student performance – and when the buck stops in the principal's office – principals have lost most control over the things that affect how well students learn' (McVicar, 2000). That is a huge problem. We are tackling it head-on, by working simultaneously to design and create small, personalized schools around the country. These learning communities are good for kids, first and foremost. But they also make it possible for the school leader to do his or her job well.

Peter McWalters, commissioner of education in Rhode Island, agrees that the job of the principal needs to change. 'You have a critical position increasingly powerless, other than through charisma. If we're going to revolutionize American education, do we think the answer is a million charismatic people, or restructuring the job?' (McVicar, 2000).

Our work carries us to the new small schools movement, where we have been embraced. Small schools are 'fragile ecosystems, often parched for political cover, eager for recourses' and concerned about 'the ability … to survive and to thrive after the departure of a powerful leader,' as participants in the program wrote (Myatt & Alexander, no date, p. 4).

Participating in the Principal Residency Network can give these schools some additional leadership hands while shoring them up in case a principal retires or the community calls for the formation of another small school. Specially designed small schools are often run democratically, with principals and teachers taking on broad leadership responsibilities for work in the school.

At Fenway High School, for example, 'every staff member … plays many roles and has multiple obligations leading toward the success of students and the smooth, yet complex, operation of the school' (Myatt & Alexander, no date, p. 8). As a consequence, many of the teachers in these small schools already have demonstrated leadership potential – and are more eager about the prospect of being a principal someday.

## The Future

The Wallace-Readers' Digest Funds has made school leadership a major funding priority. With their support, the Big Picture Company has been able to develop the Principal Residency Network, looking more deeply into residency programs that work and that can be expanded to states and countries all over the world. Our commitment is to be clear about the design principles and to develop materials that help with the training, so that people can develop more sites of the Principal Residency Network throughout the country and beyond. The Principal Residency Network is connected globally, but the work is done locally.

While we continue to develop the Principal Residency Network, we are simultaneously creating more small schools and searching for small school networks interested in partnering with us. And we search for ways to bring more people of color into school leadership positions.

Our dream is that the graduates, aspiring principals who are now principals, will in a few years' time become mentoring principals – and that they will profoundly change the design and culture of schools so that kids have better opportunities. As one African-American graduate of the program, Karen Edmonds, wrote, 'As an aspiring principal, I am looking forward to the day that I can continue the cycle and become a mentor to a young, deserving leader of color' (Edmonds & Perez, 2000, p. 3). The circle then continues, and new principals are grown.

## References

Allen, T. D. & Poteet, M. L. (1999). Developing effective mentoring relationships: Strategies from the mentor's viewpoint. *The Career Development Quarterly*, 48(1), 59–73.

Brent, B. O. (1998). Teaching in educational administration. *Newsletter of the American Educational Research Association*, 5(2), 3–4.

Daresh, J. C. & Playko, M. A. (1990). Mentor programs: Focus on the beginning principal. *NASSP Bulletin*, 74, 73–77.

Edmonds, K. & Perez, G. (2000). *Racial equity: The forgotten child of leadership reform*. Unpublished manuscript. Providence, RI: The Big Picture Company.

Gerwin, C. (2001). A matter of principal: Aspiring administrators learn school leadership by doing. *Commonwealth*, 34, 31–36.

Gray, T. (2001). Principal internships: Five tips for a successful and rewarding experience. *Phi Delta Kappan*, 82(9), 663–665.

Keller, B. (2000). Building on experience. *Education Week*, 19(34), 14–16.

Medeiros, S. (2001). *Mentoring aspiring principals to improve principal preparation*. Unpublished doctoral dissertation. Johnson & Wales University.

McVicar, M. (2000). Principal losses: Restore leadership to top job, advocates urge. *The Providence Journal*.

Myatt, L. & Alexander, K. (2000). *Growing champions*. Unpublished manuscript. Providence, RI: The Big Picture Company.

For a brochure and informational video on CD, write to:

Principal Residency Network
c/o Big Picture Company
275 Westminster Street, Suite 500
Providence, RI 02903
or visit our website: www.bigpicture.org

# 7

# Problem-based Leadership Development: Developing the Cognitive and Skill Capacities of School Leaders

Dr. Michael Aaron Copland

*Assistant Professor, Department of Educational Administration and Policy, College of Education, University of Washington, Seattle, WA, USA*

Two decades of education reform, beginning in the early 1980's have greatly expanded the expectations for school principals' leadership in understanding and solving complex educational issues. Such leadership involves encounters with myriad problems, some routine and mundane, others complex and wildly unfamiliar. Recent research on the principalship confirms that problem-solving ability is central to principals' work (Leithwood & Steinbach, 1992; Leithwood & Steinbach, 1995). Efforts to lead teams of teachers and others toward understanding and solving important educational problems comprise a significant part of what principals do.

Research suggests that some level of problem-solving skill is more than simply a necessary prerequisite for entry into the principalship. Studies indicate that greater expertise in problem-solving ability distinguishes some persons as more successful than others in this leadership role. Leithwood and his colleagues (Leithwood & Montgomery, 1982, 1984; Leithwood & Stager, 1989; Leithwood & Steinbach, 1992, 1995) have conducted a long-term research program aimed at exploring the nature, determinants and consequences of school administrators' problem-solving practices. Through this research, they came to view the problem-solving role as the core of administrative work, and crucial to an understanding of why some principals are more effective than others.

The problems that principals confront are often unpredictable. Therefore, some scholars have suggested that a focus on improving the quality of administrators' thinking and problem-solving is likely to be more productive in preparing principals than a focus on teaching specific actions or behaviors (Leithwood & Steinbach, 1992). Yet, despite a relatively extensive understanding of the importance and nature of school administrators' problem-solving behavior, the development of prospective principals' problem-solving ability remains a challenge for the field.

One strategy for preparing school leaders, problem-based learning (PBL), aims to develop the cognitive and skill capacities of prospective principals (Bridges & Hallinger, 1995). PBL originated in medical education. It is an instructional and curricular approach that employs complex, interdisciplinary problems taken from professional practice as the starting point for learning (Bridges with Hallinger, 1992). PBL in leadership preparation places students in contextualized problem scenarios, wherein they work with others to understand and solve problems associated with their future role.

Since PBL emphasizes the development of problem-solving skills (Bridges with Hallinger, 1992), research that seeks to determine whether problem-solving behavior improves among students prepared in a problem-based curriculum has a high degree of relevance for the field. This chapter reports on research that inquired about the development of prospective principals' problem-solving ability in a problem-based administrator preparation program at Stanford University.

More specifically, the chapter recounts aspects of a study that was conducted in two phases. The findings suggest that greater exposure to problem-based preparation is associated with greater problem-framing ability among prospective principals. These findings are discussed in terms of their broader implications for leadership development.

## Problem-based Learning and Problem-solving Ability

In a foundational work on medical problem solving, Elstein and colleagues (Elstein, Shulman, & Sprafka, 1978) described a theory of *hypothetico-deductive* reasoning, that was used to explain physician problem-solving process. Hypothetico-deductive reasoning suggests that when a problem is encountered, an initial hypothesis is formed. Then the problem solver seeks to locate evidence to either support or discount this hypothesis. Elstein and colleagues theorized that the hypothetico-deductive reasoning process is universally present, at some level, in the problem-solving behavior of physicians. In addition, Elstein and colleagues discovered that while novices and experts use the same basic processes to solve problems, persons get better at solving problems by building a knowledge base through repeated problem-solving efforts. This line of thinking supported the marriage of content knowledge and practical experiences in the medical model of problem-based learning (see Barrows & Tamblyn, 1980).

Perhaps due to the fact that PBL originated in medical education, empirical investigations that explore connections between problem-based preparation and students' problem-solving skill exist primarily in that field. As Berkson (1993) suggests, no one has yet been able to fully characterize, and therefore measure, all the cognitive components that make-up problem-solving. Therefore, a comprehensive answer to the question of whether PBL teaches problem-solving more effectively than other instructional approaches is unavailable. This aside, comparisons of the problem-solving ability of students from different avenues of preparation have been done using the best available theoretical models of how physicians solve problems.

There are relatively few studies in this literature that are methodologically sound; within the small group that are, results are not consistent. Kaufman and others (1989) performed a longitudinal examination of student outcomes in the University of

New Mexico's School of Medicine. This study contrasted the performance of students in a PBL track and those in a traditional track on the National Board of Medical Examiner's – Part II examination, which is designed to measure clinical competence. Findings showed PBL students scored significantly higher ($p < .01$) than conventional track peers. This finding was consistent over a period of years.

Conversely, in a study that applied the Elstein, et al., theory of hypothetico-deductive reasoning, Gordon (1978) contrasted subjects drawn from a highly traditional medical school with those from a newly accredited medical school using an innovative PBL curriculum. Comparisons of problem-solving performance between students from the PBL curriculum and those of the traditional curriculum revealed no significant differences on any of four dependent variables.

Methodological weaknesses, present in a number of other studies of PBL and problem solving, further obfuscate any general tendencies that might be present. Schmidt, Dauphinee, & Patel (1987) summarized findings from three studies (Claessen & Boshuizen, 1985; Saunders, McIntosh, McPherson, & Engel, 1990; Woodward, 1984) and detected weak evidence of higher performance among PBL students on tasks related to clinical competence. On closer examination, the Woodward's (1984) results, while favorable toward PBL graduates, were based on small comparison groups. Moreover, the ratings reflected performance ratings assigned by the students' supervisors rather than an actual test of problem-solving or clinical competence.

Saunders and colleagues (1990) contrasted PBL students with those from a traditional curriculum. Their results revealed no differences in the performance of students on questions of clinical competence, except that when scored as pass/fail rather than numerically, PBL students performed slightly better.

The Claessen et al., study (1985) of the performance of PBL medical students was essentially a measure of information recall rather than problem-solving. The resulting scores were, however, used to substantiate suggested differences in students' underlying cognitive problem-solving structures. Students from the PBL curriculum were able to recall more information than students in the conventional curriculum, a finding interpreted by the researchers as evidence of stronger problem-solving ability.

Finally, a study by Patel, Groen, and Norman (1991) contrasted students from a conventional program with those from a PBL school. The study found that PBL students demonstrated a tendency to provide better explanations and more detailed diagnoses. On the other hand, the diagnoses of the PBL students were incorrect more often than those of their conventionally-trained peers. This difference is difficult to interpret because of possibly confounding factors. For example, the academic entrance requirements at the conventional school were more rigorous than those of the PBL school. There is the possibility that the PBL students may have been more error-prone simply due to lower levels of inherent academic ability.

Within the field of educational leadership, research on outcomes associated with PBL is still quite limited. Much of the evidence to date is anecdotal, relying on student perceptions of their experiences (see Bridges w/Hallinger, 1992). The modal contributions to literature on PBL in education leadership include numerous efforts to document the development, implementation and use of problem-based curricula in the revision of educational administration programs (Martin, Chrispeels & D'emidio-Caston, 1998; Chenoweth & Everhart, 1994; Muth, et al., 1994). Frequently such reports rely on limited evidence to offer prescriptions to others in the field of education leadership endeavoring to reform preparation programs (Cordiero, 1998; Martin, et al., 1997).

A satisfactory answer to the question of whether a problem-based approach to administrative preparation creates administrators who are better problem-solvers does not leap out of this literature. Yet, there is limited evidence pointing to improvements in problem-solving ability stemming from students' engagement in problem-based instruction. Chrispeels and Martin (1998), for example, systematically collected various data over the course of one year on students in an administrative credential program that featured problem-based learning, and uncovered findings suggesting that the PBL classes allowed students to acquire knowledge and skills in group processes and problem-solving, as well as course content.

Very few scholars carefully develop and employ constructs related to administrative problem-solving to test hypotheses about specific outcomes associated with PBL. One notable exception, a study by Leithwood and Steinbach (1992), used a grounded model of administrative problem-solving to compare the problem-solving abilities of a group of school leaders who participated in a problem-based training program with those of a control group. While differences between the groups did not reach statistically significant levels, much of the data supported the claims that administrators' problem-solving abilities could be improved using a problem-based instructional approach, and that a well-designed program of preparation was a more efficient and reliable vehicle for improving problem-solving expertise than was on-the-job experience.

## Problem-based Learning at Stanford University

In the particular model of problem-based learning developed for leadership preparation at Stanford University (Bridges w/Hallinger, 1992),[1] students are expected to practice necessary job-related skills during the program. These include leadership skills, facilitation skills, interpersonal skills, and skills of consensus decision-making. At the same time, they are learning domain-specific content knowledge drawn from field believed to be relevant to the problems they encounter.

Problems are presented to students in realistic contexts modeled on professional practice. These problems are largely similar to those that students will encounter in their future professional roles as school leaders. Problems that form the central focus of PBL projects take a variety of forms. While some are relatively clear and straightforward, others are crafted to be extremely complex and 'messy.' Some problem-based curricula or projects incorporate live actors as role-players who present aspects of the problem situation.

Students work through PBL projects in extended blocks of time (generally 3–5 sessions of 3+ hours per session in length). They typically work within small groups of

[1]For a complete accounting of problem-based learning for administrators, see Bridges w/Hallinger (1992). The book includes one example of a problem-based learning project, in addition to providing a detailed account of the way in which projects are constructed, problem-selection, learning objectives, guiding questions, student role in PBL, instructor's role in PBL, and various methods of assessing student progress.

6–8 members, and use a role-based method for conducting meetings (Doyle & Straus, 1976) as the vehicle for learning how to use tools of consensus decision-making.

The content of a problem-based curriculum is interdisciplinary, and its pedagogical nature constructivist. Students are responsible for using their existing knowledge, in combination with various other resources suggested or supplied by instructors. They draw on these varied resources to understand each presented situation and to define the inherent problem(s). The student teams then generate criteria that can be used to evaluate potential solutions to problem(s), and reach consensual solution(s) with peers.

Within this context of active engagement, students assume a high degree of responsibility for deriving meaning and learning from problem-based projects. Problem-based learning instructors do not lecture. Instead, they function as guides to the group process, intervening as little as necessary in students' project work. Their primary role during the learning process is to debrief the group on process and content issues at the close of each work session. Instructors also create or select the problem-based projects, engage students in relevant questions, and contribute formative feedback on group processes and student performance throughout the project experience.

Students engage in a variety of different performance-based assessments during, and at the conclusion of PBL experiences. Many PBL projects culminate in a presentation or other product which is analyzed and critiqued by instructors and invited practitioners. These may include teachers, assistant principals, principals, staff developers, central office administrators, superintendents, school board members, and parents drawn from schools and districts. Students also assess their own learning through reflective feedback at the close of project meetings, and in writing individual reflective essays at the close of each project experience.

In the Stanford program, students enroll for three consecutive summers of problem-based coursework. This is referred to as practicum. Students receive developmentally appropriate feedback on their work, process, and performances over a series of projects. This is designed to foster their ability to know and do.

## Conceptual Framework

Problem-solving ability is a broad and complex construct. In order to narrow and define a particular approach, this research examined one aspect of problem-solving skill, specifically the ability to understand and frame problems. Little empirical evidence exists in the educational administration literature indicating a link between problem-based leadership preparation and the development of problem-solving abilities. Therefore, the warrant for studying such a connection is developed here using theoretical perspectives from cognitive and social psychology.

### Problem-framing ability

A conceptual definition of problem-framing ability developed from various sources in the literature on problem solving (c.f., Cuban, 1990; Getzels, 1977; Getzels & Csikszentmihalyi, 1976). Getzels, for example, studied the creative process in visual artists and refers to the initial formulation of a problem as 'problem-finding'. Getzels found that a skilled problem-finder would not approach a problem situation with a ready-made solution in mind. Such a person will not let past experience completely determine what is to be done. Instead, he or she will let the new challenge suggest new solutions.

This preliminary aspect of a problem-solving process is defined here as being consistent with Cuban's (1990) notion of problem-framing. Cuban notes:

> Framing a problem, then, is a subjective process. It depends upon one or more facts that show a discrepancy between what is and what ought to be. It depends upon the perceptions of the person or group that interpret the data and do the defining. What shapes (these perceptions) are (one's) previous personal and work experiences, (one's) beliefs and values, the position (one holds) within an organization, and the expected role (one) is to play within that organization. (Cuban, 1990, p. 2)

An analysis of the Leithwood and Stager (1989) expert/novice problem-solving framework, suggests that a number of various sub-skills focus directly on abilities or processes employed in the initial formulation or framing of administrative problems. For example, from Leithwood and Stager (1989), expert principals:

- have an understanding of the importance of developing a clear interpretation of a problem,
- have the ability to develop a clear interpretation of the problem and the ability to describe this interpretation to others,
- seek out and take into account the interpretation others have of the problem,
- carefully check their own assumptions relative to others' interpretation of problems,
- have less of a personal stake in any preconceived solution than non-experts,
- anticipate obstacles likely to arise during group problem solving, and
- plan in advance for how to address anticipated obstacles.

Cuban (1990) further suggests that leaders often commit the fatal error of strongly embracing a preconceived solution before a problem has been clearly defined and understood. This suggests the importance of a principal's ability to recognize a problem that has been presented with a predetermined solution, and then to reframe the problem in solution-free terms. Different and equally sound solutions to a problem may exist, particularly with regard to non-routine, complex dilemmas. Indeed, the way a problem is framed fundamentally determines whether a predetermined solution will quickly follow, or other alternative solutions will be considered.

### Theory linking PBL and problem-framing ability

Inherent in problem-based learning is the fundamental challenge of grappling with a presented problem. But what is it about PBL that leads to a hypothesis suggesting that aspects of students' problem-solving ability will improve with time and experience when using this type of instructional approach?

The theoretical rationale underpinning the relationship between problem-based instruction and students' ability to understand and frame problems of practice can be established at least two different ways. Social comparison theory (Festinger, 1954) posits that humans have an innate need to evaluate their own abilities against those of similar others. The majority of students in a principal preparation program enter with the common intention to pursue a career in school leadership. The interactive context of problem-based learning enables comparisons of the interpersonal abilities central to that work. In a PBL group, students' problem-solving skills are openly on display, and skills-based feedback between peers is encouraged.

Social psychology provides further conceptual support for the links between modes of teaching and the natural tendency to compare abilities among persons being taught. Rosenholtz and Simpson (1984) suggest, for example, that classrooms that are organized to recognize and promote multiple performance dimensions provide multiple bases for students to compare and evaluate their abilities. In a problem-based learning group, students assume multiple responsibilities in the process of completing a project. They take turns leading, plan the work, create and communicate agendas, facilitate group processes, contribute as group members, handle planned interruptions, make oral presentations, record meeting notes, and provide critical feedback to peers.

Theories of cognition also contribute to a conceptual warrant for the study. Bridges (w/Hallinger, 1992), drawing on work done on problem-based learning in medical education by Schmidt (1983), notes three conditions created within a problem-based learning environment that information theory links to subsequent retrieval and appropriate use of new information.

First, in problem-based learning *prior knowledge is activated.* Students are expected to exercise knowledge they already possess in order to understand the problem situation they face. Bridges (w/Hallinger, 1992, p. 9) notes that 'this prior knowledge and the kind of cognitive structure in which it is stored determine what is understood from the new experience and what is learned.'

Second, in PBL *new knowledge is encoded in a context modeled on practice.* Research on cognition suggests that knowledge is more likely to be remembered or recalled in the context in which it was originally learned (Godden & Baddeley, 1975). As Bridges notes (w/Hallinger, 1992, p. 9) 'encoding specificity in problem-based learning is achieved by having students acquire knowledge in a functional context, that is, in a context containing problems that closely resemble the problems they will encounter later on in their professional careers.'

Third, PBL offers students the *opportunity to elaborate on information* that is learned. Elaboration provides redundancy in memory, which in turn reduces forgetting and abets retrieval (Bridges w/Hallinger, 1992, p. 9). One might expect that the skills of problem-framing are better understood, processed and recalled through strategies such as small group discussion, peer review of ideas about how particular information applies to a given problem, and the practice of reflecting back on problem-solving processes through debriefing practices and personal essays about what was learned.

## Research Method and Findings

The research reported here first tested the hypothesis that students' with greater exposure to problem-based learning over the course of an extensive preparation program would demonstrate greater ability to understand and frame an encountered administrative problem. Three cohorts of students from the Stanford program, with varying degree of exposure to problem-based experiences, were compared on an administrative problem-framing measure developed for the study. A second phase of the research followed up with one of the student cohorts two and one-half years later. At the close of their program experience the researcher gathered additional data about the development of problem-framing skills learned in preparation.

**Phase I: Comparison between groups**

*Study design: phase I*

The subjects included in Phase I of this study were eighteen students enrolled in the Stanford University Prospective Principals Program. The specific students included in the study comprised three cohorts admitted to the program in successive years. An extensive set of background data, including age, gender, ethnicity, GRE scores, years of teaching experience, and level of prior education was collected on each student prior to beginning the study. The nature and use of the background measures is introduced in the analysis.

Assessment of students' problem-framing ability in Phase I was accomplished through the use of a quasi-experimental, post-test only design (Cook & Campbell, 1979). Given the university setting, and the cyclical nature of turnover in the Prospective Principals Program, the study is further categorized as a cohort design in a formal institution with cyclical turnover (Cook et al., p. 126). Cook and Campbell note that cohorts are useful for experimental purposes because (a) some cohorts receive a particular treatment while preceding or following cohorts do not, and (b) it is often reasonable to assume that a cohort differs only in minor ways from its contiguous cohorts (p. 127).

The purpose of the design was to test whether the three successive PPP cohorts differed in their problem-framing ability. The cohorts were tested after receiving different levels of exposure to problem-based learning. Each of the eighteen subjects was individually presented with a series of five short, written, administrative problem scenarios that were developed specifically for the purpose of this study. One of the scenarios is included as Appendix A. All represented actual problems that were faced in practice by a school principal. While not explicitly known by the subjects, each scenario featured an embedded solution as part of the problem formulation. Students were asked to respond in writing to each of the following questions about each problem scenario:

1. How has the problem been defined in this scenario?
2. Employing what you know and believe about solving problems in practice, reflect on how this problem has been framed. Be as thorough as possible.
3. If faced with this situation in practice, would you reframe the problem? If so, how? (If you choose to restate the problem, justify your reasons for doing so.)

Cohort 1 (first-year group), completed the exercise prior to participation in any problem-based learning experience, nor any program exposure to the concept of problem-framing. Cohort 2 (second-year group) completed the assessment during their second summer in the problem-based practicum, after exposure to approximately seven (7) PBL projects. Cohort 3 completed the scenarios during their final summer in the program, after prior exposure to approximately twelve (12) PBL experiences.

All three groups completed the scenario exercises under similar conditions. Students responded in writing to the five scenarios, in a consistent order, during one sitting in the same university classroom. Students were given little instruction about the exercise other than to read each scenario and respond in writing to the three questions. Students were not allowed to discuss their responses to the scenarios until after all materials had been turned in by all students. Responses were collected in a manner that insured a blind scoring process.

The sub-skills of problem-framing included ten indicators, in three categories, and formed the basis for construction of a scoring instrument used in assessing responses to the problem scenarios (Appendix B).[2] Three (3) independent raters assessed student responses to the three questions using the ten item-scoring instrument. All raters scored each subject's written responses on a 0–3 scale (0 = no evidence of indicator; 3 = strong-compelling evidence of indicator) corresponding to each indicator. The instrument was found to be highly reliable in a separate generalizability study conducted on the original data set (Copland, 2000).

*Analysis: phase I*

To attack the problem of separating the effect of the treatment from the possible effects of selection differences, a two-step statistical analysis was conducted. First, a correlation matrix was created to determine the nature of the relationships between a number of background measures and the dependent variable, problem-framing ability. A composite background measure, found to correlate strongly with the dependent variable, was identified for use as a covariate in further statistical analysis. Next, an analysis of covariance (ANCOVA) was conducted to assess differences between cohort groups on problem-framing ability, employing the composite background measure as a covariate.

Subjects' responses on the five problem-framing scenarios were scored by three independent raters, using the ten item-scoring instrument. A grand mean score, across all items (10), scenarios (5) and raters (3), was calculated for each subject. This grand mean represents the closest approximation available to a 'true' score for each subject's problem-framing ability, as measured by the instrument.

The generalizability analysis was conducted on the complete design (persons × raters × scenarios; see Copland, 2000) and reliability of the measure was established at the level of the grand mean. Therefore, analysis of the dependent variable was appropriately conducted at that same level.

An analysis of covariance (ANCOVA) was calculated on subjects' grand mean of problem-framing scores by cohort. ANCOVA uses a covariate, an individual difference variable that is highly correlated with the dependent measure, as a statistical control to reduce unexplained error variation. To do this, in ANCOVA the dependent variable is adjusted statistically to remove the effects of error variation represented by predictable differences within groups – variation due to the covariate (Ruiz-Primo, Mitchell, & Shavelson, 1996). Given the small *n*, and the fact that, for each covariate employed in the ANCOVA one degree of freedom is lost from the error term, a decision was made to include only one covariate in the model.

The research hypothesis proposed that the ability to frame administrative problems is greater among prospective principals with more exposure to problem-based learning

[2]As listed in the Appendix, the indicators derive from Leithwood and colleagues' work on expert-novice problem-solving among school administrators (Leithwood & Stager, 1989), and Cuban's conception of problem-framing skill (Cuban, 1990).

than for those who have had less exposure. The ANCOVA summary table is presented as Table 1.

Table 1. Analysis of Covariance Summary Table. Problem-framing Ability by Cohort with Covariate GREAQ.

| **Source** | **Sum of Square** | **df** | **Mean Square** | **Sag fRatio** | **Off** |
|---|---|---|---|---|---|
| Main Effect | 4.908 | 3 | 1.636 | 62.040 | 0.000 |
| COHORT | 2.399 | 2 | 1.199 | 45.486 | 0.000 |
| GREAQ (Covar) | 0.401 | 1 | 0.401 | 15.218 | 0.002 |
| Explained | 4.908 | 3 | 1.636 | 62.040 | 0.000 |
| Residual | 0.369 | 14 | 0.026 | | |
| Total | 5.277 | 17 | 0.310 | | |

*Note*: EXPERIMENTAL sums of squares: Covariates entered WITH main effects.

A significant main effect was found for Cohort ($F < .000$), representing the levels of exposure to problem-based learning. This result indicates that significant mean differences were present among the cohort groups on the dependent variable, problem-framing ability. The ANCOVA provides support for the research hypothesis, concluded that significant differences among treatment group means were found. However, which group means differ significantly cannot be determined from the initial ANCOVA table. Post hoc comparisons of the adjusted group means were conducted to make this determination.

In order to conduct post hoc comparisons between groups, Cohort means were statistically adjusted to remove the effect of the covariate.[3] It was expected in the study design that Cohort 3, the group with the most exposure to problem-based learning, would score higher on the problem-framing measure than Cohort 2, the group with the next highest level of exposure to PBL. Also, it was expected that both Cohort 3 and Cohort 2 would score higher than Cohort 1, which had no exposure to PBL. The expected pair-wise relationships were formulated as three separate null hypotheses, using the adjusted means for each cohort. In order to test these specific hypotheses, and to determine which (if any) of the observed differences were due to chance, Tukey's HSD test was used to conduct the pair-wise comparisons.[4]

All three pair-wise comparisons revealed significant differences between groups on the dependent variable, problem-framing ability. The difference in the adjusted means between Cohort 3, those students with the greatest exposure to problem-based learning, and Cohort 1, those students with no exposure, was significant at alpha $< .01$. The difference between the adjusted means for Cohort 2 and Cohort 1 was also significant at alpha $< .01$. The comparison between Cohorts 3 and 2, who had received graduated levels of exposure to PBL, revealed a significant difference at the .05 level of alpha for greater exposure to PBL.

[3]The statistical formula for calculating the adjusted means comes from Shavelson (1996, p. 516).

[4]Tukey's HSD (honestly significant difference) test calculates an observed $q$ value for each comparison, which is measured against a critical $q$ value from a generalized student range distribution (c.f., Shavelson, 1996, p. 518) to determine whether group differences are significant.

**Phase 2: Within group differences**

*Study design: phase 2*

In Phase 2, the students tested prior to exposure to PBL in Phase I were retested again twenty-seven months later at the close of their Stanford program experience. A one-group, pretest-post-test design (Borg & Gall, 1989) was employed. Following administration of the pre-test prior to exposure to PBL, students took part in a series of approximately thirteen problem-based learning projects over the course of three consecutive summers in the preparation program. In all, students spent an estimated 150+ classroom hours engaged in problem-based learning activities, and unknown additional time, individually or in small groups, outside of class. Near the conclusion of the third summer of study, after completion of nearly all the required coursework for the preparation program, students were once again tested on the identical problem-framing measure.

The scoring process was similar across both administrations, pre- and post-tests, of the problem-framing exercise. Descriptive statistics for Phase 2 are included in Table 2. The grand mean for subjects on the pre-test was 1.0067 on a 0–3 scale; for the post-test, the grand mean score was 2.0567. Slightly greater variance existed among scores on the post-test measure. As indicated by the maximum and minimum observed values in participant scores, the distributions of pre-test and post-test means were mutually exclusive. In other words, the maximum observed value on the pre-test was lower than the minimum value on the post-test.

Table 2. Descriptive Statistics: Problem-framing Scores by Test.

| Test | GMean | Stnd Err | Var | Min Val | Max Val |
|---|---|---|---|---|---|
| Pre-test | 1.0067 | 0.0509 | 0.0156 | 0.8733 | 1.1933 |
| Post-test | 2.0567 | 0.1035 | 0.0642 | 1.760 | 2.380 |

*Analysis: phase 2*

The research hypothesis for Phase 2 was concerned with the development of problem-framing skills within the same individuals over time. The hypothesis proposed that the ability to frame administrative problems is greater for prospective principals with more exposure to problem-based learning. Therefore participants' mean scores on the problem-framing post-test would be significantly higher than those on the pretest.

In this design, the same participants were observed before and after exposure to the treatment, and so scores on the post-test were dependent upon those on the pretest. It can be anticipated that in this type of design, the scores on the first measure will be correlated with the scores on the second measure. Therefore, the *t* test for dependent samples was used to test the hypothesis. The *t* test for dependent samples takes the assumed correlation into account when the *t* statistic is calculated by removing the portion of variability due to consistent individual differences from the error term (Ruiz-Primo, Mitchell, & Shavelson, 1996, p. 269).

The design requirements necessary to justify the use of a *t* test were verified to the extent possible with a small sample.

Consistent with statistical formulas detailed in Ruiz-Primo et al. (1996, p. 270–71)[5], a $t^*$ statistic was calculated to be 9.259 (see Table 3). In order to judge whether significant differences exist between the means on the two tests, the $t^*$ statistic was compared against a calculated critical value of $t$. The value of $t_{critical}$ at .01 level of alpha with 5 degrees of freedom was found to be 3.365 (Shavelson, 1996, pg. 619). Therefore, since $t^*_{obs} > t_{crit}$, the null hypothesis was rejected, indicating that the participants scored significantly higher on the problem-framing post-test than on the pretest.

Table 3. *t* Test: Problem-framing Skills.

$t^*_{obs} = 9259$
$t_{crit(01/5)} = 3365$
Therefore, since $t^*_{obs}$, Reject $H_0$: $u_1 - u_2 = 0$

## Discussion

In Phase I, the study results revealed a significant difference in problem-framing ability between subjects with varying degrees of exposure to PBL, after statistically controlling for selection bias. In Phase II, the same subjects tested before and after exposure to a rich problem-based learning curriculum showed statistically significant gains in problem-framing ability. The results suggest that greater exposure to PBL resulted in greater problem-framing ability among students preparing for the principalship. This finding suggests that a developmental learning process resulted from the PBL curriculum as differences were discovered among the same subjects.

This finding may be interpreted as promising with regard to the use of problem-based learning in preparing school leaders, as it suggests that PBL has a positive effect on problem-framing ability. Nonetheless, the results raise questions for consideration. First, the study results rest on the assumption that the change in problem-framing ability can be attributed to students' experience in problem-based learning. This may be an overly simplistic view of the learning that took place for students, inside and outside of the preparation program over the twenty-seven month period. The study does not specifically address, nor perhaps adequately untangle, the influence of other contextual and instructional factors that may be present for students, and potentially contribute to an increased ability in problem-framing. How does the extensive time in preparation factor into students' overall development? What is the influence of other experiences both in and out of the preparation program that contribute to students' 'thinking like a principal?' These are questions that the study does not answer, and point out the importance of not overstating results.

Second, this research does not answer the question of whether PBL teaches problem-framing skills more effectively than other instructional methods. Rather, it indicates only that students' problem-framing skills improve over time in this method. It is conceivable that students who experience other methods of instruction may develop in a similar way. This lack of ability to discriminate amongst instructional methods calls for further studies that might compare student learning of problem-framing skills via different instructional approaches.

[5]The actual calculation can be found on page 273 of Ruiz-Primo et al., 1996.

Third, the generalizability of findings is limited by the unique nature of students included in the study, the unique implementation of problem-based learning in this setting (40% of the total Stanford curriculum), the limited size of the sample and, in Phase 2, the possible maturation of subjects between pre-test and post-test due to other contextual factors.

Finally, the study also points out a need for further research examining whether problem-framing skills learned in preparation actually transfer into students' subsequent administrative work. A primarily qualitative study focused on two cohorts of program graduates, including the students that were the subjects of both phases of this problem-framing study, is currently underway to focus on this all-important 'so what?' question via follow-up interviews and observations of graduates in administrative practice.

The practical importance of this research stems from the assumption that successful principals must be skilled in the ability to understand, frame and solve problems encountered in practice. Moreover, as noted earlier, the research of Leithwood and others (1989, 1992, 1995) has linked problem-solving expertise, in some measure, to principals' success on the job. The findings suggest that students can develop skills in understanding and framing administrative problems can be developed during the process of preparing for the principalship.

More specifically, the findings suggest that repeated exposure to, and practice with a systematic problem-solving process, such as that which is incorporated in problem-based learning, is associated with greater student ability in framing administrative problems. While not conclusive of a link between problem-based learning and problem-framing skill, the study raises the possibility of such a link.

## Implications for the Development of Educational Leaders

Any real value inherent to this work resides in implications that it may hold for the field. The following comments are an effort to cull meaning from this research which may be useful to those who work at developing leaders for schools.

Perhaps most centrally, the study offers empirical evidence that the prospects for systematically improving aspects of the problem-solving expertise of school administrators are promising. These findings extend earlier research (see Leithwood & Steinbach, 1992) suggesting administrative problem-solving ability can be improved through instruction, and not strictly left up to chance advancement via on-the-job experiences. In practical terms, those who choose to pursue this as an instructional goal are presented with, at the very least, the possibility to make a positive difference.

Moreover, the findings move the field beyond student anecdotes as a basis for building leadership programs designed to improve problem-solving, to a more robust understanding of the possibilities for learning inherent in a problem-based approach. The findings build on the tentative support found elsewhere in the literature on problem-based learning, both within and external to educational leadership, for the particular value of investing professional preparation that includes: use of authentic, real world problems; the use of small group, collaborative work structures to tackle such problems; modeling and practice of skills within those collaborative group settings; and provision of extensive, developmentally sound feedback to students on issues of content and process.

The research also infers the importance of understanding problem-based learning as one component nested within a broader context of preparation. Other context factors apart from problem-based learning, and not focused on in this research, likely contribute to the growth in students' problem-framing skills reported in this chapter. These factors may include: the cohort structure of the preparation program, in combination with an ongoing commitment of time together, enabling the development of deep, lasting, trusting personal and professional relationships among students; a heavy commitment to student advisement on the part of core faculty; consistent attention to faculty recruitment and composition consistent with the goals of the program; and a program vision that understands problem-based learning as one aspect of a broader curriculum that also features a core course curriculum and school-based fieldwork. The findings noted herein cannot be understood as divorced or isolated from these important contextual aspects of the program. For those who create or design leadership development experiences, to simply 'do problem-based learning,' apart from consideration of and attention to other contextual and programmatic factors, may fail to generate significant student growth of the kind observed in this study.

## References

Barrows, H. & Tamblyn, R. (1980). *Problem-based learning: An approach to medical education.* New York: Springer.

Berkson, L. (1993). Problem-based learning: Have expectations been met? *Academic Medicine,* 68(10 (Supplement)), S79–S88.

Borg, W. R. & Gall, M. (1989). *Educational research: An introduction 5th edition.* New York: Longman.

Bridges, E. M. with Hallinger, P. (1992). *Problem-based learning for administrators.* Eugene, OR: ERIC/CEM.

Bridges, E. M. & Hallinger, P. (1995). *Implementing problem-based learning in leadership development.* Eugene, OR: ERIC/CEM.

Chenoweth, T. & Everhart, R. (1994). Preparing leaders to understand and facilitate change: A problem-based approach. *Journal of School Leadership,* 4, 414–431.

Chrispeels, J. & Martin, K. (1998). Becoming problem solvers: The case of three future administrators. *Journal of School Leadership,* 8, 303–331.

Claessen, H. & Boshuizen, H. (1985). Recall of medical information by students and doctors. *Medical Education,* 19, 61–67.

Cook, T. D. & Campbell, D. T. (1979). *Quasi-experimentation: Design and analysis issues for field settings.* Chicago: Rand McNally College Publishing.

Copland, M. A. (2000). Problem-based learning and prospective principals' problem-framing ability. *Educational Administration Quarterly,* 36(4), 584–606.

Cordiero, P. (1998). Problem-based learning in educational administration: Enhancing learning transfer. *Journal of School Leadership,* 8, 280–302.

Cuban, L. (1990). *Problem-finding: Problem-based learning project.* Stanford University, School of Education.

Doyle, M. & Straus, D. (1976). *The new interactive method: How to make meetings work.* New York: Jove Books.

Elstein, A., Shulman, L. & Sprafka, S. (1978). *Medical problem solving: An analysis of clinical reasoning.* Cambridge, MA: Harvard University Press.

Festinger, L. (1954). A theory of social comparison processes. *Human Relations*, 7, 117–140.

Getzels, J. (1977). Problem-finding in research in educational administration. In G. L. Immegart & W. Boyd (Eds.), *Problem-finding in educational administration* (pp. 5–22). Lexington, MA: D. C. Heath.

Getzels, J. & Csikszentmihalyi, M. (1976). *The creative vision: A longitudinal study of problem finding in art*. New York: Wiley and Sons.

Godden, D. & Baddeley, A. (1975). Context-dependent memory in two natural environments: On land and underwater. *British Journal of Psychology*, 66, 325–32.

Gordon, M. J. (1978). Use of heuristics in diagnostic problem-solving. In A. Elstein, L. Shulman, & S. Sprafka (Eds.), *Medical problem solving*. Cambridge, MA: Harvard University Press.

Kaufman, A., Mennin, S., Waterman, R., Duban, S., Hansbarger, C., Silverblatt, H., Obenshain, S., Kantrowitz, M., Becker, T., Samet, J. & Wiese, W. (1989). The New Mexico experiment: Educational innovation and institutional change. *Academic Medicine*, 64, 285–294.

Leithwood, K. & Montgomery, D. (1982). The role of the elementary school principal in program improvement. *Review of Educational Research*, 52, 309–339.

Leithwood, K. & Montgomery, D. (1984). Obstacles preventing principals from becoming more effective. *Education and Urban Society*, 17(1): 73–88.

Leithwood, K. & Stager, M. (1989). Expertise in principal's problem solving. *Educational Administration Quarterly*, 25(2), 126–161.

Leithwood, K. & Steinbach, R. (1992). Improving the problem-solving expertise of school administrators: Theory and practice. *Education and Urban Society*, 24(3), 317–345.

Leithwood, K. & Steinbach, R. (1995). *Expert problem solving: Evidence from school and district leaders*. Albany, NY, State University of New York Press.

Martin, K., Chrispeels, J. & D'emidio-Caston, M. (1998). Exploring the use of problem-based learning for developing collaborative leadership skills. *Journal of School Leadership*, 8, 470–500.

Martin, W. M., Ford, S., Murphy, M., Rehm, R. & Muth, R. (1997). Linking instructional delivery with diverse learning settings. *Journal of School Leadership*, 7, 386–408.

Muth, R., Murphy, M. & Martin, W. M. (1994). Problem-based learning at the University of Colorado at Denver. *Journal of School Leadership*, 4, 432–450.

Patel, V., Groen, G. & Norman, G. (1991). Effects of conventional and problem-based medical curricula on problem-solving. *Academic Medicine*, 66(7), 380–389.

Rosenholtz, S. J. & Simpson, C. (1984). Classroom organization and student stratification. *The Elementary School Journal*, 85(1), 21–37.

Ruiz-Primo, M. A., Mitchell, M. & Shavelson, R. J. (1996). *Student guide for Shavelson statistical reasoning for the behavioral sciences*. (Third edition). Boston: Allyn and Bacon.

Saunders, N., McIntosh, J., McPherson, J. & Engel, C. E. (1990). A Comparison between University of Newcastle and University of Sydney final-year students: Knowledge and competence. In Z. Noonan, H. Schmidt, & E. Ezzat (Eds.), *Innovation in medical education: An evaluation of its present status* (pp. 50–54). New York: Springer Publishing Company.

Schmidt, H. (1983). Problem-based learning: Rationale and description. *Medical Education*, 17, 11–16.

Schmidt, H., Dauphinee, A. & Patel, V. (1987). Comparing the effects of problem-based and conventional curricula in an international sample. *Journal of Medical Education*, 62, 305–315.

Shavelson, R. J. (1996). *Statistical reasoning for the behavioral sciences*. (Third edition). Boston, MA: Allyn and Bacon.

Woodward, C. (1984). *Summary of McMaster medical graduates' performance on the Medical Council of Canada Examination*. Hamilton, Ontario, Canada: McMaster University Faculty of Health Sciences.

## APPENDIX A: EXAMPLE PROBLEM SCENARIO

Mr. Donbee is a first-year music teacher, assigned to teach orchestral and choral music at the middle school where you serve as principal. Mr. Donbee was hired late in August to fill an open teaching position for a veteran teacher who announced her retirement late in the summer. While you felt that the pool of candidates was not particularly strong for this job opening, Mr. Donbee impressed you as mature, good-natured, and had an outstanding track record as a private music instructor in a nearby community. Although he had received his public school teaching certificate years earlier, Mr. Donbee had chosen to teach music as a private instructor, and had no previous public school teaching experience, other than his internship experience in his degree program.

In your observations of Mr. Donbee, you have noticed that he appears more comfortable leading the orchestras than the choirs, and that some minor discipline problems have erupted within the sixth grade choral class in particular. When observing the sixth grade choir, you've noticed that students don't seem very enthusiastic, that attention sometimes wanders off, and you detect a lack of respect between students and teacher. Mr. Donbee speaks with an accent that many of the sixth grade students find amusing, and you have overheard some hallway talk among students about Mr. Donbee being a 'nerd' or a 'geek.' You notice that in the sixth grade class, the students' reaction to Mr. Donbee is becoming increasingly vocal, and that he is becoming somewhat frustrated by their comments and their seeming lack of attention to music. Sensing his frustration, you counsel him on techniques for managing student behavior, and provide some resource materials for him to read and consider. In addition, you ask the sixth grade chair, a veteran teacher, if she would drop in on Mr. Donbee unobtrusively and offer collegial support. Further, you ask the district music coordinator to pay a visit to Mr. Donbee's room, to offer support and assess the situation.

One morning during second period (the sixth grade choir period), Mr. Donbee rushes into the main office appearing exasperated and upset. He has left his classroom to seek assistance from 'one of the administrators' in dealing with escalating discipline problems in the sixth grade choir. As you walk back to the classroom with him, you ask which students are creating the problem. He is not specific about the source of the disruption, but generally feels that the problem lies among four or five female students in the classroom. He mentions that he has been the brunt of some 'name-calling' by these particular girls, and that during the singing of songs, several students have chosen to whistle instead of sing. When you ask if he has previously contacted the students' parents, he is evasive and claims that it is difficult to reach them in the evening.

As you enter the room, a silenced hush falls over the class. Mr. Donbee begins to scold the class for being too loud. You calmly interrupt him and ask to see the four female students he had mentioned as the cause of the disruptions. You recognize a couple of the girls as having had some previous minor discipline problems, but the other two are not students you would expect to misbehave in class.

Back in the office, you question the girls individually about the situation in class. Without deviation, each of them tells of their dislike for Mr. Donbee, how the choir class is boring and how he often yells and stomps his feet. Each of the girls denies that they have been involved in any name-calling directed at the teacher, but admit that they were parties to the whistling. After contacting each of their parents to relay the situation and ask for their support, you assign a minor consequence (some time in detention at recess) and have the girls sit out the remainder of second period.

The next day, Julie Terry, the sixth grade counselor catches you in the hallway. She has been visited by Mr. Donbee who has expressed his feeling that several of the students in his sixth grade choir 'just weren't cut out for music' and has asked that the students be reassigned to study hall instead. Ms. Terry appears to sympathize with Mr. Donbee's situation, stating, 'You know the poor guy is just experiencing some of the first-year teaching blues.' She produces a list of six students, which includes the names of three of the four girls you spoke with yesterday. She suggests that moving the students might ease the pressure on the teacher and provide a chance for the kids to catch up on homework assignments in study hall.

## APPENDIX B: SCORING INSTRUMENT – PROBLEM FRAMING

| | | Scoring 0–3: | 0 = no evidence |
|---|---|---|---|
| | | | 1 = weak-moderate evidence |
| **Subject ID#** | Scen# | | 2 = moderate-strong evidence |
| **Scorer ID#** | | | 3 = strong-compelling evidence |

| | | | | |
|---|---|---|---|---|
| **(I) Define the stated problem** | | | | |
| 1. Clearly recognizes and defines the stated problem. (1) | 0 | 1 | 2 | 3 |
| **(II) Reflect on the stated problem** | | | | |
| 1. Identifies the importance of formulating a clear interpretation of the problem prior to considering possible solutions. (2) | 0 | 1 | 2 | 3 |
| 2. Identifies the importance of approaching a problem without holding to a preconceived solution. (2) | 0 | 1 | 2 | 3 |
| 3. Raises/checks personal assumptions about the problem scenario. (2) | 0 | 1 | 2 | 3 |
| 4. Considers the views of others in the problem scenario. (2) | 0 | 1 | 2 | 3 |
| 5. Identifies pre-existing solution(s) embedded in problem statements. (2) | 0 | 1 | 2 | 3 |
| **(III) Reformulate the problem. S** | | | | |
| 1. Restates the stated problem in solution-free terms. (3) | 0 | 1 | 2 | 3 |
| 2. Identifies and relies on personal values related to a problem-solving process in restating problem. (3) | 0 | 1 | 2 | 3 |
| 3. Anticipates obstacles likely to arise during the problem-solving process. (3) | 0 | 1 | 2 | 3 |
| 4. Anticipates ways to address obstacles should they arise. (3) | 0 | 1 | 2 | 3 |

# 8

# Research and Development in Leadership Preparation: Adapting 'Global Knowledge' for a Local Context[1]

Dr. Philip Hallinger

*Professor and Executive Director, College of Management, Mahidol University, Bangkok, Thailand*

Dr. Pornkasem Kantamara

*Management Program Leader, Asian University of Science and Technology, Chonburi, Thailand*

Over the past decade policymakers in the Asia Pacific region have conceived ambitious educational policies consistent with evolving social, political and economic aims (e.g., Abdullah, 1999; Cheng & Townsend, 2000; Gopinathan & Kam, 2000; Ministry of Education-Thailand, 1997a, 1997b; Ministry of Education-ROC, 1998; Suzuki, 2000). However, with the ever-accelerating rate and scope of global changes, governments are finding it increasingly difficult to put their new policies in practice (Caldwell, 1998; Cheng & Townsend, 2000; Dimmock & Walker, 1998; Fullan, 1990; Hallinger, 1998; Hargreaves & Fullan, 1998; Murphy & Adams, 1998). Recognition of this gap between reforms in educational policy and implementation in educational practice has contributed to an emerging global consensus on the need for more adept leadership at the school level.

This has led to a relatively new focus on the training of school leaders, especially principals. Moreover, for the first time, this trend is evident throughout the world; for example, in Europe (see Bolam, 2002; Huber, 2002; Huber & West, 2002; Reeves, Forde, Casteel, & Lynas, 1999; Tomlinson, 1999, 2002). Australia (see Caldwell, 2002; Davis, 1999, 2002). East Asia (see Chin, 2002; Chong, Stott, & Low 2002; Feng, 1999, 2002;

[1]An earlier version of this chapter appeared in Hallinger, P. & Kantamara, P. (2001). Learning to lead global changes across cultures: Designing a computer-based simulation for Thai school leaders. *Journal of Educational Administration*, 39(3), 197–220.

Fwu, 1999; Hallinger, 1999; Lam, 2002; Li, 1999; Low, 1999; Ming, 1999, 2002; Walker, Bridges, & Chan, 1996; Yang, 2001) and North America (see Bridges & Hallinger, 1995; Copland, 2002; Hallinger, 1992, 1999; Leithwood, Jantzi, & Steinbach, 1999; Murphy, 1992; Murphy & Shipman, 2002). This reflects an optimistic belief in the capacity to develop more effective school leaders as well as in the impact of leadership on school improvement (Hallinger, 1992; Marsh, 1992; Murphy, 1992; Murphy & Shipman, 2002).

Despite this optimism, the knowledge base on which to build leadership for school change remains uncertain, unevenly distributed, and poorly integrated into training programs (Hallinger, 1999). Thus even in domains that have been subject to considerable study such as educational change, Evans concludes:

> Over the past few decades the knowledge base about ... change has grown appreciably. Some scholars feel that we know more about innovation than we ever have.... But although we have surely learned much, there remain two large gaps in our knowledge: training and implementation. (Evans, 1996, p. 4)

Evan's observation is especially salient in the developing nations where the need for educational change is acute, but the knowledge base is even less mature than in the industrialized Western societies (e.g., see Bajunid, 1996; Cheng, 1995; Hallinger, 1995; Hallinger & Leithwood, 1996, 1998). When Asian school leaders receive formal administrative training, they generally learn Western-derived frameworks. This knowledge base, which is not without critics in the West, usually lacks even the mildest forms of cultural validation (Cheng, 1995; Swierczek, 1988).

This has led scholars in the Asia Pacific region to advocate steps to develop an 'indigenous knowledge base' on school leadership (Bajunid, 1996; Cheng, 1995; Dimmock & Walker, 1998; Hallinger, 1995; Hallinger & Leithwood, 1996, 1998; McDonald & Pratt, 1997; Walker et al., 1996). Calls for culturally-grounded research on school improvement set the context for our research in Thailand which has sought to understand the nature of successful school improvement in a rapidly developing Asian nation.

As our understanding of school improvement in Thailand began to grow, we became interested in finding means of transferring that knowledge into practice. A research and development (R & D) approach appeared well suited to this goal. Research and development is a strategy designed to integrate formal knowledge into products or tools for the improvement of practice (Borg & Gall, 1989).

Unlike many R & D efforts, however, we began this project with a fully-developed product: a computer-based simulation, *Making Change Happen!*™ (Network Inc., 1999). This simulation had been designed to teach leaders how to implement change in schools. The simulation was, however, grounded exclusively in theories and empirical research on educational change originating in North America and Europe.

The challenge for our R & D effort was to use knowledge of educational change and improvement in Thailand to create a Thai version of the *Making Change Happen!*™ simulation. Stated otherwise, it was our desire to adapt and validate a training tool that had been constructed on a Western knowledge base for use a non-Western context. We believe that this process and the results should be of interest in this age of global dissemination of knowledge and knowledge-rich products. Notably, these products and practices are largely derived from a Western educational, cultural context but are *exported* with little regard to the context of the importing nation.

In this chapter, we have three goals:

1. To describe the *Making Change Happen!*™ simulation (Network Inc., 1999) and its use as a tool for leadership development;
2. To describe the process of adapting the simulation for use in Thailand.
3. To reflect on the R & D process that we undertook and its implications for the emerging global industry of school leadership development.

## The *Making Change Happen!*™ Simulation

The *Making Change Happen!*™ (Network Inc.,1999) simulation has been employed in training school administrators, teachers, parents, and school improvement teams in North America, Europe, Asia, and Australia. The simulation provides a challenging and active learning environment for learning how to think *systemically* about organizational change. Its interactive design enables learners to refine their understanding of *how to apply best practices in school change and improvement to predictable problems of innovation implementation in schools.*

The *Making Change Happen!*™ was designed to provide the feel of implementing change in real schools. At the same time, the simulation is grounded in theoretical models of change that have been extensively studied in Western societies (see Hallinger, Crandall, & Ng, In press). These include the concerns-based adoption model or CBAM (Hall & Hord, 1987), change adopter types (Rogers, 1971; Rogers & Shoemaker, 1982), knowledge diffusion and dissemination (Crandall, Eiseman, & Louis, 1986) and more general change implementation and leadership (Evans, 1996; Fullan, 1990; Sarason, 1982, 1990).

Consistent with its overall purpose of teaching how to implement change in schools, the simulation has several specific learning objectives. These include:

1. To learn how to develop effective strategies for overcoming predictable obstacles to change implementation in schools;
2. To learn how to bring about change when working with different types of people in organizations;
3. To learn how to lead change efforts in ways that create a positive impact on teachers' classroom behavior and student learning;
4. To learn how to work as a team in bringing about change.

### Instructional format

The original simulation was developed as a problem-based, interactive board-game designed to be played by teams with a facilitator (Network Inc., 1988). The board game was recently redesigned as a computer-based simulation (Network Inc., 1999). While the learning objectives remain the same, the use of information technology makes facilitation of the simulation easier for the instructor. It also enables users to extend their learning since they can play the computer-based simulation on their own following use in a formal classroom setting.

The simulation requires no prior knowledge of computers. Its initial introduction is usually provided in a structured instructional session in a computer lab under the guidance of a facilitator. Learners play the simulation in teams of two to four persons at each computer. We have found this cooperative learning approach more effective at achieving the simulation's learning objectives than individuals working on their own in the classroom environment. When learning with peers, the leaders are forced to question of their assumptions and also to share knowledge and experiences.

An instructor facilitates the session in a cycle that alternates the learners' active engagement of the simulation with teacher-led debriefings. The original North American version of the computer simulation typically consumes between six and 12 hours of instructional time. The amount of instructional time allocated depends upon the depth of understanding desired, the prior experience of the learners, and the nature of the instructional setting (e.g., an in-service workshop or a masters or doctoral course).

*The problem*

The simulation employs a 'problem-based learning' approach in which learners encounter 'the problem' before they become aware of the simulation's theoretical content (Bridges & Hallinger, 1995; Hallinger & Bridges, 1997; Hallinger & McCary, 1992). The instructional design embedded in *Making Change Happen!*™ invites learners to construct the embedded conceptual frameworks out of their experience in the simulation. The actual frameworks are only presented and discussed in the final debriefing.

When learners begin the simulation, they confront the following statement of the problem.

> The Problem
>
> The new Superintendent of the Best Public School System has mandated implementation of a new learning technology system – IT 2020. The Superintendent has said, 'It's time for change. Our traditional methods of teaching and learning are inadequate to meet the needs of the global age.' IT 2020 is the Superintendent's first step in acting on his promise of change to the School Board.
>
> IT 2020 will, however, mean significant change for all who work in the system. In addition to the purchase and redesign of IT hardware and software, IT 2020 will require changing the way staff teach and share information. This will in turn affect their relationships to students and to each other.
>
> Moreover, in the Superintendent's words, 'The Best Public Schools have been slow to adopt practices and policies necessary to 21st century education.' Principals, teachers and other front-line staff are, however, already uncomfortable with the pace at which other recent changes have been forced upon them. Some veteran staff have begun to joke that the learning technology advocated by the new Superintendent just might get used by 2020.
>
> Given the scope of this change, the Superintendent has decided to proceed by pilot testing the use of IT 2020 at two schools in the Central Region of the system. Based on results of the trial implementation in these schools, IT 2020 will then roll out into other schools. Despite this step-by-step approach, the Superintendent is under pressure to show results soon. Therefore trial implementation will begin immediately.
>
> You are part of a school support team that has been selected to help manage implementation of IT 2020 in the two trial schools. Your team is comprised of people from different roles in the Central Region. You will coordinate with

Beth, the Technology Coordinator in the Central Office, and also with Al, the Regional Assistant Superintendent. Two members of the system's School Board – Carol and Dave – have been assigned by the Chairman of the School Board to monitor this project.

Your team will lead implementation of IT 2020 over a three-year period. In each year you will have a budget of money – *bits* – to spend on activities – presentations, workshops, classroom lessons, follow-up help – designed to foster use of IT 2020 in these pilot schools.

Your success will be assessed annually. At the end of three years you will be able to see how widely staff are using IT 2020 and the effects on student learning. Based upon your success you will reach of six levels of expertise in leading change: Apprentice, Novice, Manager, Leader, Expert, Master.

*The people*

Any change effort involves working with the people who will actually implement the innovation. After encountering 'The Problem' the teams find that they will work with 24 staff members to implement the new learning technology, IT 2020. The staff is distributed across two schools and the central office (see Figure 1).

Prior to beginning the actual change effort, the teams must become familiar with the staff. Thus, the next step is for each team to access short profiles of the 24 staff

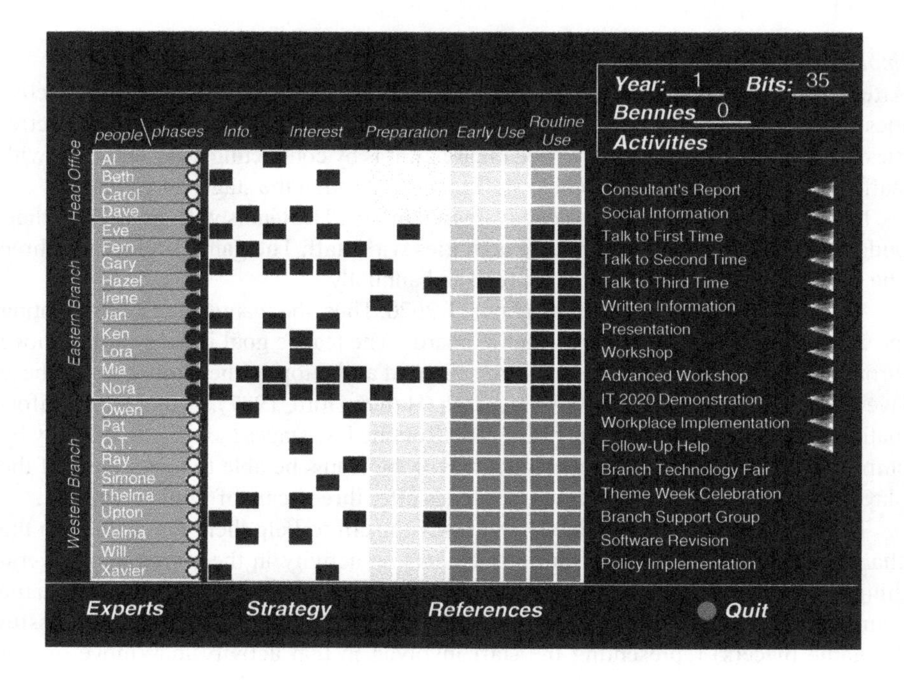


Figure 1. The making change game board.

members. These profiles were written to reflect the range of 'adopter types' typically found in schools (Rogers, 1971; Rogers & Shoemaker, 1982).

Based upon empirical studies, researchers have found a predictable breakdown among schools staffs in the U.S. on five change adopter types: Innovators (8%). Leaders (18%), Early Majority (38%), Late Majority (38%), Resistors (8%). The designers used this breakdown as a means of creating profiles for the 24 staff members.

For example, the profile for the Assistant Superintendent in charge of the Central Region reads:

> 'Al is a respected manager who is concerned with maintaining his Region's productivity. Passed over for the Superintendent's position, he has been heard to say: "The new boss may not understand how things are done around here." '

Or the description of Irene, a second grade teacher:

> 'Irene says, "When there's a job to be done, the old ways still work best." She doesn't trust technology or see a need to change her method of teaching. She will resist anything that results in more work, even in the short-term.'

The team will need to help all staff – from innovators to resistors – learn to use the new technology. Note again, however, that while these profiles were written to reflect the five different adopter types, the people are not labeled as such. The learners simply read the profiles and process the information as they would 'in the real world' as they develop and implement their strategies for change.

*Implementing change activities and receiving feedback*

After familiarizing themselves with the staff, the change teams must examine the activities they will conduct in order to foster change. These activities reflect typical activities used in school improvement (See Figure 2). It is by conducting these activities with staff that the team will begin to help staff to *move* through the stages of change.

Note that each activity has a cost expressed in *bits*. The teams will spend from their budget of 35 *bits* to implement these activities with staff. The game is played in three one year cycles and the budget is replenished annually.

At the outset, staff know nothing about IT 2020. Thus, the 'game pieces' representing the 24 staff members start 'off the game board'. The team's goal is help staff to move through stages of the change process represented at the top of the game board. These five stages are based on the CBAM research (Hall & Hord, 1987). They include Information, Interest, Preparation, Early Use, and Routine Use stages (see Figure 1). Only by employing a 'successful' change strategy, will the teams be able to move most of the players into the Early and Routine Use stages after three years of implementation.

The change team will conduct activities with staff to help them move through the change process. Each time a team implements an activity in the simulation, several things happen. Following the conduct of an activity, the team receives feedback via the computer describing what happened and clues as to why. If the activity was successful the game piece(s) representing the staff involved in that activity may move one or more spaces on the game board. If the activity was less successful, the staff member(s) will move more slowly or not at all.

CONSULTANT REPORT
Information about the schools from a recent consultant's report.
Cost: 2 bits

SOCIAL INFORMATION
Information you obtained from colleagues in the schools about the informal relationships of staff with whom you are working.
Cost: 1 bit

TALK TO
Your 1st conversation with individual people to introduce learning technology issues and IT 2020. Choose three people.
Cost: 2 bits

TALK TO AGAIN
A follow-up conversation to discuss questions about IT 2020. You must have talked to each of these people once.
Cost: 2 bits

TALK TO THIRD TIME
You go back for a 3rd conversation to discuss concerns and answer questions about IT 2020. You must have talked to each of these people 2 times before you can talk to them a 3rd time. Choose three people.
Cost: 2 bits

WRITTEN INFORMATION
A short informational brochure about IT 2020 distributed to all staff in the district (i.e. in the Central Office, and 2 schools).
Cost: 2 bits

PRESENTATION
A short presentation to all school about IT 2020 (i.e. Central Office and the schools).
Cost: 3 bits.

WORKSHOP
How to use IT 2020 in the classroom. Hands-on training designed to promote the ability to use IT 2020 in the classroom. Choose five people from one school.
Cost: 5 bits

ADVANCED WORKSHOP
Advanced strategies for applying IT 2020. Training designed to encourage discussion other applications of IT 2020 to improve learning. Choose 5 people from one school.
Cost: 6 bits

IT 2020 DEMONSTRATION
An on-site demonstration of IT 2020 for school staff. Following the demonstration, a demo model is left on display so it can also be viewed by parents and students. Designate whether the demonstration is at the Secondary or Primary School.
Cost: 3 bits

CLASSROOM LESSON
The staff that you select begin to try out IT 2020 in the classroom. Choose three people from anywhere in the Region.
Cost: 2 bits

FOLLOW-UP HELP
A conversation with staff to solve problems they have encountered in using IT 2020. Choose three people (Note: The people must have conducted a classroom lesson).
Cost: 1 bit

SCHOOL TECHNOLOGY FAIR
A staff initiated fair that shows off the advantages of IT 2020. It's open to students, staff and also to parents. Designate 1 school.
Cost: 6 bits

THEME WEEK CELEBRATION
A major event showcasing how staff in the pilot schools are using IT 2020. Staff, parents, and the media from the Region are invited to participate.
Cost: 8 bits

SCHOOL SUPPORT GROUP
A group of staff who are using IT 2020 meet weekly to help each other solve problems. Choose five people from 1 school.
Cost: 4 bits

IT 2020 SOFTWARE REVISION
Revision of the IT 2020 software to better fit the needs of the schools based on staff feedback. Form a committee of five staff.
Cost: 8 bits

POLICY IMPLEMENTATION
Change systems policies to reflect changes in curriculum and instruction resulting from adoption of learning technology. Form a committee of five staff from anywhere in the Region.
Cost: 8 bits

Figure 2. Change activities.

For example, if the team chooses to *Talk To* three people (see Figure 2), those three people may respond in a variety of different ways depending upon their backgrounds. Personalities, roles, and level of interest in learning technology. When the team *Talks To* Al, the Assistant Superintendent in charge of the pilot region, for the first time, they receive the following feedback:

> 'Al is very busy. He is involved in other projects to improve the region's productivity and doesn't have much time to talk with you today. He suggests that you coordinate with MIS staff at the Central Office. On your way out he says, 'I don't know *they* are always thinking up these new things for us to do.' Al moves one space.'

If they 'Talk to' Irene, she responds differently.

> '"I just don't like computers. They're so impersonal. How can this new system help me anyway? And what will I do when the system breaks down and I have to teach my classes? Will I be blamed when students don't learn?" Irene doesn't move at all.'

Talking to other people will generate a variety of reactions and different degrees of movement (i.e., change). Some activities also generate student benefits or *Bennies* (e.g., teaching a classroom lesson, holding a technology fair), while others do not (e.g., talking to people). If an activity generates benefits for students, this is noted in the feedback and tallied by the computer. This feature of the simulation serves to highlight the distinction between fostering interest and fostering effective use of the innovation. The teams are able to see not only their success in fostering change among staff, but also in improving learning outcomes. The instructor uses differences among the team's results on these two dimensions as a basis for the debriefing that occurs following 'each year of implementation.'

Through this process of planning, doing, getting feedback, reflecting, and acting, learners see the evolving results of their strategies for bringing the new learning technology into the schools. Yet, as becomes apparent to the learners, not all improvement strategies – the sequence of implementation of activities – are equally effective. Understanding how to implement change successfully entails the use of a 'strategic systemic approach' (Evans, 1996).

*Creating effective strategies for change*

The success of activities in the simulation depends upon two sets of factors. First, consistent with the research of Hall and Hord (1987), change activities must meet the needs or concerns of people. Consequently, in forming their strategy, the change team must match their selection of an activity *to the needs and concerns of the particular people at any given point in time.* Those needs are based on a variety of factors: their personal feelings about the innovation, their change adopter type, their role in the school, the attitudes of their peers, and most important their stage in the change process.

If staff members are in the *Interest* stage, activities that inform and increase interest meet people's needs. Activities that meet people's needs result in some level of change in attitudes and movement on the game board. In contrast, activities that focus on building skills may *not* succeed if the people are not yet interested (i.e., ready). An analogous 'decision rule' operates for people as they reach each stage represented on the game board.

Successful conduct of a given activity may also depend upon the creation of certain conditions in the school (i.e., completion of other activities). For example, the change team cannot successfully conduct a *Workshop* at a school site until they have gained support from the principal. If they conduct the *Workshop* activity before they have the principal's support, the feedback will say that they were unable to hold the workshop because they did not yet have the principal's permission.

This highlights the importance of administrative support. It also prompts the question for team members, 'How can we gain the principal's support?' In order to obtain the support or permission of the principal, the team will discover that they need to *Talk To* the principal until he or she agrees to support this initiative. This particular decision rule highlights the importance of the principal's role in implementing school-level change.

In all cases, the feedback provides not only information on the results but also provides contextualized cues as to the nature of the obstacles the change team has encountered. The team reviews this information and considers how to revise their strategy – what to do next – in order to overcome the particular obstacle (e.g., lack of principal support or lack of staff readiness).

This simulation was designed to help leaders learn how to *apply* knowledge of school change and improvement. Thus, at the end of the simulation (i.e., after three simulated years), the computer provides an assessment of the team's success. Two criteria are used: how many staff are using IT 2020 (i.e., game pieces in Early or Routine Use stages) and how many *Bennies* (i.e., student benefits) the team accumulated. Based on these results, the team is assigned to one of six levels of expertise in leading change: Apprentice, Novice, Manager, Leader, Expert, Master. Specific diagnostic feedback is provided based upon the level achieved.

As noted we use the simulation initially in a structured, team-based, cooperative learning environment. Following this initial exposure, however, we encourage individual learners to use the simulation on their own to further refine their understanding of strategic school improvement. Indeed, we use the outcome-based feature of the simulation for the purposes of grading and assessment for individual learners in classes and leadership development programs.

The simulation has been used extensively in a variety of Western industrialized countries (e.g., USA, Canada, United Kingdom, Netherlands, Belgium, Australia) with a highly positive response from practicing school leaders. Yet, both theoretical analysis and practical experience with the simulation suggested that use of the original version in Thailand would not yield the desired results. Simply stated, educational change in Thailand is based on different cultural assumptions (see Hallinger & Kantamara, 2000a, 2000b, 2001). Adaptation of the training simulation therefore would require not only translation but also cultural adaptation.

## Development of the Thai Version of *Making Change Happen!*

Borg and Gall (1989) describe research and development as, 'a cycle in which a version of the product is developed, field-tested, and revised on the basis of field-test data' (p. 781). The initial phases of the R & D cycle entail research and information collecting, planning, and developing a preliminary form of the product (Borg & Gall, 1989). Thus, our first task in approaching adaptation of the simulation was to identify the knowledge base that would underlie our Thai version. Next we developed a preliminary form of the Thai simulation. Then we finished with a cycle of field tests and further revisions of the product. We describe each of these in turn.

### Research and data collection

The authors drew upon several sources to inform adaptation of the simulation: our experience working with Thai schools, theoretical and empirical literature, advice from practitioners, results from our own case studies, field-tests and evaluations (see Hallinger & Kantamara, 2000a, 2000b, 2001).

*A cultural synthesis of Thai approaches to change*

Our research synthesis identified both similarities and differences between school improvement and change as reported in Western schools and Thailand. It is interesting to note that many of the change obstacles identified in Thailand also appear in the Western literature. These include shifting goals and policies, insufficient resources, the need for new skills among staff, staff resistance, political opposition, unclear articulation of needs, conflicting policies, traditions, lack of administrative support.

Certain 'strategic' dimensions of the change process observed in Thailand also appear similar:

- need for administrative support,
- stages in the development of new skills, attitudes and understandings related to a given innovation,
- need to engage people's commitment in order to bring about lasting change,
- importance of institutional elements in solidifying changes in the school,
- individual differences in response to the same change,
- impact of individual 'school cultures' on change efforts, and
- change as a process of development of technical skills and feelings.

Identifying these similarities in the process of school change in Thailand and the West was important. It suggested that certain fundamental dimensions of the simulation might remain more or less intact.

At the same time, however, we also found a range of differences in the response of Thai educators to change. Understanding the nature and source of these differences held the key to our R & D project. We used a cross-cultural framework developed by Hofstede to assist in analyzing the characteristics of Thai responses to change.

Hofstede defined culture as the *collective mental programming of the people in a social environment in which one grew up and collected one's life experiences* (Hofstede, 1980, 1983, 1991). His cross-national research identified four dimensions on which national cultures differ: Power Distance, Uncertainty Avoidance,

Individualism–Collectivism, and Masculinity–Femininity. The dimensions yielded a useful point of departure for comparing how Thai people respond to change.

Power distance describes the degree to which large status differences exist among people in a society and also the extent to which these differences in power are accepted. The *large power distance* characterizing Thai culture shapes the behavior of administrators, teachers, student and parents in important ways. People of lower status show much higher deference towards those of authority or senior status in social relationships than is typical in the West.

Students naturally defer to teachers, teachers to principals and principals to their superiors. This results in a pervasive, socially-legitimated expectation that decisions should be made by those holding positions of authority and reinforces the strength of hierarchical relations. Large power distance creates a cultural tendency for administrators to lead by *fiat*. There is a cultural assumption that leading change entails establishing orders – which will be followed naturally by others – and applying pressure in special cases where it is needed.

It is critical to note that large power distance describes a web of social expectations. It is not simply a matter of superordinates desiring authority, but within this culture subordinates expect them to exercise their legitimate power. Thai's refer to this cultural deference or inclination to show consideration to seniors as *greng jai*. *Greng jai* is a dominant norm that influences all social relations, not simply inside school or other formal organizations (Holmes & Tangtongtavy, 1995).

Hofstede contrasted *collectivism* and *individualism*. Collectivist societies value social relations over individual performance. People in a collectivist culture think naturally in terms of 'we' rather than 'I'.

The highly collectivist nature of Thai culture shapes the context for school improvement by locating change in the social group somewhat more than within individuals. As with other Asian societies, Thai's look primarily to their referent social groups in order to 'make sense' of events (Herbig & Dunphy, 1998; Holmes & Tangtongtavy, 1995; McDonald & Pratt, 1998). Consequently, staff are more likely to 'move in the direction of change' as a group than as individuals.

Hofstede refers to a dimension of *high uncertainty avoidance*. In cultures with a high degree of uncertainty avoidance, there is a low cultural tolerance for ambiguity and non-conformity. In Thailand, which ranks moderately high on uncertainty avoidance. People tend to avoid risks. Place a high value on conformity of opinion and behavior, and seek a high level of control over their environment (Hofstede, 1980). Thai's are strongly socialized to conform to group norms, traditions, rules and regulations. They find change more disruptive and disturbing than in 'lower uncertainty avoidance' cultures.

People who innovate by definition tend to stand out from the group. In some countries innovators are admired, but Thailand's heroes are not great individual achievers. Rather they tend to people who quietly represent the traditional aspirations of the group. This dimension suggests that Thai schools represent an even less fertile ground for innovation and change than the much criticized schools of Western nations.

The fourth dimension of Hofstede's framework contrasts femininity and masculinity. Feminine cultures place a high value on the maintenance of harmonious social relations. Masculine cultures focus on achievement and performance.

The *feminine* dimension of their culture leads Thai's to place a high value on social relationships, to seek harmony, and to avoid conflict. Thai's place great emphasis on living and working in a pleasurable atmosphere and on fostering a strong spirit of community. Anything that threatens the harmonious balance of the social group (e.g., change) creates natural resistance.

In contrast, masculine cultures such as the U.S. emphasize results, performance, and productivity (Herbig & Dunphy, 1998; Hofstede, 1980). This dimension has implications for a variety of factors often associated with school change and improvement including responses to pressure, the use of accountability, measurement of performance outcomes, and the role of informal social relationships during change.

We employed this conceptual framework to analyze the process of change in Thai schools (Hallinger & Kantamara, 2000b, 2001). We also conducted empirical case studies of selected 'successful change schools' in order to fill in the outlines that emerged from the literature review (Hallinger & Kantamara, 2000a). We then synthesized these data to generate propositions about the nature of leadership and change in Thai schools (Hallinger & Kantamara, 2001). These included the following.

1. Target formal leaders and obtain their support early in the change process.
2. Formal leaders should use strategies that deemphasize traditional norms of deference to authority and bring staff concerns to the surface so they can understand and address causes of potential staff resistance.
3. Change leaders should pay special attention to creating group consensus around the nature of the change.
4. Leaders should take more time and effort to inform and interest staff during the initial stages of change.
5. Leaders should not assume that a policy adopted is a policy implemented. Implementation must be viewed as a long-term process that requires ongoing support for the staff as a whole and as individuals.
6. Obtain and cultivate the support of informal leaders and leverage resources of the social network to create pressure and support for change.
7. Use formal authority and policies selectively to reinforce expectations and standards consistent with implementation of the innovation.
8. Find ways to inject fun, encourage group spirit, and celebrate shared accomplishments in the workplace while maintaining accountability.

On the surface this list appears similar to recommendations that might be offered to an American, British or Australian staff. This reflects several factors. Thai society is in a process of integration into a global culture. While the process of *cultural* change is slow, it is taking place nonetheless. Thus, certain global norms and values (e.g., regarding participation in decision making) are gradually filtering into all societies.

In addition, as noted above, certain dimensions of the change process appear to carry over across cultures. Thus, even some of the differences observed in Thailand are essentially differences of degree. For example, it has become a *sine qua non* in the Western school improvement literature that the principal is a key gatekeeper in the process of school improvement. Obtaining principal support is an important ingredient in successful educational change (Evans, 1996; Fullan, 1990; Hall & Hord, 1987).

In Thai culture, the 'large power distance' associated with social relations makes support from the principal even more crucial. Thai staff simply cannot move towards implementation of an innovation until their principal has signaled active support. Moreover, because decision-making in the Thai school is more centralized than in the West, the Thai principal plays a similarly critical role at each stage of implementation.

In selected cases, these differences in degree attain a level where the cultural distinctions are quite dramatic. For example, we asserted that the *collectivist* nature of Thai culture makes the group the central locus of movement during change. In combination with the *uncertainty avoidance* characteristic of Thai culture, this leads staff to avoid actions that would make them stand out from the group or disturb the status quo.

The combination of *femininity* and large *power distance* all combine to create and interesting contrast with the West. Even when Thai's disagree with a proposal, they will seek to avoid saying so. The cultural emphases on politeness and moderation blend with the need to *greng jai* or defer to those of higher status.

As noted earlier, the R & D process also entailed conducting a set of case studies of schools that had successfully implemented long-term innovations in the recent past. The case studies were designed to begin to test and elaborate on the propositions that had emerged form the literature reviews. Space limitations prevent the presentation of these data here (see Hallinger & Kantamara, 2000a). Table 1, however, displays how we translated findings from the literature and case studies into changes in the simulation.

**Planning and preliminary development of the Thai simulation**

Initial revision of the simulation involved consideration of differences in the institutional and cultural contexts of education in Thailand. Changing the institutional context to reflect the Thai educational system was not difficult. This involved small changes in the titles of positions, the problem description, and the nature of the school organization.

These revisions were far less significant than changes resulting from differences arising from the social culture of Thai schools. The linkages between cultural characteristics, their effects on change in Thai school organizations, the implications for leading change, and the resulting revisions to our change simulation are detailed in Table 1. Weaving these features into the simulation in a way that would seem realistic to Thai educators and accurately model the process of change in Thai schools would prove to be the real challenge of adaptation.

In terms of change strategies embedded in the Thai version, we concluded that learners would need to develop a change strategy that differs in at least three important ways from the original version.

1. The Thai version of the simulation would require the change team to pay even greater attention to building interest among the staff prior to actual implementation of the new learning technology.
2. The change team must pay greater attention to leading change as a group process.
3. There is an even greater need for support from the principal than in the original version.

Space limitations preclude us from describing all of the changes made to the simulation. Instead we focus on providing *representative* types of changes made to reflect the

Table 1. Cultural translation of the change simulation (from Hallinger & Kantamara, 2001).

| **Findings from Literature & Research** | | **Conclusions and Strategies** | |
|---|---|---|---|
| **Cultural Dimension** | **Effect on Change** | **Implication for Leading Change** | **Simulation Revisions** |
| **1. Large power distance** | | | |
| • Highly hierarchical and bureaucratic society<br>• Implicit belief that power differences are 'natural'<br>• There is an emphasis and high value placed on being 'polite'; in practice this means avoiding creating situations that<br>• Deference to authority and to seniors in age and rank<br>• People 'expect' to be told what to do; participation is not viewed as a right or as something to be sought<br>• 'Just do it' mentality prevails throughout society<br>• Managers have more power but also greater obligations to their subordinates<br>• Real leadership is earned | • People believe status differences are natural and accept their position in society from birth<br>• Reluctance to question – why? – at all levels<br>• Reluctance to make a decision on one's own at all levels<br>• Greater authority vested in people holding formal administrative positions such as principals<br>• Staff tend to follow orders at least at a surface level<br>• Compliance culture; people at implementation level often lack commitment to change<br>• Assume adoption represents real change; insufficient focus on supporting implementation<br>• Informal relationships heavily shape expectations | • Focus on articulating the moral purpose behind change rather than just the institutional orders<br>• Use non-public as well as strategies to uncover the varying perspectives of people<br>• *Listen* to people more; sell them on the change less. Do not mistake polite acceptance as commitment<br>• Take a long-term approach to reducing power distance and fostering genuine participation<br>• Foster individual and group initiative; ensure leaders' actions are consistent with words<br>• Use power selectively<br>• Support the practical tasks of implementation<br>• Gain support of informal leaders; share responsibility for implementing change | • Negative responses are softened; opposition is more indirect and couched in polite phrasing<br>• Early and Late Majority 'adopter types' respond neutrally or positively when the team Talks To them the first time. Unlike the Western version, they *do not move even though they are polite and positive*<br>• Persistence in Talking To the Thai staff pays off. The Early and Late Majority 'adopter types' who do *not* move the first time the team Talks To them, move one space when they Talk To them for a second time<br>• Staff support for IT 2020 is linked more directly to the support of their superior for IT 2020 than in the original version<br>• The leaders' support directly influences the benefits. This does not happen in the original |

*(Continued)*

Table 1. *(Continued)*

| Findings from Literature & Research | | Conclusions and Strategies | |
|---|---|---|---|
| **Cultural Dimension** | **Effect on Change** | **Implication for Leading Change** | **Simulation Revisions** |
| **2. High collectivism**<br>• 'We' consciousness prevails rather than 'I'<br>• Change is moderated through the 'eyes' of the group<br>• Group spirit is a fundamental prerequisite to individuals gaining confidence<br>• Fear of not meeting the group's expectations tends to be greater than fear of individual failure | • Almost all people move through change with the group<br>• Group consensus and spirit moderate efforts to change more than individual decisions<br>• People are highly sensitive to social acceptance and sanctions to direct their behavior during change<br>• Actions which make one stand out from the group are avoided<br>• Conflict with others in the group is avoided and disagreement with the direction of change is hidden | • Use more change activities that focus on the group and allow development of a group consensus<br>• Focus especially on obtaining support from the group's informal leaders<br>• Allow the group to 'make sense' of the change both inside and outside of the formal school setting<br>• Use team-building, synergistic activities that enhance the group's spirit even as they address technical aspects of the innovation<br>• Connect innovations to important values of the school and society | • Responses in Thai text feedback on activities are changed to show the teachers' greater concern about what others think about IT 2020<br>• An activity – 'School Visit' – is added to the Thai version. This is an overnight trip to view IT 2020 in action at another school. It gives staff opinion a chance to coalesce around the change in an informal atmosphere<br>• A group lunch is mentioned in feedback response to a couple of activities to reflect the importance of informal social relationships and group activities.<br>• The Thai text in several responses emphasizes the importance of the innovation to the children's opportunities in the future |

*(Continued)*

Table 1. *(Continued)*

| Findings from Literature & Research | | Conclusions and Strategies | |
|---|---|---|---|
| **Cultural Dimension** | **Effect on Change** | **Implication for Leading Change** | **Simulation Revisions** |
| **3. High uncertainty avoidance** | | | |
| • Cultural norms foster stability and continuity more so than in low UA cultures on which the simulation was based | • Change is often slower in high UA cultures such as Thailand | • Demonstrate clarity and seriousness of purpose in words and actions | • Innovators and Early Majority Types move one space fewer than in original version, the first time the team Talks To them |
| • Thai culture tends neither to seek innovation nor to reward innovators<br>• Innovation and 'being different' more generally are regarded as undesirable and disruptive qualities to the stability of social relations<br>• Traditions and rules exert a stronger reign on individual and group behavior | • High level of discomfort with uncertainty, ambiguity, and complexity, all of which come as a part of change<br>• 'Innovators' are marginalized<br>• Accept rules and traditions as 'natural' and with a logic of their own even when they cease to make sense<br>• Once a change is made, it is very difficult to change<br>• Reluctance to make decisions that depart from status quo<br>• Focus on guidelines rather than purposes of change<br>• Strong bureaucratic emphasis creates 'order-taking' or 'wait and see' mentality at all levels | • Gain support of the group behind the purposes and practices associated with the change<br>• Gain support of formal and informal leaders<br>• Connect the purposes of the change to past policies and traditions<br>• Expect change to be slow and persist in the face of opposition<br>• Use public activities to signal changes and create new vision of what is possible<br>• Celebrate incremental successes<br>• Provide consistent and visible support for implementation new practices | • Late Majority adopters do not move at all on the first Talk To<br>• Text changes in feedback to activities reflects tendency of staff to wait and see principal's response first<br>• Text changes in feedback to activities reflects tendency of principal to wait and see supervisor's response first<br>• Text changes in feedback to activities reflects tendency of staff to respond to moral purposes articulated for the change |

*(Continued)*

Table 1. *(Continued)*

| Findings from Literature & Research | | Conclusions and Strategies | |
|---|---|---|---|
| **Cultural Dimension** | **Effect on Change** | **Implication for Leading Change** | **Simulation Revisions** |
| **4. Feminine Culture**<br>• Caring for other people and the preservation of harmonious social relationships is emphasized in the society and the workplace<br>• Social relations are valued more than productivity or performance at both the individual and group levels<br>• Harmony between individuals and among groups is sought and conflicts are avoided as much as possible<br>• People act on feelings more than on logic; in Thai to 'understand' each other is to 'enter each other's hearts'<br>• All relationships entail reciprocity; those with largest power distance carry the greatest obligation on the part of the senior member | • People seek to maintain social harmony, even if it means foregoing potential benefits of change<br>• Open disagreement over goals or procedures is avoided<br>• Resistance to change remains passive, covert and 'underground'<br>• Tendency to view lack of dissent as support. Leaders proceed without real support resulting in partial implementation<br>• 'Group processes' popular in Western cultures fail to obtain the desired results<br>• Logical arguments for change carry less weight<br>• People mix work and play; work without fun achieves fewer results | • Demonstrate moral leadership; connect change to the needs of people<br>• Show sincerity in words and actions<br>• Demonstrate seriousness of purpose<br>• Try to resolve conflicts by compromise and negotiation; look for win-win solutions<br>• Show sincere personal interest in people as individuals; demonstrate caring as people struggle to change<br>• Create opportunities for staff to have fun and develop team spirit during the change<br>• Celebrate success and provide moral support | • Staff respond neutrally or positively, but do not move the first time the team Talks To them<br>• Negative responses to the change are softened in the text; opposition is more indirect and couched in appropriate cultural phraseology<br>• 'Having a major fight' at a key Board Meeting related to Policy Change is taken out<br>• Text refers to cultural and organizational traditions and rituals to foster group spirit<br>• The Site Visit activity description and text feedback emphasizes the process by which the staff mix work and play to achieve consensus on the innovation |

cultural adaptation of the simulation (see Table 1). Specific modifications to the simulation fell into several categories:

1. Revision of the descriptions of text descriptions and activity feedback;
2. Revision of the change activities;
3. Revision of the decision rules underlying player movement through the stages of the change process and in the student benefits accruing from activities.

*Descriptions of staff and feedback dialogue*

The original version used Rogers' (1971) *adopter types* to classify staff's attitudes towards change. Given the absence of similar data on Thai schools, we stayed with the same breakdown. We only changed the descriptions of people to reflect differences the more 'polite' and conservative nature of Thai people.

Considerable revision was made in the feedback and dialogue provided in response to activities. For example, when the team *Talks To* staff in the original version, there are many questions a fair amount of overt resistance is expressed. In the Thai version staff ask no questions, and the tone of resistance is softened considerably. Their responses reflect the cultural tendency towards overt. Polite compliance (i.e., *greng jai)* even in the absence of any change in behavior. This type of revision was carried out as deemed appropriate throughout the simulation text.

*Activities*

The change activities represent the vehicle by which the team fosters interest, acceptance, learning, and long-term use of IT 2020. The activities included in the original version of the simulation (see Figure 2) represent the same activities Thai schools typically use to foster change. However, our research suggested a need to add one additional activity to the Thai simulation: an overnight visit to observe the use of IT in another school.

Typically such visits involve the staff traveling together to another school some distance away from home. Teachers will observe in classrooms and talk with other teachers. In the evening they will typically eat, talk, and perhaps sing together.

This activity provides an opportunity for *the group* to make sense of the change outside of the formal school setting. Consistent with the importance of *sanook* (fun) in Thai culture, the trip builds a bond among the group members and set the stage for building support back at the school. Like another of the activities, the *Demonstration of IT 2020* at the school site, this activity is an important stimulus for creating interest and making the abstract notion of IT 2020 more real. Given the more passive orientation of Thai staffs, it is even more critical for leaders to create opportunities where teachers can ask questions and find personal meaning in the early stages of the change process.

*Decision rules*

When revising the decision rules to reflect the Thai context, we needed to maintain the theoretical integrity and internal coherence of the simulation. Revisions in one decision rule could have an unintended but potentially important impact on another dimension of the simulation. Again, however, revision was informed by three general differences observed in Thai schools.

By way of example, one significant change entailed the *Talk To* activity. In the original version of the simulation, it is critical that the team take time to *Talk To* people as a means of informing them about IT 2020, but also as a means of finding out staff perspectives on the change. When the team *Talks To* individuals their responses and subsequent movement are linked to their adopter types; the staff member may move 3 spaces (*Innovators*), two spaces (*Leaders*) one space (*Early and Late Majority*) or not at all (*Resistors*).

Based on the *large power distance* observed in Thai culture, we made two relevant changes on this activity. We changed the programming so that staff falling into the *Early Majority* and *Late Majority* Adopter Types respond politely and/or positively the first time the change team Ta*lk To* them. They ask no questions, and evince no negative opinions. However, instead of moving a single space as in the original version, they do not move at all.

This reflects the tension between the cultural need to show polite deference and the underlying uncertainties that still accompany change. This norm of overt compliance and passive resistance is an important *pattern* that school leaders in Thailand must recognize and address if real change is to take place.

Another decision rule adaptation involved the role of the school principals. In the original version, the principal's support is necessary in order to conduct activities in the schools. To reflect the even greater importance of the Thai school leader in the change process, we increased the *Bennies* accruing from school-level activities (e.g., *Workshops*) if the team has obtained strong support from the principals.

These are just a few examples of the revisions made to the simulation. See Table 1 for a fuller but still incomplete list of the revisions.

**Field tests and further revision of the Thai change simulation**

Field testing of the simulation proceeded through several phases. Four separate field trials were conducted with the simulation. Each field trial consisted of using the simulation in a computer lab setting with between 25 and 45 school leaders in a full-day workshop. Between each field trial, revisions were incorporated into the simulation based on formative and summative evaluation results.

*Formative evaluation*

Formative and summative evaluation of the simulation were conducted using a variety of instruments including:

1. direct observation by the authors,
2. a talk-back sheet soliciting formative feedback on strengths and weaknesses of the simulation and the accompanying instructional process,
3. verbal debriefings with the workshop participants.

The formative evaluation data informed the further adaptation of the simulation and the instructional process. Revisions included a variety of minor revisions to the game's decision rules to maintain its internal consistency.

*Summative evaluation*

Summative evaluation was conducted using two main data sources:

1. pre-post test on relevant concepts derived from the learning objectives of the simulation,

2. short (two page) essays in which the learners focused on key learnings they acquired from the simulation.

Taken together the summative evaluation results yielded several conclusions.

First, the simulation met the goal of introducing important strategic concepts of change leadership. It was useful at stimulating the learners to think more deeply about change in their own schools. The results suggested improvement on the primary goals of understanding obstacles to change and the elements of effective change strategies.

At the same time, the degree of understanding of change strategies did not meet the authors' desired level of mastery. The dramatic change in the nature of instruction led the authors to underestimate the amount of time needed to solidify the learning. Thai school leaders are accustomed to a lecture format. Few had ever worked in either a formal cooperative learning or computer-based learning environment.

It took them longer than North American educators to adapt to the computer-based instructional design. However, once they got over the initial confusion, they enjoyed it and remain highly engaged. In the fourth field trial we allocated eight hours instead of six hours and obtained better results on the summative evaluation. Thus, we concluded that eight hours of instruction would be needed to meet the learning objectives at a high level of mastery in Thailand.

Second, we observed an unanticipated outcome of the simulation. It appeared to have a significant impact on the learners' attitudes towards the use of learning technology. Learning through the computer-based simulation appeared to stimulate new attitudes towards both technology and change. It also changed the perspective of numerous participants towards the value of learning technology.

*Future research*

The evaluation program undertaken to date with the Thai version of the simulation has focused on ensuring a high level of face validity. Through several rounds of data collection with primary and secondary school administrators and teachers, the Thai school leaders concurred that the context, the characters and process of change as they unfold in the simulation 'feel real' to them. The embedded change strategies also made sense to them, despite the fact that conceptualizing change as a systemic strategic process was new to them.

At the same time, we do not yet have data that shed light on the external validity of the simulation program in terms of its use as a training tool with school leaders in Thailand. This phase of research and development will entail using the program with leaders engaged in the change process and subsequently observing the extent to which their leadership strategies and behaviors have changed. A program of validation could also compare more systematically the degree to which the strategies conceptualized as effective in the simulation result in change in a set of real schools. We view this as an important extension of the current research and development project.

## Reflection on the Role of Research and Development in an Era of Global Knowledge Dissemination

The rapid pace of globalization makes it critical that we become more conscious of the limitations of our own cultural contexts. Too often 'what we don't know we don't know' is our greatest handicap (Hallinger, 1995; Hallinger & Leithwood, 1996, 1998). Future leadership development efforts must be embedded in a 'knowledge base' that is relevant not only to global trends in education but also grounded in the norms of local cultures (Bajunid, 1996; Cheng, 1995; Chin, 2002; Hallinger & Leithwood, 1996, 1998; Lam, 2002; Walker et al., 1996).

We would emphasize that we started this project with the explicit assumption that the knowledge base on which we were building needed to be adapted and validated in the local context. Undoubtedly there are educational practices that are not culturally bound. If that is the case, however, we suggest that it remains the responsibility of those who would disseminate knowledge to confirm that assumption. We find too many staff involved in school leadership development all too ready to *carry their training packages* several thousand miles from home under the easy assumption that the end-users will make the necessary adjustments.

The process of research and development that we have described in this chapter is time-consuming, but we believe worth the investment. Indeed, we would view this type of process of knowledge transformation and validation to be the prime responsibility of tertiary instructions in developing nations. The many leadership development centers and institutes springing up globally seem primed to provide training to scores of prospective and practicing school leaders. We would suggest, however, that the key role to be played by universities in this process should be knowledge generation, knowledge transformation, and the development of valid tools for training school leaders.

The findings from this project highlight the inherent limitations of applying knowledge gained in one cultural context to another. While we have only begun to understand elements of successful school improvement in Thailand, there is no question that substantial culturally-derived differences exist when compared with Western nations. We believe that many of these differences are shared by other Asian nations, though this awaits empirical verification.

Despite our confidence in the efficacy of this type of cultural analysis, we would also caution against the reification of indigenous knowledge during this global era. We agree with McDonald and Pratt's assertion that training programs: 'need to be directed at educating tomorrow's professionals and leaders, and therefore we should be including in curricula not only extant knowledge, but also academic fundamentals in support of future scenarios' (1997, p. 55).

Globalization will continue to influence the 'future scenarios' that shape education in all societies. Therefore, an emerging challenge for scholars and practitioners in school improvement is to generate, interpret and balance knowledge gained from global and indigenous sources. Our experience suggests that this challenge not only holds potential for improving educational practice, but also for breathing new life into the academic enterprise of higher education.

## References

Abdullah, A. S. (1999). *The school management and leadership directions in Malaysia for the 21st century*. Paper presented at the 3rd annual Asian Symposium on Educational Management and Leadership. Penang, Malaysia.

Bajunid, I. A. (1996). Preliminary explorations of indigenous perspectives of educational management: The evolving Malaysian experience. *Journal of Educational Administration*, 34(5), 50–73.

Borg, W. & Gall, M. (1989). *Educational research: An introduction, 5th Edition* (781–804). White Plains, New York: Longman.

Bolam, R. (2002). The changing roles and training of headteachers: The recent experience in England and Wales. In P. Hallinger (Ed.), *Reshaping the landscape of school leadership development: A global perspective*. Lisse, Netherlands: Swets & Zeitlinger.

Caldwell, B. (1998). Strategic leadership, resource management and effective school reform. *Journal of Educational Administration*, 36(5), 445–461.

Caldwell, B. (2002). A blueprint for successful leadership in an era of globalization in learning. In P. Hallinger (Ed.), *Reshaping the landscape of school leadership development: A global perspective*. Lisse, Netherlands: Swets & Zeitlinger.

Cheng, Kai-Ming. (1995). The neglected dimension: Cultural comparison in educational administration. In Wong, K.C. & Cheng, K.M. (Eds.), *Educational leadership and change: An international perspective* (87–104). Hong Kong University Press, Hong Kong.

Cheng, Y. C. & Townsend, T. (2000). Educational change and development in the Asia Pacific region: Trends and issues. In T. Townsend & Y. C. Cheng (Eds.), *Educational change and development in the Asia Pacific: Challenges for the future* (317–344). Lisse, Netherlands: Swets & Zeitlinger.

Chin, J. (2002). Reconceptualizing administrative preparation of principals: Epistemological Issues and perspectives. In P. Hallinger (Ed.), *Reshaping the landscape of school leadership development: A global perspective*. Lisse, Netherlands: Swets & Zeitlinger.

Chong, K. C., Stott, K. & Low, G. T. (2002). Developing Singapore school Leaders for a learning nation. In P. Hallinger (Ed.), *Reshaping the landscape of school leadership development: A global perspective*. Lisse, Netherlands: Swets & Zeitlinger.

Copland, M. (2002). Problem-based leadership development: Developing the cognitive and skill capacities of school leaders. In P. Hallinger (Ed.), *Reshaping the landscape of school leadership development: A global perspective*. Lisse, Netherlands: Swets & Zeitlinger.

Crandall, D., Eiseman, J. & Louis, K. S. (1986). Strategic planning issues that bear on the success of school improvement efforts, *Educational Administration Quarterly*, 22(3), 21–53.

Davis, B. (1999). *Credit where credit is due: The professional accreditation and continuing education of school principals in Victoria*. Paper presented at the Conference on Professional Development of School Leaders, Centre for Educational Leadership, Hong Kong University, Hong Kong.

Dimmock, C. & Walker, A. (1998). Transforming Hong Kong's schools: Trends and emerging issues. *Journal of Educational Administration*, 36(5), 476–491.

Evans, R. (1996). *The human side of change*. San Francisco: Jossey Bass.

Feng, D. (1999). *China's principal training: Reviewing and looking forward*. Paper presented at the Conference on Professional Development of School Leaders, Centre for Educational Leadership, Hong Kong University, Hong Kong.

Feng, D. (1999). China's principal training: Retrospect and Prospect. In P. Hallinger (Ed.), *Reshaping the landscape of school leadership development: A global perspective*. Lisse, Netherlands: Swets & Zeitlinger.

Fullan, M. (1990). *The new meaning of educational change*, Teachers College Press, New York.

Fwu, Bih-jen, & Wang, Hsiou-huai. (2001). *Principals at the crossroads: Profiles, preparation and role perception of secondary school principals in Taiwan*. Paper presented at the International Conference on School Leader Preparation, Licensure, Certification, Selection, Evaluation and Professional Development, Taipei, ROC.

Gopinathan, S. & Kam, H. W. (2000). Educational change and development in Singapore. In T. Townsend & Y. C. Cheng (Eds.), *Educational change and development in the Asia Pacific: Challenges for the future* (163–184). Lisse, Netherlands: Swets & Zeitlinger.
Hall, G. & Hord, S. (1987). *Change in schools: Facilitating the process*. Albany, NY: State University of New York Press.
Hallinger, P. (1992). School leadership development: Evaluating a decade of reform. *Education and Urban Society*, 24(3), 300–316.
Hallinger, P. (1995). Culture and leadership: Developing an international perspective in educational administration. *UCEA Review*, 36(1), 3–7.
Hallinger, P. (1998). Educational change in Southeast Asia: The challenge of creating learning systems. *Journal of Educational Administration*, 36(5), 492–509.
Hallinger, P. (1999). School leadership development: State of the art at the turn of the century. *Orbit*, 30(1), 46–48.
Hallinger, P. & Bridges, E. (1997). Problem-based leadership development: Preparing educational leaders for changing times. *Journal of School Leadership*, 7, 1–15.
Hallinger, P., Crandall, D. & Ng Foo Seong, D. (In press). Making change happen: A simulation for learning to lead change. *The Learning Organization*.
Hallinger, P. & Kantamara, P. (2000a). Educational change in Thailand: Opening a window onto leadership as a cultural process, *School Leadership and Management*, 20(1), 189–206.
Hallinger, P. & Kantamara, P. (2000b). Leading at the confluence of tradition and globalization: The challenge of change in Thai schools, *Asia Pacific Journal of Education*, 20(2), 46–57.
Hallinger, P. & Kantamara, P. (2001). Learning to lead global changes across cultures: Designing a computer-based simulation for Thai school leaders. *Journal of Educational Administration*, 39(3), 197–220.
Hallinger, P. & Leithwood, K. (1996). Culture and educational administration: A case of finding out what you don't know you don't know, *Journal of Educational Administration*, 34(5), 98–119.
Hallinger, P. & Leithwood, K. (1998). Unseen forces: The impact of social culture on leadership. *Peabody Journal of Education*, 73(2), 126–151.
Hallinger, P. & McCary, M. (1990). Developing the strategic thinking of instructional leaders. *Elementary School Journal*, 91(2), 90–108.
Hargreaves, A. & Fullan, M. (1998). *What's worth fighting for out there*. New York: Teachers College Press.
Herbig, P. & Dunphy, S. (1998). Culture and innovation, *Journal of Management Development*. 5(4), 13–21.
Hofstede, G. (1980). *Culture's consequences: International differences in work-related values*. Beverly Hills, CA: Sage.
Hofstede, G. (1983). The cultural relativity of organizational practices and theories, *Journal of Business Studies*, 13(3), 75–89.
Hofstede, G. (1991). *Culture and organizations: Software of the mind*. Berkshire, England: McGraw-Hill Books.
Holmes, H. & Tangtongtavy, S. (1995). *Working with the Thais: A guide to managing in Thailand*. Bangkok, Thailand; White Lotus.
Huber, S. G. & West, M. (2002). Developing school leaders – A critical review of current practices, approaches and issues, and some directions for the future. In K. Leithwood & P. Hallinger (Eds.), *Second international handbook of educational leadership and administration*. New York: Kluwer Academic Press.
Lam, J. (1999). *Balancing stability and change: Implications for professional preparation and development of principals in Hong Kong*. Paper presented at the Conference on

Professional Development of School Leaders, Centre for Educational Leadership, Hong Kong University, Hong Kong.

Lee, C. (1990). Determinants of national innovativeness and international market segments. *International Marketing Review*, 7(5), 39–49.

Leithwood, K. (1994). Leadership for school restructuring. *Educational Administration Quarterly*, 30(4), 498–518.

Leithwood, K., Jantzi, D. & Steinbach, R. (1999). *Changing leaders for changing schools.* Buckingham, UK: Open University Press.

Li, Wenchang. (1999). *Organizing principal training in China: models, problems and prospects.* Paper presented at the Conference on Professional Development of School Leaders, Centre for Educational Leadership, Hong Kong University, Hong Kong.

Low, Guat-Tin. (1999). *Preparation of aspiring principals in Singapore: A partnership model.* Paper presented at the Conference on Professional Development of School Leaders, Centre for Educational Leadership, Hong Kong University, Hong Kong.

Marsh, D. (1992). School Principals as instructional leaders: The impact of the California School Leadership Academy. *Education and Urban Society*, 24(3), 386–410.

McDonald, P. & Pratt, G. (1997). Management education within a cultural confluence: Twinning programmes in Malaysia, *Malaysian Management Journal*, 2(1), 43–57.

Ming-dih Lin. (2002). Professional development for principals in Taiwan: The status quo and future needs. In P. Hallinger (Ed.), *Reshaping the landscape of school leadership development: A global perspective.* Lisse, Netherlands: Swets & Zeitlinger.

Ministry of Education-R.O.C. (1998). *Towards a learning society.* Taipei, Republic of China: Ministry of Education.

Ministry of Education-Thailand. (1997a). *Introducing the Office of the National Primary Education Commission.* Bangkok, Thailand: Ministry of Education.

Ministry of Education-Thailand. (1997b). *The experience from the Basic and Occupational Education and Training Programme.* Bangkok, Thailand: Ministry of Education.

Murphy, J. (1992). *The Landscape of leadership preparation: Reframing the education of school administrators.* Newbury Park, CA: Corwin Press.

Murphy, J. & Adams, J. (1998). Reforming America's schools, 1980–2000. *Journal of Educational Administration*, 36(5), 426–444.

Murphy, J. & Shipman, N. (2002). Developing standards for school leadership development: A process. In P. Hallinger (Ed.), *Reshaping the landscape of school leadership development: A global perspective.* Lisse, Netherlands: Swets & Zeitlinger.

Network Inc. (1988, 1999). *Making change happen!*™ Rowley, MA: The Network Inc.

Reeves, J., Forde, C., Casteel, V. & Lynas, R. (1999). Developing a model of practice: designing a framework for the professional development of school leaders and managers. *School Leadership and Management*, 18(2), 185–196.

Rogers, E. (1971). *Diffusion of innovations.* New York, NY: The Free Press.

Rogers, E. & Shoemaker, F. (1982). *Communication of innovations: A cross culture approach.* New York: The Free Press.

Sarason, S. (1982). *The culture of the school and the problem of change, (rev. ed.).* Boston: Allyn and Bacon.

Sarason, S. (1990). *The predictable failure of educational reform: Can we change course before it's too late?* San Francisco: Jossey Bass.

Swierczek, F. (1988). Culture and training: How do they play away from home? *Training and Development Journal*, 42(11), 74–80.

Suzuki, S. (2000). Japanese education for the 21st century: Educational issues, policy choice, and perspectives. In T. Townsend & Y. C. Cheng (Eds.), *Educational change and development in the Asia Pacific: Challenges for the future* (57–82). Lisse, Netherlands: Swets & Zeitlinger.

Tomlinson, H. (1999). *Recent developments in England and Wales: The Professional Qualification for Headship (NPQH) and the Leadership Programme for Serving Headteachers (LPSH)*. Paper presented at the Conference on Professional Development of School Leaders, Centre for Educational Leadership, Hong Kong University, Hong Kong.

Tomlinson, H. (2002). Supporting school leaders in an era of accountability: The National College for School Leadership in England. In P. Hallinger (Ed.), *Reshaping the landscape of school leadership development: A global perspective*. Lisse, Netherlands: Swets & Zeitlinger.

Walker, A., Bridges, E. & Chan, B. (1996). Wisdom gained, wisdom given: Instituting PBL in a Chinese culture. *Journal of Educational Administration*, 34(5), 98–119.

Yang, C. L. (2001). *The changing principalship and its implications for preparing, selecting and evaluating principals*. Paper presented at the International Conference on School Leader Preparation, Licensure, Certification, Selection, Evaluation and Professional Development, Taipei, ROC.

# 9

# Developing Leaders for Self-Managing Schools: The Role of a Principal Center in Accreditation and Professional Learning

Bruce Davis

*Educational Consultant, Flinders, Victoria, Australia*

The redistribution of power between education bureaucracies and individual schools caused profound educational change during much of the last decade in many countries around the world. In most cases, professional educators were not the initiators of these decentralizing educational reforms. Rather, the redistribution of authority among educational agencies has been a political response led by governments and business communities.

Throughout developing and industrialized societies alike, concerns about international competitiveness and uncertainty about the future have fueled discontent with the educational status quo. The international response has increasingly focused on the self-management of schools, the introduction of common curriculum standards, and the overhaul of student assessment processes. The implementation of this wide-ranging agenda has impacted on the work of hundreds of thousands of teachers and principals.

The political restructuring of school education in Australia followed patterns already developed in New Zealand. In Australia, reforms of this type first appeared in 1989 in New South Wales and Tasmania. Other Australian jurisdictions quickly followed suit. By 1993 all eight government education systems in Australia were somewhere along the change continuum. Although Victoria was one of the last jurisdictions to initiate school restructuring, by 2000 it had undertaken the most radical and far-ranging reform scheme of all Australian states.

The professional growth of principals[1] was quickly viewed by Australia's educational policymakers as a strategic necessity. Indeed, funds have generally been provided for

---

[1] 'Principal' is here meant to include deputy principals, assistant principals, the heads of schools in large educational establishments and the heads of campuses of multi-campus schools.

the professional education of school principals. Yet, with the exception of Victoria (and later Tasmania), few of the professional development resources allocated to principals were under the control of the professionals themselves.

This is not a surprise; professional development in Australia has always been a somewhat unpredictable commodity. In one notorious case, professional development funds were bartered for a pay-rise. This event exposed the topsy-turvy values that an employer and industrial organisations can bring to even the most serious professional issue.

As state governments throughout Australia embraced self-management, no set pattern emerged in the distribution of professional development funds. Some systems held the dollars centrally, exempting the majority of professional development funds from the self-managed basket of goodies. Others devolved the whole professional development budget. Some systems believed they knew best how to employ a limited resource; others believed that control over professional development funds was fundamental to principals' success in leading self-managed schools.

With hindsight, the *give-it-all-away approach* seems to have been inappropriate. Systems rarely succeeded in convincing principals to give priority to *systemically* important professional development. At the same time, many principals seemed unable to provide appropriately for their own professional education. Control over the professional development funds did not seem to help them spend money effectively on processes and activities that would have a significant effect on their professional effectiveness.

This pattern of practice raises questions salient to system providers in Australia as well as other nations. What is the appropriate balance between system priorities and personal, professional needs and goals in professional development? Have principals as a profession lagged behind other professions (e.g., doctors, scientists, private sector managers) in developing the habits of lifelong learning? If so, why and how could principals' centers be a positive influence towards increasing the professionalism of school leaders?

This chapter focuses on issues related to the professional development of Australian principals. The author draws on experience gained during his tenure as Chief Executive of the Australian Principals' Centre in Melbourne, Victoria from 1996–1999. The chapter discusses the Centre's approach to professional development of principals and reflects on the challenges facing those who would employ training and development as vehicles for professionalization of the principalship.

## The Australian Principals Centre: Leadership Accreditation and Competencies

In October 1992, Victorians elected a new government – one committed to restructuring the State's economy. This included restructuring the cost to the taxpayer of public services, especially education. The government was anxious to make changes quickly, even if it risked alienating public servants and their industrial associations[2].

---

[2]I am not able to go into all the details. I was not living in Victoria at the time, nor have I researched the history of events. Refer to Caldwell, B. J. & Heywood D. K. (1998). *The Future of Schools: Lessons from the Reform of Public Education.* London: The Falmer Press. Mr. Heywood was Minister for Education in the Kennett government during the years of most change and Professor Caldwell was chief among those who provided advice to the minister about self-managed schooling.

With regard to schools, the new policymakers held the view that the principal of a school should be *the* school leader at a separate leadership level from the teaching staff. They viewed schools as overly democratized institutions that needed a push towards adoption of management practices commonly found in business and industry. The government, therefore, gave particular attention to the role, status, remuneration and power of the school principal.

New policies required the principal to become 'boss' and to carry that whole-enterprise accountability which is the responsibility of every chief executive. The emphasis was important because, at the same time, the government was establishing the definitions, standards and outcomes that would define its commitment to the public accountability of state schools. The government had decided the extent to which principals would be held accountable under its self-management initiatives well before it planned the restructuring of school management.

Not surprisingly, negotiations between the government and the principals' associations were earnest, vigorous and professional. Agreement was reached on a wide range of issues and a number of initiatives were put in place. It was this negotiation in the context of reshaping the role of school principals that led to the creation of the Australian Principals Centre Limited (APC) in 1995.

The Australian Principals Centre is a company limited by guarantee and is owned by four incorporated bodies:

- Victorian Government Department of Education Employment and Training (In 1995 the department was titled Department of Education),
- University of Melbourne,
- Victorian Association of Secondary School Principals, and
- Victorian Primary Principals' Association.

The APC exists because these four organisations believed that there were two particular needs that a jointly directed body could best fulfill:

- establishment of the principalship as a profession distinct from teaching,
- provision of continuing education adequate to the changing professional needs of the principalship.

With these goals uppermost, the APC was established as an independent corporation in 1995.

**Professional accreditation**

In order to advance the establishment of the principalship as a distinct profession, the APC created a structure of accredited membership using the frameworks and language common to other professions in Australia. In 1999 there were three existing categories of accredited membership[3] and one category in the planning stage:

- *Affiliate* (AFAPC): for those who aspire to the principalship, or who are in the principalship but have insufficient experience to be awarded membership as Associate Fellow;

[3]There are two other membership categories that are of little concern here: Life Fellow (for those APC members whom the APC wants particularly to exalt) and Honorary Fellow (for those 'friends of the APC' who would like to maintain a close association with the Centre, but who are professionally ineligible to join as accredited members).

- *Associate Member* (AMAPC): for affiliates who have completed pre-appointment training (the proposed new category referred to above);
- *Associate Fellow* (AFAPC): for those who are already competent principals (which notion includes the assumption that this requires a minimum of three years of practice);
- *Fellow* (FAPC): for those who have achieved substantial leadership in the profession (which notion includes the assumption that this requires a minimum of six years of practice).

**APC accreditation processes**

Establishing membership through accreditation is not an easy task. Issues of definition are difficult to sort out. For example, what are the real differences assessors apply when distinguishing between the award of Associate Fellow and Fellow? Moreover, the introduction of Associate Member as a level of membership related to pre-appointment training has made progress a little slower than preferred. Nevertheless, accreditation is beginning to work in a properly accountable fashion.

The fundamental tenet of APC accreditation is that of *accreditation of principals by principals*. Not everyone is, however, comfortable with this. For example, employers do not automatically accept that principals will exercise a sufficient degree of rigor in the professional assessment of their colleagues. This would be an unwelcome criticism for any profession.

For the APC and its member principals, this criticism is difficult to confront, especially when it is inferred rather than vocalized. It is, though, a criticism that the APC treats very seriously. It accepts that the only way to deal with innuendo is a consistent application of high ethical principles and demonstration that disparagements of this sort are unfounded. It is especially important to the APC that its accreditation processes exclude any propensity for professionals to *scratch each others' backs*.

In this context, APC accreditation progressed smoothly enough and with employers' approval. At the same time, the Department of Education Employment and Training continued to run its own accreditation process. This is a *one-off assessment* associated specifically with contract employment and salary remuneration. However, this practice is likely to be reviewed. The Department of Education Employment and Training is considering how its own certification needs can be accommodated in a broader accreditation process. This would likely include the option of using the APC as an accreditation agency[4].

Until now, accreditation has been limited to membership as Affiliate, Associate Fellow or Fellow. The process has relied heavily on the formal references of others in the work place and recognition of an applicant's prior learning (RPL). This conforms with current vocational educational practices in Australia. Prior learning is usually taken into account when assessing vocational competence in Australian business and industry.

[4]At the time of writing, many education authorities in Australia (and, indeed, other parts of the world) are examining the need for new registration procedures for teachers and school heads. In Australia, the accreditation of principals raises as many political issues as it does professional ones. This is particularly true of re-accreditation, where the need for the reassessment of practitioners is well outside the existing culture of the education industry.

The methodology of assessment is obviously important. In this regard, the APC sees a need to clearly achieve that level of rigor which employers and the general public expect from a highly ethical professional body. The APC has been looking at two particular options:

- assessment using assessment centers, like those of the National Association of Secondary School Principals (NASSP) in the United States of America
- assessment through the analysis of professional portfolios[5].

Experience in other places suggests that assessment center methodologies have the best record of predicting professional competence. However, they are expensive to operate. For this reason, the APC is giving detailed attention to the use of portfolios. Initially, it is the APC's intention to use portfolios for the upward accreditation of members from Affiliate to Associate Fellow and from Associate Fellow to Fellow. At the time of writing, the move to portfolios continues as work in progress and is not expected immediately to affect accreditation for membership. It has been important to work out ahead of change how portfolios can best be used. The concept of portfolios is new to Victorian principals in respect of their own professional and career planning. However, portfolios are widely used by teaching staff in their continuing education and career management. Thus, most principals are experienced in the use of portfolios by others.

---

[5]'A portfolio provides concrete evidence of the complexities of a job and allows for the identification of core values and beliefs as evidenced by a leader's decisions and practices. The portfolio allows positive, comprehensive and authentic documentation of what the school leader does, how it is done and what the results are. It is not reliant on a checklist of dos, but illustrates the dynamic and fluid nature of schools and school leadership. Above all, a portfolio enables a record to be kept of a leader's focus on student outcomes.'

'The establishment of a professional portfolio both increases the focus on goal setting for school improvement and realistically evaluates what worked well and what didn't. It is the on-going reflection and application of learning using a portfolio which promotes leadership capable of improving student outcomes.'

'Portfolios are constructed from a collection of materials (artefacts) – from events, programs and activities in which the school leader has participated. The APC portfolio is used to evaluate elevation in membership status, which requires demonstration of leadership competencies in the workplace and application of on-going professional learning. A portfolio should be developed with this in mind.'

'Portfolios should be limited to approximately twenty pages, therefore it is important that the selection of artefacts be done with care. The portfolio should reflect the breadth of leadership experiences and demonstrate leadership competencies in an effective and concise manner. Most artefacts will represent more than one competency and often more than one leadership dimension. Portfolios are not a collection of everything done by the leader, but rather an edited selection from a bigger folio that efficiently demonstrates the leader's knowledge, skills and practices.'

This is an extract from the APC's draft policy paper: Australian Principals Centre. (1999), *Guidelines in the Use of Portfolios*, Melbourne: APC.

The APC's thinking on portfolios has benefited much from the work of Genevieve Brown and Beverley Irby, whose *The Principal Portfolio* provides a most practical guide to the effective use of portfolios by principals. (Brown, G. & Irby, B.J. (1997). *The Principal Portfolio*. Thousand Oaks: Corwin Press).

The APC expects that existing accreditation assessment (relying on references and RPL) will continue for some time for the accreditation of affiliate members, who make up about half the APC's accredited membership. For a number of reasons, this simpler process is appropriate to a level of membership established for those who are newly appointed or are yet to enter the principalship. Every professional organisation needs

- an easily accessed beginner level from which mature membership can grow, and
- a level of membership to which people can belong while they work out whether or not their career future lies in the mainstream of the profession or in another leadership direction altogether.

Assessment by references and RPL seems to adequately meet these circumstances.

**Dimensions and competencies of leadership**

The Australian Principals Centre uses a set of competencies to describe the essential skills of the school principal. It was developed at the request of the APC by the Management and Research Centre (MARC) of the South Australian Training and Education Centre Inc., which is associated with the University of South Australia. MARC's work for the APC was based on a set of competencies already in use by business and industry. For more than three years it seemed appropriate to both circumstances and the times. In 1999, however, the APC came to the view that its competency framework needed review. There are a number of reasons why:

- the original set was always a little light on competencies with an ethical dimension,
- the whole notion of competencies as 'skills that you need to do the job' no longer carried the conviction that it did in 1995,
- the notion that a competency is something you either have or don't have was increasingly viewed as simplistic.

Nevertheless, it has not been easy to produce a substitute. For its own reasons, the Department of Education Employment and Training revisited the notion of leadership competencies during the second half of 1999. The outcome of their review is set out in a report, *Excellence in School Leadership: Creating a First Class Leading Environment* (Report), prepared by the department's consultant, HayGroup[6]. The APC was asked to contribute to the review. This was a very timely opportunity for the APC, as it has no commitment to inventing its own unique set of professional competencies – indeed, quite the opposite.

As a consequence of the review, the APC now has access to solidly researched material on attributes of quality leadership. This knowledge base has changed the focus from elementary notions of competency to a more flexible collection of capabilities useful in assessing professional expertise. It is now possible to reduce the number of indicators (originally there were 20 competencies) and to accommodate the notion that 'degrees' of capability are needed to describe the roles of school leaders at different

[6]In 1999, the Hay Group researched the leadership characteristics of principals and senior staff of Victorian schools. They reported their findings to the Department of Education Employment and Training. (Refer: Hay Group (1999). *Excellence in School Leadership: Creating a First Class Leading Environment.*)

stages in their careers and in different environments, locations and educational and social frameworks.

For example, the 20 competencies included attributes like:

- *Team building* (where the leader develops cohesive teams focused on the achievement of the organization's goals),
- *Influencing* (through which a leader contributes to the achievement of organizational goals by the empowerment of others), and
- *Organizational Culture* (where the leader oversees the establishment of the cultural values, behaviors and systems which best contribute to the achievement of the organization's goals).

Instead of establishing new generic statements of leadership competence, the Hay-Group researched the way school leaders go about their jobs. Thirteen behavioral characteristics (called *capabilities* in the Report) were observed as definitive of the leadership required in schools. The Report treats capabilities in a markedly different way to the traditional have/have not application of professional competencies. Instead, expectations for each capability can be separately established for principals, assistant principals and leading teachers. Moreover, each capability may be ordered by complexity or sophistication as well as scaled in a way that indicates increased skill.

For example, about the capability *contextual know-how*, the Report says that principals use *contextual know-how* to build the sort of commitment which comes through an awareness and understanding of the internal and external forces that exert an impact on the school. For leading teachers, however, *contextual know-how* has a narrower focus, one specific to the informal structure of the school and its complementary culture.

Most interestingly, the Report lists the capabilities in a series of charts called *Paths to Excellence*. In these tables, *contextual know-how* is rated at the highest level of importance for principals, at the mid-level of importance for assistant principals and at a lower level of importance for leading teachers. In the capability *analytical thinking*, however, no differences were seen to apply and *analytical thinking* is rated at the mid-level importance to all three groups of educators.

Just how the APC will use these data is not clear at the time of writing. They are not as easy to apply as yes/no competencies. Yet this, perhaps, will prove to be the richness of their strength. Using capabilities in the manner envisaged in the Report cannot help but give better direction to the definition of the profession and the design of relevant continuing professional education.

In the meantime, the APC continues to use its competency model to define professional proficiency, knowing it is possible for up to 80 per cent of a principal's skill to be described in terms of agreed-upon statements about leadership competence. The other 20 per cent is subject to numerous factors and is important for this very reason. In particular, it is in this 20 per cent that the freedom to express religious, cultural and philosophical differences can flourish and thereby distinguish a school and its leader from its neighbors.

Staff of the APC have always been a little reluctant to use competencies in any public description of the principalship as they are too hard to explain definitively in every circumstance of debate. In time, the adoption of capabilities instead of competencies may overcome this difficulty. But until then, the APC will rely on a simpler set of criteria it developed, called *professional leadership dimensions*. These have provided a helpful flexibility when presenting the profession to other constituencies in education, business and government. There are four:

*The dimension of educational leadership that leads a school community in its learning activity*
Creating an environment that is conducive to high levels of student learning and achievement is the core purpose of educational leadership. Implicit in this educational purpose is a set of beliefs about what constitutes effective schooling and appropriate student achievement. Quality leadership in schools is about making these beliefs explicit in the school environment. This incorporates curriculum planning, policy development and the delivery of learning and teaching programs that enable students to develop their capacities as fully functioning and responsible citizens.

*The interpersonal and personal dimension of leadership*
People-centered leadership is concerned with gaining the respect and cooperation of diverse individuals and groups in the community and building effective and purposeful relations between them. In this dimension, leadership is focused on the development of self and others in an environment where individuals are valued and cared for. Building constructive and purposeful relationships between them is integral to effective schooling. Fostering positive group dynamics and building cohesive and competent teams are fundamental leadership skills in any school community.

*The ethical dimension of leadership that provides predictable decision-making and a morally responsible social environment*
Quality schooling presupposes fundamental moral and ethical imperatives. It is essential to lead in such a way that the learning and welfare of students is sustained in a context of ethical clarity and moral predictability. The personal qualities, ethics, vision and values of the leader are integral to the creation of a principle-centered culture and value system within the school community. It is the leader's responsibility to constantly communicate those values and to model them in daily practice.

*The strategic and managerial dimension of leadership which makes things work today, tomorrow and the day after*
Effective strategic leadership is responsive to the wider external environment. This includes the alignment of internal school priorities with the mandates and pressures imposed upon the school from outside. Within the context of a changing educational and social world, school direction and outcomes must be planned strategically. Change management and the development of a culture of achievement are as important to strategic leadership as is the effective management of the school's financial, physical and human resources.

## Professional Development for the Principalship

Everyone in education and government seems to have personal convictions about professional development for school principals. I am no exception. I nurture a bucketful of ideals, expectations and successes, as well as some not very attractive prejudices, disappointments and failures. During my time at the APC, I had ample opportunity to re-examine my views. I also had the chance to experience first-hand the difficulties that confront providers of professional development for principals – those who want high quality professional education offered to every school leader.

**Professional development challenges for the education profession**

Nothing in my time at the APC altered my conviction that there is expanding work to be done in the professional development of school heads. I hold the view that much of the currently provided professional development is unlikely to prepare principals to be the successful leaders in schools-of-tomorrow. If this situation is to be changed, there are some deep-seated disorders that require attention as soon as possible. There are challenges here for all segments of the education profession.

*Head-teachers or head-learners?*

There is reluctance among many school leaders (and teachers) to demonstrate their belief in the efficacy of lifelong learning. While it is a cliché in this era to profess support for lifelong learning for students, there is an equally urgent need for school leaders to proactively apply lifelong learning to their own careers. In Australia, the number of teachers and school heads involved in relevant continuing education is disturbingly small.

At one time, the APC estimated that in Victoria more than 70 per cent of its quality professional development was undertaken by less than 30 per cent of the principalship. How can so many profess a belief in lifelong learning for others and yet find so little need to extend their own professional knowledge? Changing this professional norm among teachers in general and school leaders in particular is an urgent priority for the education profession.

*Institutional rhetoric or support for professional development?*

There is a critical failure among employers to 'walk the talk'. Why are governments and employers so reluctant to generously fund professional development for school leaders? In Australia, the rhetoric about school improvement has never been matched with the vocational training needed to produce the specified outcomes.

Moreover, this attitude seems peculiar to the education sector. Australian governments have been quite purposeful in their provision of leadership training for other government administrations (e.g., foreign affairs, land management, motor vehicle registration). Why is there an apparent indifference to the identical need in schooling? Is it simply a matter of funding? Or, is there a quiet prejudice in government treasuries against providing funds on such a large scale for educators? Does this signify a true scarcity of educational resources or an institutional skepticism about the impact of budget allocations for teacher and principal learning?

*Industrial strength through professionalism or more rhetoric?*

There is a professional indifference within some teachers' and principals' associations when it comes to professional development. Many of the organizations still view their primary role as looking after employees' employment and working conditions. They do not see themselves as part of the educational system or part of the solution to the complex educational problems faced by society.

The professional development of their members ought to be at the center of their agenda as it is the ultimate route to enhancing the credibility, status (and pay) of the professional. Why, therefore, is there a reluctance among many teachers' and principals' associations to promote the need for continuing career development and professional education? Proclamations of the importance of professional education are at the core of most professional associations' accreditation and re-accreditation processes. In engineering, architecture, pharmacy and accounting, professional institutes work hard to educate their members to higher professional standards and then to ensure that the public is aware that their members' learning achievements lead to higher levels of service. Why is this not the case in education? Why do we take ourselves for granted?

*Why isn't quality at the core of professional development?*
Finally, indifferent teaching quality is the hallmark of too many professional providers of vocational education. This includes the broad array of institutions involved in this significant industry including universities, institutes, and private sector providers. Why do schools of education have so little impact in the industry, especially when education *per se* is their declared expertise?

**Suppliers: the professional development industry**
Any examination of principals' continuing professional education must acknowledge that, for many years, remarkably little professional development was designed or supplied in direct response to principals' own declared needs. In Australia, a disturbingly large number of principals seemed content to rely on the initiatives of others to upgrade their knowledge and skills. They preferred an infinite extension of the *status quo*, where challenge is avoided and experimentation in professional development methodologies is discouraged. They cherished past experience and resisted all but the most simple evolutionary change.

To some extent this attitude still prevails in the profession. This does not characterize all principals of course, but it is true for far too many. The profession as a body is still not leading the expansion of its expertise. By default, the leadership role in conceptualizing, designing, and delivering professional education falls to others. As a consequence we still have a 'supply-driven' professional development industry. That is, professional development exists because people want to teach courses or because policymakers want courses taught. Learning is not engaged on the basis of requests from school leaders or specific identified learning needs of school leaders.

It is commonplace, therefore, for professional development to be a proliferation of courses based on what presenters are able to teach. University personnel teach their 'subjects'. Retired principals describe their experience. Over time both groups can easily become irrelevant in the work place.

Having said that, assisting providers with a better view of principals' continuing professional education requirements is really quite difficult. Assessing professional development needs is politically complex. It is rarely a simple matter of curriculum design, teaching and learning. Getting a view on what really is needed can be as wearying as it is frustrating. In Australia, a number of factors contribute to this difficulty.

*The pipers call the tunes*

In general, in Australia, the providers of funds tend to define the professional development curriculum. Commonly, providers are also the employing authorities. Consequently, professional development about systemic policy is the first to be resourced from available professional development funds. When this professional development does little more than explain changes required by policy or practice, there is little improvement in the effectiveness of school leaders in the wider contexts of their task. Disappointingly, even good professional development of this sort is usually applied indiscriminately to all. Rarely is there any acknowledgement of individuals' prior learning. Moreover, it is all too often delivered by in-house experts rather than by presenters skilled in adult learning.

*Design by committee leads to mediocrity*

When consultative processes are used to establish continuing education programs (and this is a common requirement of professional development funders) the professional development committees are often too large. Moreover, they too frequently bring together 'professional representatives' rather than professional development educators. Attempts to represent everyone's interests are common and agreement can be difficult to achieve on anything.

The resulting compromises on content and design result in mediocrity. Policies and precedent obscure vision and progress. And poorly established professional development committees too often tend to extremes – they either seek answers through yesterday's experience or become *gung-ho* with uncontrolled experimentation.

*Lack of a grounding framework of professional school leadership*

As earlier stated, the profession has not been good at establishing its own understanding of professional proficiency from which would flow industry-wide frameworks for continuing vocational education. The lack of a 'capital-P' profession for the principalship has allowed the administratively and industrially strong to dominate the definition of the principalship and its continuing educational needs. This has not necessarily proved helpful in assisting principals to lead their schools into the future.

In the face of the difficulty of getting agreement on what continuing education is really necessary, it is not surprising that available professional development has become a collection of what traditional deliverers are able to provide. With academics and retired principals dominating the supply of professional development, it has become a bit of a closed shop – disturbingly comfortable to both teachers and learners. There has been little emphasis on the development of private-sector expertise. It is of more than passing interest that the consulting firms that have done so much leadership training for commercial business and industry have had very little impact on school leadership. There are a number of reasons for this:

- The culture of schools is different and a bit of a shock to presenters used to business and industry (and *vice versa*). Significant differences exist in the self-images of executive leaders in schools and their counterparts in business. In each case, professional maturity seems to be assessed by different sets of principles and values.

- Available professional development funds are meanly provided compared with the norms consultants find in business and industry – very meanly, in fact. Schools rarely have enough to pay for the better, let alone the best. When they do, there is a guilt associated with the spending of money on the education of school leaders.
- Principals have high (often rather naïve) expectations of teaching and presentation quality. They do not hesitate to discount the content of courses if presenters lack teaching talent. This can be an unsettling, even devastating, experience for a presenter of average skill. Few will persevere in the face of continued disparagement.

For private consulting firms, delivering professional development to school educators can be singularly unrewarding. Firms have frequently started out providing state-of-the-art professional development (often at the invitation of employers) only to lose money, suffer indignity and feel irrelevant in an industry they find difficult to understand. Some persevere, find their niche and make very worthwhile contributions to school leadership. But others retire hurt from a commitment they were ill prepared to discharge. They treat any further engagements with reluctance, if not outright rejection. This is not an outcome which benefits the leadership of our schools.

## An APC Framework for Continuing Education

The APC has been concerned to develop continuing professional education that both encourages participation in better learning processes and discriminates against processes that give lesser results. With this in mind, it drafted a framework to direct APC members towards continuing education most likely to provide effective learning experience. The framework was designed as an assessment tool for determining whether or not accredited members of the APC are doing sufficient continuing education to maintain their accreditation.

The draft schema is presented below in Figure 1. While obviously of debatable content, it does credit the benefits of superior learning processes to learners' advantage. It is a 'first attempt' to include the means of learning as a tool when measuring the quality of individual professional development commitments.

The Figure 1 does not deal specifically with content. Because the profession is innately complex, the professional development needs of principals vary widely. Content is covered by requiring members to expand their proficiency in each of the four dimensions of leadership referred to earlier. In retrospect, it is perhaps rather *laissez-faire*. But the alternative produces the other extreme – a detailed prescription of courses where prior learning is too often set aside and workplace realities are too frequently ignored.

The Figure 1 itself includes in the left-hand column the common ways a principal might broaden professional learning. These are broken down into sub-strands in the adjacent column so that points can reflect expectations of quality. The right hand columns indicate the points individual professional development activities 'earn' and the maximum that can be accumulated in each category in any one year. The quantum of professional development required to sustain membership was proposed at 300 points over a four-year period.

| Activity | Detail | Points per Event | Max Points per Year |
|---|---|---|---|
| **Attendance at meetings & briefings** | | | |
| Category A | Zone, cluster, regional meetings; professional associations | 1 | 20 |
| Category B | National or state-wide meetings, e.g., AHISA, DEET, ACEA* | 2 | 10 |
| Category C | Member of national or state-wide committees which focus on schooling or the profession, e.g., BOS, APC, ACEA† | 3 | 12 |
| **Participation in short courses** | | | |
| Category A: one-day (longer than 3 hours) | • APC accredited<br>• Non-APC accredited | 2<br>1 | 6 |
| Category B: course requiring two to four days' attendance | • APC accredited<br>• Non-APC accredited‡ | 8<br>4 | 16 |
| University course modules | Module must relate to school leadership competencies | 16 | 32 |
| **Conferences, seminars & symposia (activities greater than one day)** | | | |
| Attendance | | | |
| Category A: cluster, zone, regional, state | • APC accredited<br>• Non-APC accredited | 2<br>1 | 6 |
| Category B: national, international | | 4 | 8 |
| Presentation to Colleagues | | | |
| Category A | Presentation of not less than 1 hour at professional meeting/conference | 4 | 8 |
| Category B | Paper and presentation at a state/ national conference | 5 | 10 |
| Category C | Major research paper presented and printed in state-wide publication or international paper and presentation | 20 | 60 |
| **Self-initiated professional development and work-related learning Publication** | | | |
| Category A | Educational article published for distribution to local, zone/cluster, region | 5 | 5 |
| Category B: first author only | Publication in state-wide journal, topic related to education | 30 | 60 |
| APC Accredited Course | Extended course greater than 4 days with school-based project extending over 6-month period | 30 | 30 |

*(Continued)*

Figure 1. A proposed accounting of engagement in continuing professional education.

| Activity | Detail | Points per Event | Max Points per Year |
|---|---|---|---|
| Peer Learning Activities | A group meeting a minimum of 4 times/year of one-hour duration with structured, clearly defined learning targets. Learning journal or reflective writing required as evidence | 40 | 40 |
| | Working with a colleague in a learning partnership (similar to mentor/coaching/ peer coaching) on a specific plan/purpose. Maintenance of learning journal required | 40 | 40 |
| **Research** | | | |
| Action Research Project | Must include evidence of reading, application of theory and analysis of outcomes. | 50 | 50 |
| Professional Reading | Specific reading directly related to PD plan. Learning journal or evidence of critique of ideas and issues required | 5 | 20 |

*Association of Heads of Independent Schools of Australia (AHISA), Department of Education Employment and Training (DEET), Australian Council of Educational Administration (ACEA)
†Board of Studies (BOS) Australian Principals Centre (APC), Australian Council of Educational Administration (ACEA)
‡Special consideration may be given to awarding higher points

Figure 1. (*Continued*).

The Figure 1 was distributed for critical comment. Reaction from both APC-accredited members and others was mixed. The proposal had no soft edges and was not easily absorbed into the present culture. There was an evident tendency to shy away from any mathematically constructed schema, particularly one which so unequivocally measured individual effort. Yet in spite of its rather direct demands, it does seem to meet the criteria set by the APC:

- It has a weighting system that enables better adult learning processes to accumulate significantly more points.
- It can be applied to existing professional development programs and courses without much definitional difficulty.
- The total effort required does not force people to do what they cannot achieve (if you can't write articles or present papers you can still achieve the points required, and not necessarily with greater effort).
- It can be used in a self-monitoring recording mode.

It also has some obvious shortcomings:

- It cannot be easily interpreted into the personal professional agendas of individual school leaders.
- It needs numerous examples to sufficiently cover the diversity of experiences that make the principalship such a complex profession.

- The activities appear ambiguous to some principals – the meaning of words really matters.
- The number of points proposed and their comparative weightings appear arbitrary (which to some extent they are).

In general however, critical response was supportive. Notably, commentary focuses rather more on the way the proposition was described rather than the purpose for which it was designed. A Version II was planned to provide a more user-friendly statement – one more directly related to adult learning practices, perhaps. The Figure 1 (or its successor) is more likely to become an appendixed guide to written text – although in the end it is hard to see how the mathematics embodied in the chart can be dispensed with if accountability is to be transparently expressed, adequately measured and publicly understood.

**Pre-appointment training**

With the same commitment to good adult learning practice, the APC is planning enhanced training for teachers who aspire to the principalship through professional training and development. By introducing the membership category of Associate Member, the APC can introduce a level of accreditation to cater for this circumstance.

The desire to accommodate pre-appointment training is not a new direction for the APC. Since 1996, the APC has provided a program through which it trained leading teachers for the principalship. The program was flexible enough to recognize prior learning where appropriate and broad enough to cover the capabilities needed for effective principalship. It included one compulsory course, *Being a Principal*, and four other courses run by university faculties on behalf of the APC. These included:

- *Leading a Learning Community*
- *Curriculum Leadership and Management*
- *Finance and Business Management*
- *Education, Technology and Change*

The APC has since concluded that these courses, in their original form, no longer deliver a coherent program suited to potential principals seeking pre-appointment training. It believed that attempts to tweak the original courses were unlikely to provide sufficient change. The program itself needed to be reinvented. Reinvention was planned around three levels of certification: course certification from the APC, advanced certification with vocational education training (VET) status[7] (also from the APC) and regular higher education graduate study. The program embraced a number of quite tight constraints:

- The framework of the courses must substantiate the APC's dimensions of leadership describing the principalship and also cover the proficiencies that employers and the general public associate with professional competence.
- APC-certificated courses must articulate to VET status.
- VET-status courses must give awardees credits towards university graduate courses.

[7]The introduction of certification at the level of vocational training is new and stems from the desire to externally validate the APC's own certification, particularly of those courses that relate to pre-appointment training.

- Assessment for VET and university courses must be the normal assessments associated with these courses.
- The courses must stand alone as useful professional development as well as combine into an APC certificate of professional proficiency.
- Provision must be made to recognize appropriate prior learning.

Not surprisingly, getting all this together is difficult. It will force the APC to become a registered provider of vocational education training – not in order to present courses with its own staff, but to be able to negotiate quality with university and other providers. Even more difficult will be the construction of progressive articulation. Connecting a non-VET APC course to a VET-certified APC course to university post-graduate learning is far from straightforward and may take a long time to negotiate satisfactorily.

Nevertheless, the APC believes it must be done. The high cost to students of post-graduate university training in Australia is a very motivating consideration. Additionally, as participants need to be encouraged into formal continuing education, people need the opportunity to start with a suck-it-and-see approach. For example:

- Do one day and get a non-VET APC certificate.
- Do more with assessment and get a VET certificate.
- Negotiate this achievement as a credit towards a university post-graduate qualification.
- Indeed, get three or four VET-accredited certificates and negotiate a sizeable credit towards a graduate diploma or masters degree.

This proposal is even now in its infancy. But it does seem to hold a promising future. Best of all, it enhances the possibility of bringing people back into university post-graduate study. In Australia the numbers have fallen away markedly since the introduction of full-fee payment in post-graduate teaching institutions.

Completion of a pre-appointment program will give an applicant status in the APC as an Associate Member. At this stage it is not proposed to expect more than VET-level achievement for accreditation as Associate Member. High levels of participation in this venture will depend on the extent to which the APC and others are prepared to recognize prior learning. This is an absolutely important issue for potential principals. They do not expect to be asked to learn skills they have already mastered. Experience to date suggests their expectation is justified. School educators who have had opportunity for real leadership experience usually fulfill more than half the assessment criteria currently applied by the APC.

What will be the advantage of being an Associate Member of the APC? Initially, the APC expects Associate Membership will provide members with a competitive edge when applying for jobs in the principalship. The APC will be in a position to proclaim the proven commitment of its Associate Members to quality leadership in the principalship and continuing professional education. In time, pre-appointment training is

---

[8]It is interesting to note that the Australian tradition of requiring no pre-appointment training and the tradition of the United States of America which insists on it seem to produce a very similar level of professional competence – an observation that should concern those involved in the professional development of school leaders.

almost certain to become a standard prerequisite when applying for a principal's position – as it is in some other countries, notably the United States of America[8].

At the moment, Australian principals remain somewhat skeptical about pre-appointment training. However, even if the education profession doesn't immediately take all this to heart, the legal implications of untrained professional chief executives will eventually force a change of view. One large successful common law claim against an individual principal will change the culture overnight.

**Post-appointment training**

The range of courses provided through or by the APC varies from year to year. Some are sponsored by employers (e.g., the Department of Education Employment and Training). Some are entirely paid for by those who attend. Attendance is not limited to APC-accredited members, although they enjoy lower registration rates and preferential booking for some courses. The range of courses can be viewed on the APC's website, www.apcentre.edu.au.

## Governance and Administration of the Centre

It is not my intention to describe the administration of the Australian Principals Centre Limited in detail. However, it will help in assessing the work of the APC if a little information about governance and management is provided. In many ways it is the governance structure of the APC that makes it unique among principals' institutes.

The APC was established in 1995 as an independent corporation, governed by a board of six directors nominated by and representative of the four owners. In 1999 two additional directorships were added to the board – one from the Catholic Education Office[9] to directly represent the needs of non-government school leaders and one from the APC membership itself (a serious previous omission).

The Board has appointed a Professional Accreditation Council to advise it on matters of accreditation and continuing education. The Council is drawn from the accredited membership of the APC. Members of the Council are always principals or assistant principals.

A Policy Advisory Council provided advice to the APC chief executive and the staff in the drafting of APC policy. This was particularly important in the early years. The council was made up of invited educators from principals' organisations, universities, employing authorities, teacher industrial organisations and, of course, schools from the government, Catholic and independent sectors of Victorian schooling.

When I directed the APC, five staff were employed by the Centre. In addition to the chief executive, there were two professional officers (Project Directors who were

[9]Of the 2,322 schools in Victoria, 490 (21.1%) are systemic schools of the Catholic Education Office. Of the remainder, 1,631 (70.2%) are part of the Victorian Government education system administered directly by the Department of Employment, Education and Training and 201 (8.7%) are independent schools either in very small systems or free from systemic oversight. (Source: Department of Education Employment and Training. (2001). *Department of Education Employment and Training Annual Report* (1999–2000: Melbourne. DEET)

recently school principals, one secondary and one primary) and two administrative staff. Minor changes to the staffing structure have occurred since.

**Finances**

The APC's income relies on an annual grant from the owners and annual subscriptions from members and those who subscribe to its publications. It is this funding that makes quality accreditation and quality professional development possible; it could not otherwise be attempted. Unfortunately, this funding is not sufficient to make professional development free to members. Equally unfortunate, there is no historic expectation that school principals will pay the cost of their professional education or career development. This is in marked contrast to most other professions. A strong expectation of and dependence on subsidy overshadows all professional development planning.

There is no limit to the demands that society can place on schools and no likelihood that society will turn elsewhere for the life-preparation of its young people. Continually increasing expectations of schools are a certainty, so leadership capabilities will have to grow commensurately. Success in a climate of expanding expectations is utterly dependant on the leadership quality of principals and their ability to achieve responsible leadership in others.

Continuing professional education is the foundation from which success can be planned. For principals in Victoria, the Australian Principals Centre is well placed to continue to ensure that the continuing education for principals meet both their own professional needs and the expectations of the wider Australian community.

# 10

# Developing Singapore School Leaders for a Learning Nation

Dr. Chong Keng Choy

*Associate Professor, Department of Educational Administration, National Institute of Education, Singapore*

Dr. Kenneth Stott

*Associate Professor, Department of Educational Administration, National Institute of Education, Singapore*

Dr. Low Guat Tin

*Associate Professor, Department of Educational Administration, National Institute of Education, Singapore*

> 'Mom tells me you have enrolled for a course in stealth and cheese acquisition,' said Father Mouse to his youngest son.
> 'That's right,' replied Malcolm Mouse.
> 'But why do you need to go on a course? You seem to steal your fair share of cheese without too many problems.'
> 'I know, but things are changing. We are facing many discontinuities and uncertainties, and I am not too sure that my present understanding of the environment and its increasing complexity will enable me to guarantee the same degree of success in the future.'
> 'You mean the cat's getting smarter?' said Father.
> 'Precisely!'

Our natural inclination is take the attitude: if it ain't broke, don't fix it. However, arguably the best time to change is when things are going well. That is the approach Singapore has adopted almost across the board. With its declared intent to be a 'learning nation' – the nation's vision is 'Thinking Schools Learning Nation' (Goh, 1977, p. 20) – it has been going through a process of questioning a whole range of systems, processes, operations and programmes in a search for improvement.

Part of this questioning process has centred on how we manage and lead our schools, and, consequently, how we prepare leaders for their changing roles. In this chapter, therefore, we shall examine the Singapore context for education and highlight some of the significant changes that are having an impact on the way in which school leadership is conceptualised and practised. Essentially, the new context is best explained by the term 'learning nation'.

We shall explain in some detail what this means and the implications for people working in the system. We shall then describe some of the programmes the National Institute of Education runs for school leaders and explain how these, despite their immense success, have been and are currently being subjected to scrutiny in an effort to make them more relevant and effective. Our final thoughts focus on the importance of the partnership between the National Institute of Education, the Ministry of Education and the schools. It is the quality of this collaborative endeavour, we believe, that will take us forward in producing leaders for a dynamic and unpredictable context of education.

## A Changing Context

The speed of change in the external environment dictates the need for a nation's citizens to be equipped to face the daunting challenges ahead. People naturally look to schools to do this equipping. This has placed the teacher as one who not only socialises and teaches but who also prepares students for a changing world, and explains in part why the demands on teachers are escalating.

But it is not only teachers who have borne the brunt of unprecedented change. The demands on and expectations of leaders are considerable. Not only must they be knowledgeable and skillful; they must have the capability to be involved in school and policy design. They may need to become 'organisational architects' (Simon, 1990: 299).

Much of the rhetoric in Singapore in recent years has dwelt on the importance of learning. Probably every educator in the country is familiar with the acronym TSLN: Thinking Schools, Learning Nation. The notion of a learning nation involves a network of learning organisations, where lifelong learning is the norm. And schools are expected to play a critical role in the realisation of this desired future. School leaders are meant to design, lead and manage what in Singapore are called 'thinking schools'. They have to be the architects of institutions that represent a major paradigmatic shift from the narrowly-focused, compliant, examination-driven schools of former years. For some, this expedition into foreign territory has been fraught with difficulty; for others, it has represented an agenda that is intensely stimulating. At the National Institute of Education, our part in this scene is to prepare Singapore's future school leaders for 'thinking schools' and for a context characterised more by the unfamiliar than the known.

What does the concept of the learning nation mean to the individual? The implications are spelled out by Chong (1998) in his explanation of what people will gain in a lifelong learning mode. They will learn to:

- use the tools of information technology;
- think creatively and critically; and
- live and work globally in a networked world over their life-span.

How can this scenario be realised? The schools are being designed to prepare young people for the realisation of a desired future of work and workplace for Singapore. The Prime Minister has called for the application of the Thinking School, Learning Nation 'formula to enable Singapore to compete and stay ahead' (Goh, 1997). School leaders are being identified to attend emerging learning programmes that will prepare them to design, lead and manage thinking schools in a learning nation.

### Using information technology

Lifelong learning for a nation can be supported only by providing an appropriate infrastructure for people to pursue it and by giving ready access to this infrastructure. With a nation made up of organisations, this infrastructure can link them all into a network for pursuing lifelong learning. Information and communication technology can give life to such a network. Singapore has a plan for a national information and communication network that will connect all organisations, including schools, into one network.

Schools of the future, like all organisations, must be held accountable to the external environment for their interactions with other organisations. Indeed, with the increasing application of information technology (IT) in the management and curriculum of future schools (Chong & Leong, 1995), students will be learning to use the tools of information technology in the emerging e-learning mode within the e-government infrastructure of Singapore.

In a networked world, an individual can be in all 'places' at the same time. This makes it possible for people to pursue global living. One can be away from home and yet be at home, both at the same time. Also, an employee can be at home and yet be elsewhere at the same time. An employee can be working with teams of people that are not in one geographical location. Easy and speedy access to information worldwide also requires employees to package information creatively for application in their work. Employees could access workplace knowledge on demand while on the job. Both students in formal educational institutions and employees could have individualised learning programmes on demand. Employees' ability to learn over their life span will likely be a competitive advantage for nations and organisations. This is particularly so when we think of assigning greater value to the mass participation of citizens in packaging knowledge for the creation of a nation's wealth.

### Thinking creatively and critically

Lifelong learning should include the three concepts of:

- individual learning;
- innovation through learning; and
- workplace learning.

Individual learning will be enhanced when employees are able to make friends with others for the purpose of learning together, and when employees are able to seek feedback on their performance and then reflect on that feedback. Seeking to learn from one another and learning for continuous improvement are new competencies for employees.

Learning today tends to be equated with attending classes (participants are either willing or unwilling) rather than with bringing new skills to the workplace. Employees could learn something new that they want to try out in their workplace. To apply a new approach for the first time could be termed 'innovation through learning'. This approach is needed, because learning that employees try to use in their work could be foreign to workplace practice.

In order to innovate through learning, employees have to collaborate with their supervisors to ensure they are able to use what they have learned in the workplace. They have to think about ways that their innovative practice could affect other employees. Innovating through learning in the workplace is a new form of capability for employees, and this is needed more and more in an increasingly rapidly changing and competitive environment of work.

Workplace learning includes a process of clarifying work problems, by which employees might discern symptoms from problems that require solutions. Since employees work with others most of the time, they must have the ability to build consensus among the people they work with in defining problems and finding solutions. Learning while doing is another new form of capability for employees. Employees are not automatons or appendages to machines.

The concept of the learning organisation, as popularised by Senge (1990), focuses the attention of executives on employees as human beings who can learn and who want to learn. When employees are given a chance to learn, there is convincing evidence that they address the learning process positively, and this has been witnessed in many organisations using such vehicles as quality circles and work improvement teams.

#### Living and working globally

A networked world increases contact among people from all corners of the world. With rapid modes of travel, people may have the desire to meet people whom they have contacted through the IT network. While people could travel much and gain a cosmopolitan outlook, they may still want to feel at home somewhere. Singapore must be a place where they could come home to rest from their weariness and migration, where peace and security reign, where they may put down their guard, where they are able to surround themselves with familiar and pleasant things, where all their needs are met, and where they enjoy the comfort of having their families with them.

In the networked world of the future, sometimes called the 'global village', there must be a place to call home. An integrated concept of home and family gives meaning to life-long learning. People could keep on learning to use IT, think creatively, and live globally, only if they can hope to build a better home and bring up a more successful family.

### Leaders for Thinking Schools

The learning nation we have described above requires a new educational agenda, and this in turn demands a new type of school leader, one who is innovative and proactive in a dynamic, complex and sometimes uncertain context. The new corporate leader in education has an expanded and more intellectually demanding role. That is why we need to educate principals who can 'think' their way creatively and globally through complex, sometimes unique, and often persistent issues. The new leaders need to guarantee high degrees of quality in teaching and learning, orchestrate strategic innovations and influence the school's public at the school-community interface. Principals' roles are becoming identifiable with – though not strictly identical to – those of chief executive officers (CEOs). This being so, they need to be supported by teams of middle managers, who must assume increased responsibility for leadership in learning.

Since the mid-1980s, the National Institute of Education (NIE), which is part of Nanyang Technological University, has offered a number of leadership programmes. While the flagship programme is the one that prepares talented individuals to assume positions of principalship, the importance of other programmes, which prepare people for leadership positions in other parts of the school, is becoming increasingly evident. This is because we cannot view leadership as an isolated activity to be enacted by the principal alone. Indeed, Hord, Hall and Stiegelbauer (1983) claimed some twenty years ago that 'this rhetoric, abundant in literature, quite obviously hangs like a heavy mantle on the principal. However, what is becoming increasingly certain and abundantly clear is that the principal does not bear the weight of leadership responsibilities alone.' Their study showed that other individuals emerged to work alongside the principal.

In the Singapore context, this has not been left to chance. The Ministry of Education has drawn up a framework where promising teachers are selected for various leadership or managerial positions in the school. Whilst most teachers remain in the classroom throughout their careers, those with leadership abilities may progress to other positions: senior teachers, subject heads, level heads and heads of department.

NIE has over the years worked in partnership with the Ministry to provide training for individuals selected to fill these leadership positions. In the past, some of these training programmes have been full-time over a period of one year. That represents an enormous resource commitment on the part of the Ministry, but it has proved to be a worthwhile investment, since the quality of leadership and management in Singapore schools is seen by international visitors to be unparalleled. However, in recent years, there has been a scaling down of programme length, partly in order to cater for more people still waiting to be trained. In real terms, there has been no reduction in the Ministry's resource commitment.

Lest one should think that the training is merely functional in order to meet existing skills and knowledge needs, the ambitions of NIE and its approach to leadership development need to be made clear. NIE has chosen a different route to many training providers, which traditionally have responded to identified needs. Rather, NIE has taken the course of proactively effecting change in the practice of leadership and management by establishing a new agenda for training. Such ambition is best summarised by a short extract from the handbook for heads of department training.

It states that NIE contributes to advances by 'helping such leaders to confront the cutting edge of leadership knowledge in education, so that they can heighten corporate capability in schools and take their operations into new realms of excellence' (DDM handbook, 2001). Lofty ideals they may be, but there is a belief that this can be achieved.

In the following section, we describe the leadership programmes offered at NIE. Participants in these programmes are selected by the Ministry of Education (MOE) and, if they belong to government schools, they are fully sponsored by the MOE. That means they are on full pay while attending the programme. We start with a comparatively recent initiative and what we describe as our 'flagship' programme, the Leaders in Education Programme (LEP), since this is one that has generated considerable international attention due to its radically new approach to leadership preparation. Then we shall describe the other programmes that form the leadership course portfolio.

## Leaders in Education Programme (LEP): A New National Initiative

For over fifteen years, the Diploma in Educational Administration (DEA) was known for its excellence in training school leaders. Indeed, many of Singapore's senior educators, including superintendents and directors have passed through this prestigious programme. It was a programme characterised by executive skills training and learning from excellent principals through a mentoring process. This programme and the mentoring process have been documented and published (e.g., Low, 1995; Low & Chew, 1997).

It was from a position of strength that NIE, in partnership with the MOE, decided that changes should be made to the programme. The changing context for school leadership demanded a rethink of the nature and orientation of the programme. What was this 'new context' that precipitated change?

That has been partly answered in the earlier part of this chapter, where we outlined the thinking behind the TSLN vision. The educational landscape in Singapore was giving rise to new and escalating challenges. A series of landmark initiatives – these included 'ability driven education paradigm', an IT master plan, and a focus on national education – increasing levels of autonomy for schools, and calls for quality improvements, wider accountability and raised levels of achievement, all pointed to the fact that the preparation of school leaders needed to be thought a new. Whereas previously the compliant and efficient manager was valued in a system almost completely controlled from the centre, the demand for differences between schools to become more conspicuous and for clusters of schools to take responsibility for their own affairs shifted the search for more independent-minded principals.

Indeed, the new educational agenda demanded a new type of school leader, one who could cope proactively with a dynamic, complex and sometimes uncertain context. The old leadership thrived on conformity. The new leadership had to be ambitious and independent, innovative, and able to succeed in conditions that were less clearly defined.

It was also clear as we talked to educators both in Singapore and abroad that the new principal would have an expanded and more intellectually demanding role. We needed to train principals who could 'think' their way through complex, sometimes unique, and often persistent issues in schools. Such individuals would need – as we said earlier – to guarantee high degrees of quality in teaching and learning, orchestrate the strategic agenda and direct operations at the school-community interface. It was in this context that, in March 2001, the Leaders in Education Programme (LEP) was launched at the National Institute of Education.

The programme aims to produce chief executive officers for schools, who can demonstrate extraordinary performance operating in an emerging learning nation environment. We are no longer mandated with producing simply 'good' principals, but leaders who have the capability to transform schools and even move beyond principalship. But this needs to be done quickly. Thus, the new programme has an executive orientation – similar in scope and intensity to some of the highly rated executive courses in leading business schools, but with the prime focus on education – and it is short and intensive, so that its graduates can be returned to the system in just six months. To achieve this, it has a diverse and powerful learning agenda that we believe is different from what has been attempted in comparable executive programmes across the world.

Such an intensive learning programme, designed to give leaders the capability to operate successfully in a rapidly changing future requires contextual learning. If the new programme is to produce innovative principals who can take their schools to new heights of performance, learning experiences need to be intensified by locating them in the real setting of schools. Thus, much of the learning is in the school workplace, and is supported by learning in the university class- and tutorial-room, in business environments, and in educational institutions both in Singapore and overseas.

For that reason, participants – as innovators in a knowledge-based economy – are attached to a school throughout the programme and they spend regular weekly time in that school carrying out a major innovation project. They receive support and guidance from the principal of that school (whom we call a 'steward principal'), the cluster superintendent and the NIE tutor. The project is expected to yield rich benefits to the school in terms of needed improvements and is meant to be a profound learning experience for the participant.

These principals-in-training are exposed to leadership in the business environment and to ideas from various sources, including government organisations. To further enhance such influences, key officers are invited from the education service to engage in dialogue with participants and to observe some of the work undertaken on the programme.

There is also an international component to the programme. Participants investigate successful innovative practice overseas, undertake critical analyses, and gain significant insights into how educational innovation in Singapore might be managed. The inclusion of this international component in the course, while not unique, adds prestige and raises the programme's profile on the international stage. In its first year of operation, the participants visited institutions located in the USA and Canada.

**Delivery architecture of the programme**

While the interest of most parties is directed at the 'content' of a programme, in the LEP, it is the 'delivery' that is our prime concern. The content is there as learning support, but the delivery architecture is what sets it apart from other programmes and enables it to achieve its intended outcomes. Basically, we use the concept of action learning. In this concept, participants know what they are taught, but they do not know what they will learn. They have to create their own knowledge through team learning, and this takes place in what we call 'syndicates', a group of about six people meeting weekly and facilitated by a university professor. They know what knowledge they have created only when they come to the end of the programme.

It is useful at this stage to give a brief account of our understanding of action learning, since it is so central to the delivery process. For our purposes, action learning is group learning among people who are committed to action by using acquired learning for obtaining systems-wide outcomes, which might result in customer benefits. Its original formulation by Revans (Marquardt, 1999, p. 19) is $L = P + Q$, where $L$ = learning, $P$ = programmed knowledge, and $Q$ = questioning insight. In the LEP, programmed knowledge (P) refers to what is taught in all the seven modules, what is read, presentations by guest speakers, and all other opinions, theories and know-how shared. The LEP syndicates serve to encourage questioning insights.

Learning (L) in action learning is different from the traditional formulation, which equates learning and programmed knowledge. In our approach, the seven modules are relegated to a support role. Despite this, it is still of interest to know the areas we cover, since these give some indication of what we consider important agenda items for the future principal. We therefore provide below a brief outline of the support modules' content.

- Managing competitive learning school organisations: Optimising corporate capability through knowledge and innovation; organisation design and management; creating the future and sustaining competitive advantage; scenario planning and mental models; open systems and systems thinking; changing trends in the external environment.
- Marketing and strategic choice: Values, quality and innovation; corporate vision and identity; strategic agendas; school-community interaction to make a difference to teaching and learning; marketing theory and practice; partnerships with parents.
- Applying the new technology in managing learning: Technology and innovation; the e-learning environment and its learning culture; wireless learning environment; pedagogical models; new paradigm of learning with technology; assessing new learning; creating borderless and international arenas of learning; managing and promoting innovation.
- Achieving excellence in teaching and learning: Curriculum design for a new work environment; process curriculum; new paradigms of assessment; monitoring and assessing teaching quality; high achievement cultures; working with data and evidence; packaging knowledge.
- Building human and intellectual capital: The 'war for talent'; retention and rewards; deployment; optimising professional capability; building networks of extraordinary teams; strategic financial issues; performance management.
- Leadership for the new millennium: The learning organisation; working with teachers to use the tools of the five disciplines (Senge, 1990); leading the learning team; the leader as coach, steward, and designer, team learning and shared vision.
- Personal mastery and development for principals: Leading schools in a networked e-learning environment; influencing and motivating; philosophy, values and passion; generative conversation in dialogue and discussion; written and oral information transfer; dissensus and consensus generation; creativity and innovation cultures; accounting for performance.

While we believe that the training of extraordinary school leaders is paramount, they can operate effectively only if they are supported by teams of people who are able to lead significant developments in teaching and learning. We therefore now turn our attention to a range of programmes designed to develop leaders who will complement and support school principals. We take a brief look at these programmes and indicate some of the developments that are in the pipeline.

## The Diploma in Departmental Management (DDM)

The main professional support role in our schools is the head of department. Most schools have heads of department to take responsibility for the major subject areas (e.g., Mathematics, English) and for cross-school support (e.g., Discipline, Information

Technology, Pastoral Care). These are the curriculum or instructional leaders. The roles are multifaceted. They teach; they undertake administrative duties; they set the direction for their respective departments; they plan, implement and evaluate; and they deal with professional development for their teachers.

The Diploma in Departmental Management (DDM) is a seventeen weeks full-time programme that prepares educators to take on these roles and carry them out effectively. We say 'take on', but in fact most of the participants are already heads of department, so it is more of a question of enhancing their capability.

What is the programme's focus? Essentially, it aims to develop the participants' knowledge, skills and competence to deal expertly with the work situation. This work situation is one of managing curriculum and implementing curriculum improvement projects. It is thus taking on a more innovative nature. Heads of department are also involved in working with teachers to help them improve their teaching competence, and this necessitates going into the classroom to observe and give feedback. Perhaps the major thrust of the programme is best explained through a set of intended outcomes, which are delineated in the programme's handbook. It explains that, by the end of the programme, participants should be able to:

- Demonstrate an up-to-date knowledge of theory relevant to departmental management, and apply it to departmental personnel and activities effectively;
- Reflect, think and reason independently about complex curriculum and instructional issues, and understand how innovative practice leads to gains in students' educational achievement;
- Design, develop, implement and evaluate curriculum activities in their field of expertise;
- Formulate strategies to support teachers' motivation and satisfaction, and foster a climate of collaboration in the department;
- Identify and facilitate appropriate professional development activities that support departmental, school and national priorities;
- Assess student learning and teacher effectiveness validly and reliably, and apply appropriate assessment processes within the department; and
- Understand the wider educational context, which includes national priorities and constraints in a multi-racial society.

At present, there is a major focus on departmental management. A large part of the programme is devoted to managerial skills, managing the departmental team, instructional leadership, and staff development and appraisal. Some of the topics covered include the skills of delegation, influencing, teamwork, conflict management and the management of funds. Also covered is resource management, including financial management, and participants are given an understanding of how this is linked to the overall vision of the school.

Since each head of department generally has a span of control of about 10–20 teachers, he or she needs a range of people skills. Thus, heads of department need to learn about motivation, stress and working with small groups, and more importantly, how to develop an awareness of self so that they can manage self. The programme covers material on professional development, including planning and implementing a range of professional development strategies that will lead to real gains in the classroom. Participants learn about direction setting, and designing and implementing comprehensive programmes of instruction. They learn about leadership theories, leadership behaviours, and cover a number of issues relating to the ways in which they can introduce needed change.

It is clear from this that the head of department is the main curriculum leader in the school. Unlike schools in other systems, our schools are extremely large. Some of them have over 2,000 pupils. To have the principal as the key curriculum leader would be an unrealistic expectation, and so heads of department take on this critical role. With that in mind, the programme places some emphasis on curriculum design, development and evaluation.

There is also material on assessment and curriculum evaluation. Heads of department need to be familiar with alternative assessment modes and evaluation. Thus, they cover the issues of assessment planning, norm-referenced and criteria-referenced assessment, and group and individual accountability.

As part of the programme, participants are required to visit schools in order to view exemplary departments in their own subject areas. In doing this, they gain insights into how others run departments and, armed with a list of questions devised by the programme team, they develop new ideas on how they might improve things in their own schools.

## Other Leadership Programmes

Over the years, with schools becoming larger and the multidimensional job of head of department becoming more expansive, the system decided to create additional responsibility roles to share the burden. Thus, subject heads and level heads became posts in which enthusiastic individuals could take on development responsibility in a limited area. For example, a head of science may have subject heads for chemistry, physics and biology working with him or her. This has been a useful initiative, because it puts people in position, who have an in-depth knowledge of their specialist subject areas.

The post of level head, as the name implies, carries a responsibility for working with teachers across a level (e.g., Secondary 1). Accordingly, NIE now runs programmes for these educators with responsibility, though on a much smaller scale than programmes for heads of department. They undertake studies in mentoring and supervision for staff development; human relationship skills; and curriculum implementation and co-ordination.

Another programme we run is for teacher-mentors. These are senior teachers who are charged with coaching and mentoring newly recruited teachers. They also work with teachers who might need some assistance with their classroom teaching. The senior teacher is generally thought of as a skillful classroom teacher, who can model good practice for others in the school.

In this programme for teacher-mentors, there is a strong focus on mentoring. The mentoring process is one that has been highly valued over the years in Singapore education. In our system, we define a mentor and mentoring as 'a senior person who undertakes to guide a younger person's development, both in their personal growth and where their career is concerned … we view mentoring as a developmental process within more formal training programs' (Chong Low & Walker, 1989). Thus, these senior teachers are taught about the roles and functions of mentoring and the mentoring process, alongside other people-related skills.

## Rethinking Leadership Programmes

While all the programmes are based in NIE, relevance is a major consideration. Thus, participants are encouraged to use issues and problems they face in their schools as their basis for discussion, study and assignment work. It would be wrong to suggest, however, that relevance dictates everything we do. Indeed, while it may sound paradoxical, irrelevance may be more relevant to the future than an undeviating focus on what we know now.

It is true that our programmes, especially those for middle managers, have to give them the competence to deal with the here and now, but we also have to give them the capability to thrive in a different future. Things are changing. The past and the present are no longer an accurate indication of how things will be even in the foreseeable future. We know, for example, that our schools are facing increasing competition and that they are subjected to global influences. We know that the expectations of stakeholders are changing. And we know that the new breed of innovative and visionary principals are creating new expectations of how they will work with their teams of professionals in schools. To train merely for the present is futile.

As we move forward, then, we have to constantly look and re-look at what we are doing. Our programmes used to have a ten-year life cycle. Now it is nearer two years. We have already introduced the Leaders in Education Programme as a response to a vastly changing future. Now, we are working on our heads of department programme, and shall soon review the other programmes.

Central to much of our development in leadership programmes has been a true sense of partnership between NIE, the schools and the Ministry of Education. For example, the Ministry has been involved in helping us think through the issues in planning new programmes and, as a client, has provided useful feedback on the perceived impact of our training. Similarly, the schools have worked with us to help develop a deeper understanding of how change is manifesting itself at the operational level. These are important contributions to building professional development experiences that will make a real difference.

The partnership is especially important at a time when the social and educational landscape has emphasised the 'self-assertive' values of expansion, competition, quantity and domination (Capra, 1996). What is needed is 'a dynamic balance of the self-assertive and the integrative. From this perspective, partnership takes on a new importance, since it represents a shift away from the hierarchical to networks, from authority to influence' (Stott & Trafford, 2000, p. 2).

Through our networks of influence, therefore, we have experienced how schools are fast changing to prepare young people to live and work in the emerging learning nation of the future. We have felt the impact of pressures to update and upgrade school leaders on a continuing basis throughout their career spans in order to expand their capability to foster competitive learning school organisations. And now, the onus is on NIE to bring about change in the way we prepare professional educators for a new, challenging, but immensely exciting, future.

## References

Capra, F. (1996). *The web of life*. London: Harper Collins.

Chong, K. C. (1998). Singapore: Learning nation. *STADA Annual – 1998*, 6–16.

Chong, K. C. & Leong, W. F. (1995). Toward an IT-driven education system in Singapore – an interpretation of Delphi data. *Journal of Educational Technology Systems*, 23(3), 241–250.

Chong, K. C., Low, G. T. & Walker, A. (1989). *Mentoring – A Singapore contribution*. Singapore Educational Administration Society, Monograph, 3.

NIE. (2001). Diploma in Departmental Management Programme Handbook. Singapore: National Institute of Education.

Goh, C. T. (1997). Shaping our future: 'Thinking Schools' and a 'Learning Nation'. *Speeches 97*, 21(3), 12–20.

Hord, S. M., Hall, G. E. & Stiegelbauer, S. (1983). *Principals don't do it alone: The role of the consigliere*. Paper presented at the annual conference of the American Educational Research Association, Montreal, Canada.

Low, G. T. (1995) Mentoring, what and how do protégés learn? *International Studies in Educational Administration*, 23(2), 19–27.

Low, G. T. & Chew, J. (1997). Preparing the next generation of school principals. *Journal of the International Society for Teacher Education,* 1(2), 129–134.

Marquardt, M. J. (1999). *Action learning in action*. Palo Alto: California: Davies-Black Publishing.

Senge, P. M. (1990). *The fifth discipline: The art and practise of the learning organization*. New York: Doubleday/Currency.

Simon, E. In P. M. Senge, (1990). *The fifth discipline: The art and practise of the learning organization*. New York: Doubleday/Currency.

Stott, K. & Trafford, V. (2000). Introduction. In K. Stott and V. Trafford (Eds.), *Partnerships: Shaping the future of education*. London: Middlesex University Press.

# 11

# Balancing Stability and Change: Implications for Professional Preparation and Development of Principals in Hong Kong

Dr. Y. L. Jack Lam

*Chair Professor and Head, Department of Educational Administration, Chinese University of Hong Kong, Shatin, Hong Kong*

The powerful role of school leaders in effecting changes and improvement is universally recognized and accepted. Engulfed in the sweeping changes associated with education reform, the vibrancy and effectiveness of the school principalship is, more than ever before, being closely scrutinized. This is no exception in Hong Kong.

The role of the school principal in Hong Kong is caught between the forces of stability and change. Stability is derived from institutional traditions, societal culture, and the psychological orientation of educators. The forces of change are based in radical political, economic and social environmental transformation. School administrators in Hong Kong seem traumatized by the dilemmas involved in finding a balance between these opposing forces.

This chapter explores the delicate psyche of the school principal in Hong Kong as they attempt to adapt to externally imposed changes. The author will assess the success of the government's effort in tapping principals' synergy toward revitalizing the school system. Finally, the chapter will advance a framework for guiding principals in their professional preparation, selection and development. Given Hong Kong's location at the intersection of Eastern and Western cultures, the author hopes that this discussion both resonates with readers elsewhere and perhaps illuminates their own local issues.

## Forces of Stability and Change

As noted Hong Kong's school leaders are poised between forces of stability and change. Although change is the norm in education worldwide, it is hard to imagine a city in which more change has come more rapidly. School-based management, curriculum integration, IT in teaching, Chinese as the medium of instruction are but a few of the significant system-wide changes that have arrived in a matter of years. At the same time, conservatism and suspicion of change are rampant, not only in schools but also throughout the society as the city-state seeks to accommodate to its new status within China. This section reviews sources of tension and discusses their impact on schools and their leaders in Hong Kong.

**Forces of stability**

In sharp contrast to their Western counterparts, the career patterns of most principals in Hong Kong are typically stable and life-long, punctuated infrequently by resignation arising from ill health and mandatory retirement. Such a pattern remains intact, irrespective of the long British colonial rule and occidental influence. Deeply ingrained as common convictions are cultural beliefs that the headship of a school should be stable in order to provide continuity. Similarly, there is a deep-seated belief that school heads should be accountable only to their superiors in the school system.

Job endurance is, therefore, the rule rather than an exception. This stable career pattern is further solidified by characteristics of Eastern culture. As others have noted, education is an essentially human activity and is heavily intertwined with the societal culture it professes to serve (Dimmock, 2000). This is also true in Hong Kong where Chinese culture is overlaid on institutional traditions built into the educational system during the years of British colonial rule.

In this context, Hofstede's cultural comparative framework (1991) is most illuminating on Eastern perspectives on leadership. The framework gives insight into how societal and organizational expectations and psychological orientations influence the role of leaders and followers in Asian cultures. In societies with large *power distance*, as is the case in Hong Kong, greater inequalities of power distribution are taken for granted by people (Hofstede, 1991). Leaders and followers in organizations, including schools, tend to accept authority associated with positions and roles as 'natural'.

The effects of large power distance are evident throughout the educational system. Some effects are positive; others are negative. Large power distance breeds a 'natural' respect for authority, discipline within a system structure, and an ability to mobilize people from a central source. These characteristics have served Hong Kong and other Asian cultures well during the recent period of rapid economic ascent.

However, the same characteristics may also have dysfunctional effects under certain circumstances. In situations where proper management structure and processes are lacking, the same cultural traits can lead to 'unnatural' situations. When the organizational structure does not place appropriate limits on authority, principals can become 'little emperors' with unrestricted dictatorial powers. The school becomes their fiefdom and they rule without any evident constraints (Education & Manpower Branch & Education Department, 1991, p. 14).

This tendency has been reinforced by the *collectivist* orientation of Chinese cultures in which the group exerts a stronger influence on the behavior of individuals than is typically found in Western cultures (Hofstede, 1991). In the context of Hong Kong schools, this produces an interesting interaction effect when considered in combination with large power distance. It tends to be accepted as natural that the personal goals of staff should be subordinated to the goals of the organization. Moreover, the school's goals are often a reflection of the personal philosophy of the principal rather than a shared vision reflecting the aspirations of other stakeholder groups. Tradition and harmony prevail out of a need to maintain group solidarity and to preserve 'face'.

These factors have shaped the evolution of the principal's role in Hong Kong schools. As noted, traditionally there have been few internal threats to job security. Performance or lack performance was seldom an issue. In this context of stability, position-oriented power, and tradition there have been little incentives for most principals to prepare themselves for change. Principals have been guardians of the status quo. They are among the role groups who perceive themselves with the most to lose from recent school reforms.

Principals' individual inertia and that of their schools seem further reinforced by their previous encounters with changes. Up until 1997, prior to the return of Hong Kong to the sovereignty of China, official educational plans and ideas had come and gone, without leaving much in the way of permanent marks on the system. The obvious centralization of decision-making power in the hand of the Colonial government reconciled paradoxically with a strong reverence for traditional school autonomy. Consequently, there is a long tradition in Hong Kong in which schools are run virtually without external interference.

Mutual respect between principals and their staffs is readily translated into mutual tolerance and a clear-cut division of labor within the school. As noted above, this reflects the predominant cultural norms (i.e., power distance, collectivism) to seek group harmony and avoid conflict. Despite their profound interest in their children's schooling, parents were usually kept at a respectful distance and had no significant role to play in school affairs. Community interest groups were likewise kept at bay so as not to 'complicate' school governance.

**Forces of change**

The political transition of Hong Kong in 1997 changed the educational landscape abruptly when the government made it a priority to overhaul the entire school system (Education Commission, 1996). There were no more safe havens, even for practices that might be deemed sacred by the community of professional educators. Past emphasis on quantity increase has been refocused towards quality improvement.

This was reflected in a barrage of new initiatives undertaken by the Education Department:

- Curriculum changes,
- Upgrading of teachers in language proficiency,
- System-wide implementation of new information technology,
- Regrouping of schools by the medium of instruction (i.e., mainly Chinese mother-tongue),
- Adoption of school-based management.

These changes arrived in simultaneous wave rushing through the system, leaving principals and teachers gasping for air. The pace of reform has been such that, as some principals noted, over a period of three months, more than 600 directives reached their offices from the Department of Education. Notably, most of these changes originated outside of the educational system. The impetus for reform, and even the specific types of reform (e.g., Chinese as medium of instruction), were initiated largely by political and business leaders, not educators.

The public was discontented with a fossilized system of education. The traditional approach to education that focused on rigidly selecting limited students for academic or career advancement created a favorable social condition for change in this new more open era. Calls from the business community for more skilled and knowledgeable manpower to transform economic infrastructure necessitated a reorientation from elite to mass education. It also provided impetus for a shift away from an emphasis on general curriculum to one with greater focus on information technology. The increase in complex social problems, such as youth crimes and rising school violence, has similarly rekindled popular demands for more life-skilled education. New political orders and economic ties with China have compelled the education system to provide parallel emphasis on dual language (English/Chinese) proficiencies.

In searching for new directions for altering the educational system, Western models have tended to be fully transplanted into Hong Kong. These changes are implemented with almost the same speed as in the U.S., Canada and England (Lam, 2001a). Probably because of its concern for losing momentum and competitiveness, the government of Hong Kong has reacted by trying to direct all changes in education. On more than one occasion, education policies have been implemented without prior public input, or at best, with only a few rounds of token consultation. The spirit of transformation that so bewildered and horrified UK educators in the form of 'economic rationalism' (Burke, 1997; Dudley & Vidovich, 1995) suddenly descended upon Hong Kong schools. The spell of tranquility (and stagnation) was suddenly broken.

For scholars interested in the external environment of schools (e.g., Milliken, 1990; Rowan, 1993; Scott, 1992), there has been a general realization that context of schools has undergone radical change. In the past, public school educators perceived their organizations as social systems packaged with social technology, and operated with rules establishing identities and activity schema that comply with societal expectations. To the extent that they conform to the designated roles in societies, they are guaranteed on-going support and legitimacy. In other words, by serving faithfully the needs of its 'institutional environment', schools become 'domesticated enterprises' that do not need to further justify their existence. Patterns of incumbents' behaviors become routine and school organizational forms are 'rationalized' (Fennell & Alexander, 1987).

Recent systemic reforms, so deeply colored by conservative political ideologies, have converted the accustomed 'institutional environments' into market-like competitive 'task environments'. These demand goal setting, goal achievement and effectiveness. Political, social, economic, cultural and technical contextual factors have certainly raised the stronger pressure on public schools to increase productivity and performance. The education community, laboring under high stress, has certainly felt under siege. General resistance against such an abrupt change by institutional forces is not surprising (Hoy & Miskel, 1996).

**Current dilemmas**

Bewildered by abrupt changes in the external environment, and with few past precedents to guide behavior, principals of Hong Kong schools have felt conflicted in dealing with school reforms. Factors associated with the cross-cultural transfer of reforms from the West have accentuated this sense of dissonance. The educational authorities in Hong Kong have not carefully assessed the outcome of these changes in their homelands, never mind crafted them to suit Hong Kong.

There are many examples. Although inclusive education has not fared well in the West, it was introduced indiscriminately in Hong Kong. The hastily conceived target-oriented curriculum (outcomes-based education) became entangled with the integrated curricular approach. The adoption of information technology has blended oddly with the not so successful school-based management. Extremely labor-intensive undertakings like quality assurance assessment have been implemented while more sensible self-assessment initiatives were neglected. Under these circumstances, school leaders find their time consumed by the implementation of an ever-changing list of new priorities even while old problems (and priorities) still persist. Public school educators feel completely overwhelmed as they attempt to navigate a sensible path through this political minefield!

## Patterns of Adaptation

In the tortured mind of Hong Kong principals, reactions to school reform have been highly idiosyncratic. Principals seemed to be influenced by some combination of prior presumptions, current situational assessment and projected cost-benefit analysis. Their patterns of adaptation are more ambivalent than consensual, more opportunistic than rational, and more utilitarian than moral.

### Do the minimum

To a small group of principals coming from the 'prestigious' schools, that have enjoyed 'success' by conventional criteria (i.e., students achieving outstanding performance in public examinations) the present education reforms are viewed as a threat to their ideal world. Inherent in this rather apathetic perspective is a strong conviction that if they hold fast to the existing system, the wind of changes will blow and pass on. It is not a surprise to note, when visiting their schools that all operations go on almost unchanged despite numerous directives from the Department of Education to do otherwise.

Staffs of these schools likewise seem oblivious to what is happening beyond the school walls. When challenged by government officials to comply with the mandated changes, these principals even take on a high profile of open defiance, questioning the need and legitimacy for altering the system. This group simply sees no need to change. Their schools have been successful and they expect that to continue.

### Select the essential

To other school administrators, the snowballing effects of change have greatly stressed themselves and their staffs. Their coping strategy, therefore, is to consciously select those directives arriving from the Department of Education that match their school needs. Much care has been taken in terms of time and scope of changes so as to ensure that modification is taking place in good order and their staffs are capable of absorbing the changes into their routine. While the transformation had occurred incrementally, the changes in these schools are individualistic and highly personal.

Chief among the concerns is the fact that the technical level of the school operation should be sheltered from constant external disturbance. The subsequent unparalleled development is perhaps not unfamiliar to their counterparts in North America where external constraints have been selectively filtered on their way to the classroom level (Lam, 1997). The key role of the principal in these schools has been to moderate the externally imposed changes to achieve a reasonable fit to the schools.

**Change with the flow**

There are also a substantial number of principals who tend to follow directives from the Department of Education sheepishly. Conformity to the authority has always been the mode of operation for the majority of principals in Hong Kong. However, silence from the educational leadership should not be equated to support or consent. Lack of enthusiasm increases the time lag between receiving the marching orders from the Education Department and the actual implementation of new policies.

Misinterpretation of the nature of change has often led to confusion and frustration. Under such circumstances, principals find themselves having difficulty in motivating and assisting their staffs, crucial ingredient for effective schools (e.g., Teschke, 1996). For this group of principals, a combination of half-hearted compliance with Department policies results in the appearance of reform without any real changes that benefit students.

**Tapping fresh resources**

For some principals, reform has provided them with rare opportunities to improve school resources at a time when revenue for education is closely scrutinized. Incentives are normally available when mandated changes are enforced. The establishment of a Quality Education Fund by the Government, for instance, allows them to install air-conditioning units, to renovate multi-media rooms and to upgrade school buildings.

The expansion of information technology provides schools with additional technicians to ensure accessibility to technical support in teaching and learning. Curriculum modification and integration gives rise to free in-service training for principals and their staffs. Devolution of decision-making power offers school leaders some flexibility in fiscal management. In short, by participating in reform, some principals ensure that their schools reap the benefits of technical, curriculum, human and financial resources, which under normal circumstances are difficulty to come by.

**Active involvement in reforms**

To a handful of principals, current reforms allow them glimpses of hope in rectifying system-embedded problems that they have sensed, but felt helpless to address. The present overhaul of the school system offers these school leaders chances to rectify undesirable features in order to realize long-sought educational goals. Their personal commitment and enthusiasm for change solidify them into a small core of true supporters of the present reforms.

**Summary**

Based on these patterns of adjustment, principals in Hong Kong are, on the whole, not acting in full synchrony with government-initiated changes, even though some may realize that change is essential for the system improvement. The principals initiate few proactive measures, individually or as a group. Most tend to drift along, cooperating only reluctantly with the directives from above.

Very likely, the wave of quite abrupt reforms caught them completely off-guarded. Principals are wrestling with the difficult task of balancing school organizational stability and steering along the mandated courses of change. Ambivalence and agonizing doubts merge with a diversity of personal motives. These create constant dilemmas that render the change efforts in disarray. In face of the growing politicization of school changes (Lam, 2001b), constant shift in the rhythm and direction of reform from above further adds confusion and frustration to the process of implementation.

## Critical Leadership Issues for Principals in Hong Kong

Arising from the above scenarios is a key issue concerning the quality of school leadership in Hong Kong. There is little question that this will ultimately determine the reform outcomes in schools. In terms of their personal and professional backgrounds, the majority of principals in the systems are products of the past decades. Most of them have steady, uninterrupted teaching career patterns of over twenty years and some administrative experiences as middle managers of the schools.

Rising through the teaching rank and file, most did not receive formal training in administration or leadership. Most will have attended a few workshops provided by the Department of Education. These ad hoc sessions would have focused primarily on clarifying education policies, curricular matters and foreign educational concepts. What we think of as true professional development opportunities for them are few and far between. Further studies beyond the basic degree are not required; therefore, only a handful have pursued graduate studies on their own.

Consequently, the administrative expertise that these principals possess comes primarily from their experience. New principals look towards more experienced colleagues for guidance, and practice their administrative skills through trial-and-error. Experienced principals count on previous precedents to resolve current crises. Both groups, however, have been completely disoriented during this era as their roles and public expectations have undergone major modification.

Indeed, the educational reforms of the past decade have brought forth a new challenge for these principals. They have redefined a new set of roles for the principals (e.g., Daresh, 1999) that they are expected to execute and on which they will be assessed.

These emerging responsibilities include:

- Transformational leader which entails restructuring and power sharing in collective governance.
- Transparent managers who are accountable for their actions to their constituencies.
- Instructional leader both in curriculum reorganization and in improving the quality of teaching and learning for different streams of students.
- Guardian angels looking after the welfare of their staffs and students.
- Social workers dealing with growing complexity of students' social and family problems so as to contain unwanted disturbance or violence in school.
- Law enforcement officers preventing, monitoring and dealing with criminal offences that might occur in schools.
- Liaison officers forging closer ties with parents and local communities.

- Negotiators soliciting resources from the government officials.
- Resource explorers searching for and securing 'new sources' of revenues to support innovative programs or simply updating school facilities.

This list would likely sound familiar to principals in many Western countries as well. The points worthy of emphasis in the Hong Kong context are twofold. First, this represents a sharp even radical break with the past in terms of role expectations. Second, the cultural context of Hong Kong runs counter to the assumptions that underlie many of these new policies and trends.

## Recent Efforts for Addressing Leadership Problems in Hong Kong

Evidently, principals in Hong Kong are working under a new set of expectations. There is also a growing realization among the Government officials that if the current reforms are to have any chance of success, vibrant leadership in the schools is essential. Policymakers are contemplating two general thrusts to rejuvenate an aging and somewhat ineffective system of preserving leadership quality: elimination of the 'old corps' and enhancing the professionalization of the principalship. A quick review of these attempts is in order.

### Elimination of the 'Old Corps'

Conventional wisdom points to the fact that a complacent and aging leadership team will not inspire school staffs to actively engage in change. To make room for a fresh and energetic group of principals, the government is revising existing education statutes to enforce their retirement age at 60. This is a measure that has long been practiced for all civil servants and teachers, but not for principals.

This policy change is being implemented despite anticipated opposition from the principals and a likely court challenge. Amidst the outburst of hostility and criticism, a compromise solution is being put into practice. The prescribed retirement age of 60 remains unchanged. However, the sponsoring bodies of individual schools can apply, on behalf of the concerned principals, for a one or two year extension if the principal has demonstrated a high degree of effectiveness.

Observers with less self-interest in the Government's position would likely reach the conclusion that mandatory retirement for principals is timely and appropriate. The sweeping enforcement on this policy registers the determination of the Government to remove the old guard, paving way for more fresh blood to enter into leadership positions. What has not been carefully scrutinized is the basis upon which requests for extension from the schools' sponsoring bodies is made. Commonly, such recommendations are made either for 'humane' or other 'relationship' reasons, rather than a rigorous review of their competencies and effectiveness. In fact, few assessment tools are available, and even fewer have been applied to job performance appraisal.

Three serious consequences arise from this situation:

1. Unavoidable misuse of the extension for 'exceptional cases' perpetuates ineffectual leadership, which tends to defeat the very purpose of prescribing the retirement age.
2. Permission of principals to stay on beyond the retirement age regardless of their job performance does not provide a good model for the school in light of

the current reforms where efficiency, effectiveness and accountability are being emphasized.

3. Creation of exceptional cases reinforces discriminatory practices and heightens misgivings among colleagues. Complicated internal politicking and appeals will multiply.

What is needed, therefore is for the government to promptly re-examine these loopholes and provide a more comprehensive guideline for the school sponsoring bodies to follow before they launch appeals for extending the services of their principals. Continued neglect and persistent procrastination will defeat the very purpose for change.

### Professionalization of the principalship

As most principals advance to their role through teaching, many are deficient in basic administrative knowledge, attitudes and skills. In the face of the new challenges that come with reform, many are totally unprepared. The government has attempted to rectify the situation by making professional training mandatory for principals at different stages of their careers. In the blueprint, which is being developed and refined, three categories of individuals are recognized:

- 'aspiring' principals,
- 'newly appointed' principals, and
- 'experienced' principals.

The aspiring principals are middle managers of the schools – vice-principals, panel chairs, department heads – who aspire to become principals. Newly appointed principals, as the name suggests, are individuals who were appointed principals by their sponsoring bodies within the past one to two years. Experienced principals refer to those who have been more or less established as school leaders for some time.

The blue print also contains some specific programs to be developed for these three groups. The Department of Education envisions working closely with the universities for program delivery. In brief, the program encompasses a model based on need assessment, a leadership seminar series, and professional components that cover the full spectrum of school administrative responsibilities. Based on the strong recommendation of the universities, the notion of 'multi-track' and 'multi-level' approaches is reflected in the program design. This will more adequately accommodate the divergent needs of educators at different stages of their career development.

This is indeed a laudable attempt to professionalize the whole group of school leaders in Hong Kong. For the first time, perhaps, the Government has realized that leadership, more than anything else, will determine the success of the implementation of its reform programs. Recognizing that there may be backlashes against such a move, the Government has taken a cautious approach, consulting with different principals' groups and with concerned units at the universities. A special task force consisting of all stakeholders was created to examine the operational details. Experienced principals and some selected newly appointed principals have been asked to take part in the trial run of need assessment tools.

### Reasons for lack of success

Despite all these cautious measures, the plan has already suffered some unexpected setbacks. A great deal of the difficulty has to do with the hesitation and diffused attention shown by the government in its planning efforts. The blueprint was, in fact, unexpectedly shelved due to the militant opposition of teachers. To avoid confronting the battle on both fronts, the government has unfortunately, signaled to the educational community that it is again wavering on its position. Its sincerity and credibility in upgrading the leadership corps are suddenly in question.

The second plausible reason for the delay, perhaps, has to do with the magnitude of training expenses. Initially, the Government hoped to finance the whole project by tendering it out to the Post-secondary institutions. The institution that offered the cheapest (not the best) way of conducting the in-service activities would have been awarded the contract. Unfortunately, re-estimation of the cost showed that the projected budget for upgrading some 1,200 experienced principals, 120 annual newly appointed principals and close to 4,500 aspiring principals plus subsequent follow-up professional activities far exceeds what the government is willing to pay. The modified course of action is that only newly appointed principals will receive partial financial support. All the rest must pay on their own. This raises an interesting question. As mandatory retirement is in effect and as more new blood enters into principalship, will the government further retreat from such a commitment? Time will tell!

A related concern, based on the initial trial run of need assessments of newly appointed principals suggests that the quality of leadership potential is marginal. This points to the necessity of a critical review of established procedures and criteria in the recruitment and selection of new principals. A major challenge is to persuade the sponsoring bodies of schools to develop, with some unified guidelines, more defensible criteria for choosing principals. Without this step, great variations in terms of leadership will result and leadership development will produce weak effects.

**Some possible remedies**

It is evident that any superficial bandaging of the existing system of leadership selection and training is bound to fall short of the ambitious goal of overhauling the quality of the principalship in Hong Kong. It is equally clear that any half-hearted measure of change only sends the wrong message to die-hards who, by intensifying the politicization process, hope to maintain the present status quo. A thorough clean up of the situation calls for courage and a radical departure from past practices. Let me suggest possible directions.

*Expansion of selection criteria*

To avoid the current hit-and-miss approach, the present principal's recruitment process should be replaced with a more systematic and reliable means of identifying potentially good candidates. To this extent, the interview team should not be restricted to the lay board members of the sponsoring bodies, but should also include retiring principals and teacher representatives. In this fashion, multiple perspectives in assessing the candidates can be ensured and more professional inputs in selection are assured.

Along the same line of thinking, criteria for selection should also be made more comprehensive (e.g., Murphy, 1983; Scannell, 1988). Sole dependence on performance of the interview, focusing frequently on educational philosophy and vision alone, is inadequate. Profiles of past school performance, evidence of further studies, frequency of attending in-service and professional training activities, documentation of past initiatives taken in school, compatibility with the staff, track record on public relations work done to date and personal qualities such as honesty, diligence, ability to inspire (Richardson, Lane & Flanigan, 1996) should all be used in assessing candidates. In other words, belief, vision, achievement, and appropriate personality are all important characteristics for selecting future leaders for schools. By pooling all relevant information together, chances for selection errors will be reduced.

*Orientation and preliminary training*

There is a common but erroneous assumption that once an individual is assigned to the leadership position, he/she will become effective over time. The general preference for choosing the vice-principal from the same school to be the successor reflects this assumption. In reality, as roles and responsibilities change, vision and working relationships with fellow-teachers also change. There are more challenges and problems on the job. The tradition of isolation that comes with the top position compounds the problems facing new administrators.

Early confusion on the job can be overcome when retiring principals are engaged to provide initial orientation and on-the-spot consultation. Such mentorship is consistent with the Eastern culture where respect for experience and elders are the norm. Advice given in such a context should be considered as 'reference' rather than binding guidelines so that newly appointed principals would still have room for making their own decisions.

Additional professional training sponsored by the Department of Education should also be arranged to provide support to newly appointed principals. There should be dual intervention for new principals: orientation and on-going professional training. This could shorten their learning curves and reduce confusion as they (and their staffs) adjust to new leadership roles.

*District support group*

Throughout a principal's administrative term, it is likely that he/she will continue to encounter non-routine personnel problems that require detailed review, careful identification of problem sources and comprehensive solutions. Involvement of teaching staff in those situations may, at times, be inappropriate. An informal support group consisting of fellow-principals from the same district may provide a useful forum for the deliberation of these delicate matters. This is not dissimilar to the formalized leadership centers (e.g., Barth, 1984) that have sprung up in the USA over the past 20 years. A non-threatening environment and high collegial support can bring forth fruitful discussion and collective decision-making.

If issues are too complex for the group to handle, an invitation for expert opinion can be sought from faculty members of institute of higher learning. Quite often, the expansion of membership to university professors can be done without incurring much expense, as professional services are part of faculty members' job when they apply for tenure and promotion. The combined input from the professional educators and academicians should improve the quality of decisions and enhance the growth of principals during their early years.

*On-going monitoring and assessment*

Too often, when a principal is comfortably on the job for some time, complacency will set in and the school under him/her will once again enter another period of stagnation.

To ensure that a principal will remain alert on the job, there should build into the system a monitoring and assessment procedure. If the district support group is in place, we might ask some members from the team to undertake the function of on-going supervision and each principal will operate under the watchful eyes of his/her colleague(s). Under such an arrangement, mutual learning will be maximized.

Likewise, the provision of feedback in the form of formative assessment and performance outcome based assessment (Cairns-Danald, 1995; Valentine, 1986) will provide useful self-reflective data for principals themselves. To ensure that the formative assessment is fair and comprehensive, it should be done by people who work closely with the concerned principals. Indeed, data can be supplied by representatives on the student council, teaching and support staff, lay members from the Parent-School Association and colleagues supervising the principals on the job. The resultant portfolio will serve several important purposes:

*Basis of self-improvement*
School leadership, like any other position in the school organizational hierarchy, requires continued feedback to help guide their own perspective on the job done. As the present system does not provide such a feedback mechanism, most tend to muddle through from day to day with a great deal of uncertainty and ambiguity. This can be directly detrimental to the morale of the principals and indirectly to that of their staffs. The significance of having an assessment system as some basis for self-improvement cannot be de-emphasized.

*Accountability*
Formative assessment refocuses our attention on accountability; a term, which symbolizes the spirit of the current reform and a term that has been talked a great deal but seldom put into practice. Up to the present, improvement on student achievement, curriculum reappraisal, teachers' language proficiency testing, school performance indicators construction, have all been discussed in different education forums. Ironically, what is missing, as a central piece, is principals' evaluation. Logic and empirical evidence (e.g., Teschke, 1996) suggest that when leadership is not effective, all other aspects of school operation will not be successful. The installation of principals' assessment should certainly eliminate ambiguity and elevate school performance to the next level and makes leadership more accountable for its action and more responsive to the school's constituent needs.

*Direction and momentum for change*
Too often, educators jump on the bandwagon without fully understanding why change is essential. Consequently, their efforts are spent going in circles. Formative assessment on leadership performance, on the other hand, zeros in on crucial internal problems that require urgent attention. It helps unearth many manifested or latent school organizational problems as well as the effectiveness of the adopted coping strategies. In this context, periodic appraisal of leadership should provide a clearer direction for change. At the same time, it further legitimizes change and provides momentum for keeping the transformation on track.

*Model for others in school to follow*

Up to the present, educators at all levels in the school system have been subject to close scrutiny except the principals themselves. This begs the question as to why a critical area such as school leadership, should be immune from examination. If leadership is viewed as 'effective', there will be a greater tolerance for such a discriminatory practice. On the other hand, if leadership quality becomes a 'suspect', its immunity from formal appraisal is demoralizing to the rest of the staff. The installation of principals' assessment, even if it is formative in nature, should fill in a vital missing gap and should pacify common rumbling discontent in Hong Kong schools.

*Efficiency*

When there is a review process for all, efficiency of the school operation should be increased. When accountability for all positions in the school system is reinstated, there are fewer opportunities for anyone to take irresponsible action. When there is a clearer direction for change and greater momentum for sustaining such a change, resources can be put to maximum use while restarting the engine for change becomes unnecessary. When assessment applies to all positions in the school organization's hierarchy, the morale of the subordinate will be elevated. All these guarantee greater efficiency for initiating and maintaining change.

**In-service training for principals**

There is an inaccurate assumption and practice in Hong Kong that professional training is a terminal activity. Once a professional (doctor, lawyer, teacher alike) receives his/her initial formal training, he/she becomes certified for life and can practice his/her trade forever unless violations of serious professional ethics or criminal codes occur.

In the present conceptualization of principal training, the Hong Kong government is still stuck to this out-dated framework. Western professional models rest on the assumption that change is constant, that professional development is an on-going process. Therefore life-long education should be the basis for preventing professionals from becoming obsolete. In maintaining the quality of education leadership in Hong Kong, on-going training should be made compulsory. In other words, once principals complete their 'formalized training', they should be asked to attend on regular annual basis, workshops, seminars, and/or conferences. Evidence of their attendance or participation should be verified, and some credits associated with such activities should be assigned to the record to ensure the validity of their certificates. Those failing to comply with the in-service regulation will be required to do extra make-up work (such as attendance of University course work in educational administration etc.) or their licenses will be suspended. Continued professional development should keep principals abreast of development in the education field and should provide more up-to-date information for them to guide their schools.

**Summative evaluation**

For a long time in Hong Kong, few are willing to challenge the tradition of life-long appointment for principals despite the fact that their qualities may not match with the changing needs of the time. Neither do we see that many can keep up with the pace and magnitude of transformation. Revamping of the school system indeed is incomplete when the current system of principals' retention remains intact (e.g., Reed, 1989). Furthermore, if formative assessment can be argued to be beneficial to the school system, it is logical to stress that reinstatement of summative evaluation cannot be delayed.

When a systematic summative evaluation scheme is in place, there is a more rational basis to retain or dismiss principalship. The government has realized its significance, and is pushing hard to change the civil service system from a guaranteed life-long appointment to contract employment. Yet, such a concept has yet to take root in principalship, partly because only a small number of schools are directly under its control and partly because of the political ramification this may arouse from the powerful principals' group. And yet, this is a critical step that must be taken. Lack of political will and moral courage to adopt this approach will only dishearten those sincere in bringing about change but also demoralize the rest of the education community.

## Conclusion

Confronted with the globalization of educational reform, many school principals in Hong Kong display a strong inclination to hang on to the status quo. In part, this is directly related to their misinterpretation of the changing nature of the external context of the school (i.e., from one of institutional environment to that of task environment). With such a radical transformation comes a new set of role expectations to which many Hong Kong principals are finding it difficult to adjust. In part, the globalization of educational reform tends to ignore the cultural characteristics of a given region.

Any top-down imposition of such a change posts direct psychological threats to those in power and tends to trigger much inner resentment or outright animosity. Not surprisingly, the overall response pattern of Hong Kong school principals has been reactive rather than proactive, lukewarm rather than enthusiastic, calculative and utilitarian rather than moral, and ritualistic rather than genuinely committed. With such a strong dissonance among principals concerning fundamental changes to the context, failure of the Government's efforts may be a foregone conclusion.

Beneath the more obvious behavioral and psychological make-up of Hong Kong principals is a more worrisome issue: quality of the headship corps that was nurtured or tolerated by the past lasseiz-faire policy and practice in leadership selection, training and development. A fundamental overhaul of the entire process is in order if the quality and viability of the schools for this new millennium are to be assured. In undertaking such a monumental task, the government should not be sidetracked by the apparent lack of consent, nor retreat under intensified politicization of opposing force.

Rather, it should have the dogged determination to complete the reform agenda. But in the process of implementation, it should envision some of the existing problems encountered by schools, and provide necessary resources. It should be sensitive to the localized culture so as to determine the priority and the scope of modification in order to ease the transition. Knowing the significance of the caliber of leadership to school improvement, it should reconceptualize the entire process of principal selection, preparation, in-servicing needs, and performance assessment.

Hong Kong's policymakers and school leaders should not view this process as a one-shot deal to clear the slate. Instead they should frame it as an on-going, life-long learning opportunity for system leaders. This is the only approach that has a chance of enhancing the effectiveness of leadership quality in Hong Kong in the years to come.

## References


Barth, R. S. (1984). *The principal's center at Harvard University.* A paper presented at the Annual meeting the National Catholic Educational Association, Boston, MA: April 23–26.

Burke, C. (1997). *Leading schools through the ethics thicket in the new era of educational reform.* Melbourne: Australian Council for Educational Administration. ACEA monograph No. 22.

Cairns-Danald, V. (1995). *Research Report: Critical skills necessary for Montana principals.* Institution: Montana State Board of Education, Helena. (ERIC Document Reproduction Service ED 390 822).

Daresh, J. C. (1999). Preparing school leaders to 'break ranks'. Connections: *NASSP*, 74(3), 30–31.

Dimmock, C. (2000). Hong Kong's school reform: Importing Western policy into an Asian culture. In C. Dimmock & A. Walker (Eds.), *Future school administration: Western and Asian perspectives* (pp. 191–222). Hong Kong: The Chinese University Press.

Dudley, J. & Vidovich, L. (1995). *The politics of education: Commonwealth schools policy 1973–1995.* Melbourne: Australian Council for Educational Research.

Education Commission. (1996). *Quality school education* (EC report No. 7). Hong Kong: Government Printer.

Education & Manpower Branch & Education Department. (1991). *The school management initiative: Setting the framework for quality in Hong Kong schools.* Hong Kong: Government Printer.

Fennell, M. L. & Alexander, J. A. (1987). Organizational boundary spanning in institutionalized environments. *Academy of Management Journal*, 30(3), 456–476.

Hofstede, G. H. (1991). *Cultures and organizations: Software of the mind.* London: McGraw Hill.

Hoy, Wayne K. & Miskel, C. (1996). *Educational administration: Theory, research, and practice* (5th Ed.). London: McGraw-Hill.

Lam, Y. L. J. (1997). Loose-coupled responses to external constraints: An analysis of public educators' coping strategies. *The Alberta Journal of Educational Research*, 43(1), 37–50.

Lam, Y. L. J. (2001a). Economic rationalism and education reforms in developed countries. *Journal of Educational Administration*, 39(4).

Lam, Y. L. J. (2001b). *Language proficiency controversy: An anatomical analysis of teacher militancy amidst school reform.* Manuscript submitted for publication.

Murphy, C. (1983). *Effective principals: Knowledge, talent, spirit of inquiry.* San Francisco, CA: Far West Lab. for Educational Research and Development. (ERIC Document Reproduction Service No. ED 234 521).

Milliken, E. J. (1990). Perceiving and interpreting environmental change: An examination of college administrators' interpretation of changing demographics. *Academy of Management Journal*, 33(1), 42–63.

Persell, C. H. & Cookson, J. (1982). *Effective principals: What do we know from various educational literatures*? National Institute of Education, Washington, DC. (ERIC Document Reproduction Service No. ED 224 177).

Reed, R. J. (May 1989). *The selection of elementary and secondary school principals: Process and promise.* (ERIC Document Reproduction Service ED 316 913).

Richardson, M. D., Lane, K. E. & Flanigan, J. L. (1996). Teachers' perceptions of principals' attributes. *Clearing House*, 69(5), 290–92.

Rowan, B. (1982). Organizational structure and the institutional environment: The case of public schools. *Administrative Science Quarterly*, 27, 259–279.

Scannell, W. (1988). *The leadership of effective principals. Insights on educational policy and practice*. (Report No. 5) Austin, Tex: Southwest Educational Development Lab. (ERIC Document Reproduction Service No. ED 330 049).

Scott, W. R. (1992). *Organizations: Rational, natural and open systems* (3rd ed.). Englewood Cliffs, NJ: Prentice Hall.

Teschke, S. P. (1996). Becoming a leader of leaders. *Thrust for Educational Leadership*, 26(2), 10–17.

Valentine, J. W. (1986). *Performance/Outcome based principal evaluation: A summary of procedural considerations*. Paper presented at the Annual Meeting of the National Middle School Association, Atlanta, GA, October 22–25.

# 12

# Professional Development for Principals in Taiwan: The Status Quo and Future Needs

Dr. Ming-Dih Lin

*Associate Professor, Institute of Education, Chung Cheng National University, Taiwan, Republic of China*

> Because the environment is so dynamic, to stand still is to fall behind. (Ruohotie, 1996, p. 442)

> The most difficult people to reach for staff development are superintendents and principals. (Dunn, 1999, p. 118)

Nobody expects a doctor, accountant or lawyer to rely for decades solely on the knowledge, understanding and approach that were available at the time when they began their career. Good professionals are engaged in a journey of self-improvement, always ready to reflect on their own practice in the light of other approaches and to contribute to the development of others by sharing their best practice and insights. They learn from what works. (Department for Education and Employment [D*f*EE], 2000b, p. 1, David Blunkett, Secretary of State for Education and Employment of UK.)

> [M]uch still needs to be done to develop a coherent, sequential, and pedagogically effective programs (*sic*) for principals at different stages in their careers. (Kelley & Peterson, 2000a, p. 1)

Although professional development for principals and their schools has been called an 'education disaster area' (Wagstaff & Collough, 1973 cited in Kelley & Peterson, 2000a, p. 1), in recent years, awareness of its importance has increased around the world (Beaumont, 1997; D*f*EE, 2000a; Lin, M., 2000b; National College for School Leadership [NCSL], 2001). The movement of making educational leadership preparation programs internationally has generally followed trends originating in the U.S. The focus over the past two decades has been on 'the content of preservice programs ("what" future leaders should know)' and 'the delivery of preservice programs ("how" people should learn what they need to know)' (Daresh, 1997, p. 3). Numerous associations, institutes and organizations that originally concentrated their efforts on principal preparation have incorporated in-service professional development for principals into their practices (Kelley & Peterson, 2000a).

Similar to other countries, in Taiwan efforts are taking place to improve practicing principals' knowledge, understanding of work, skills, and future performance (Huang, 1990; Lin, W., 1999). Several forces have facilitated this movement. These forces include:

1. Recent educational reform initiatives have promoted adoption of school-based management (SBM). This allows school principals to have more latitudes in decision-making and at the same time places more obligations and pressures on the principalship. In order to fulfill the requirements of SBM, principals need more opportunities for personal growth and development.
2. In recent years, curriculum and instruction reforms are becoming hot topics in Taiwan. According to the schedule of Ministry of Education (MOE) R.O.C. (2001), beginning in September of the year 2001 all elementary and secondary schools will begin the gradual implementation of a nine-year integrated curriculum. Collaborative teaching strategies, school-based curriculum development, parent involvement, school-community relations, and selective courses, to name a few related reforms, will be introduced. All of these require the principals to have new and more in-depth knowledge, understanding of tasks, skills, and actions concerning instructional and curriculum leadership. Because of this emerging redefinition of the principalship, principals need to develop and renew themselves both personally and professionally.
3. The method of school principal selection has been replaced. In the past, after principals passed exams and took their positions, the full responsibility of the principals' appointment and transference (demotion or promotion) belonged with the Bureau of Education. As long as principals made no serious mistakes they could stay in the position as long as they wished. Now, according to newly issued regulations, after serving a four-year term of principalship, principals need to apply for another four-year term. The decision of approving the application will be made by a committee consisting of representatives of teachers, parents, administrators of the Bureau of Education, and community members. The principalship is no longer a 'life-long career' for every principal who has obtained the required qualifications. Candidates whose performance is below standard are less likely to be reappointed.
4. Recently, the opportunities and sources of professional development for principals have increased. These include new established departments in the fields of educational policy and administration in universities, more graduate programs for practicing principals, and more workshops for principals.
5. The inauguration of related professional groups is also having an impact in this domain. Within the year of 1999, two important associations for school and educational administrators were formed, the Association of Educational Administration of ROC (AEA-ROC) and the School Administrators Research Association, ROC (SARA-ROC). These organizations will provide professional development opportunities for principals, including biannual forums,

conferences, networking, and visits to allied institutes and organizations abroad.

6. More scholars in Taiwan are taking an interest in the fields of leadership and the principalship. Local research findings indicate that the principals are very influential both to the school and students' performance as well as the quality of social interactions among members of the school (Chang, D., 2000; Chin, 1997; Hsieh, 1993; Wu, 1999). Leading this way, it is important that principals are engaged in a journey of self-improvement and try to contribute to the development of the profession and others by sharing their best practices and insights.
7. The ideas of life-long learning and a learning society promoted by the MOE have recently evoked the awareness of the importance of educators' continuous professional improvement including school principals. Principals are the symbols of learning for the schools. They have to take the initiative to participate in the movement of individual and organization learning.

In this chapter, the author provides a concise description of the status quo of principals' professional development in Taiwan, particularly focusing on issues of source, content, and instructional strategies of professional development. Based on the findings of related literature, the future needs and trends in this area will be discussed. Finally, the author argues that making a genuine change in principals themselves and in their schools should be the top priority of any professional development programs.

## The Status Quo

In Taiwan, with regard to principalship, most efforts had focused on issues of selection, pre-service training, and transference (Ko & Lin, 1996). Up until the year 2001 there were no specific preparation programs for principals in universities, no required licensure, no requirements for renewal, almost no formal evaluation, and no Principal Center or Academy for the purpose of principal's professional development. Professional development opportunities for principals in Taiwan were sparse, unplanned, incoherent, spontaneous, and without sequence. However, there are efforts that have the potential for impact on improving the practice of principal leadership.

Since September 2001, there are two principals' training centers in Taiwan. They are located in Taipei City. One is in Taipei Municipal Teachers College. The other one is located in Taipei Provincial Teachers College. These centers will provide pre-service training for principals and offer licenses and certificates.

### Sources of professional development for principals in Taiwan

There are several formal ways that principals can develop personally and professionally. They are described as follows.

1. Universities such as National Chung Cheng University offers on-the-job (in-service) Master degree programs with a specific focus on school-related issues. Several universities and colleges also have doctoral programs. Some Teachers Colleges provide specific programs called 'School Administration' and 'Educational Administration' for teachers and administrators pursuing Master degrees. In a few programs only school principals are eligible to enroll.

2. Universities such as National Taiwan Normal University provide on-the-job non-degree graduate level programs. Most of these programs require the principals to take at least 40 credits in order to earn a certificate. According to the author's observation, a large proportion of the elementary and secondary principals have taken these kinds of courses.
3. The Ministry of Education and Bureaus of Education sponsor workshops and conferences that focus on particular needs or management and leadership skills. Newly revised policies, orders and regulations, which are very important to school leadership and school development, are announced and explained on these occasions.
4. Reform projects such as 'the Spirit of Small Class Instruction' and 'the Nine-year Sequent and Coherent Curriculum' require focused training on leading and implementing the reforms.
5. National professional associations such as AEA, ROC and SARA, ROC regularly hold conferences with keynote speakers, workshops, and discussions. These conferences also provide opportunities for networking, which is helpful for principals' personal and professional improvement.
6. School-based professional development for teachers, especially the topics of curriculum design and teaching and learning, also allows the principal to become informed regarding the core purposes of schooling. Principals become familiar with the state of the art of teaching and learning.
7. Teacher Centers such as the Taipei In-service Teacher Training Center invite principals who have finished the program of the pre-service training and has assumed the principalship to participate in several days of topical training. This helps practicing principals' to continue their life-long learning.

In addition to the formal ways of professional development for principals described above, there are also informal ones that help principals to improve their leadership practices. Some of them are as follows.

1. Study Circles or Reading Groups for principals and school administrators provide opportunities to reach updated research findings and related policies. These circles or groups also make the actions of sharing, proposing and networking among principals possible. It also contributes to establishing 'a community of leadership practice', which is essential for their professional development (Lin, M., 1999).
2. Informal sharing of experience*s* among peers also helps principals to overcome difficulties, problems and to meet challenges. This is particularly helpful to principals who are in their early career stages.
3. Finally, principals can develop personally and professionally through reading related information and recent research findings. They can also learn from occasional observations of others' practices, and from self-refection, dialogues and conversations with colleagues in and out of the school.

**Contents of professional development programs for principals**

The content of professional development programs for principals are related to the sources described above. Different types of programs have different purposes. Basically, different programs vary in focus, curriculum design, and specific course content.

Normally, Master degree and non-degree graduate programs provided by the universities have more planned content. In contrast, keynote speeches, workshops, conferences, programs initiated by reform initiatives, and informal ways of development address a much wider range of topics and contents. The author offers two examples of formal sources of programs and describes their purposes, curriculum designs, and courses in the following paragraphs.

*Program one: National Chung Cheng University*

The first one is an on-the-job Master degree program offered by the Institute of Education, National Chung Cheng University (2001). Fulfillment of 36 credits, including a Master thesis, are required for students to get the degree. The degree lasts from two to five years. Classes meet in the evenings each Wednesday and Thursday. A few courses are offered on-line.

The purposes of this program include:

1. Promoting the professional knowledge and capabilities for different levels of educational administrators.
2. Improving the professional knowledge and capabilities of principals, department heads, and related administrators in the areas of educational administration and school administration.
3. Developing elementary and secondary teachers' professional knowledge in the area of classroom management, curriculum and instruction, guidance, and counseling.

In order to fulfill these purposes, a variety of courses are needed. This program includes three curricula categories: theoretical fundamental courses, research methodological courses, and educational professional courses.

Based on the curriculum design above, the theoretical fundamental courses consist of Moral Education Theory and Implementation, Study on Philosophy of Schooling, and Educational Ethics.

Research methodology courses include Educational Statistics and Computer Application, Research Methods of Education, Psychological and Educational Measurement, and Educational Action Research.

Finally, the educational professional courses encompass Research on Reforms in Schools, Research on Educational Leadership, Research on Instruction System, Curriculum Design Studies, Staff Development in Open Flexible Education, Cognitive Sciences and Instruction, Teachers' Professional Development, Educational Law, School and Community Relations, Educational Evaluation, Comparative Studies in Education, Research on Educational Technology, Educational Economics, and Readings of Educational Theories and Practices.

*Program two: Taipei Municipal Teachers College*

The second example is an on-the-job Master degree program offered by the Taipei Municipal Teachers College (2001). The difference is that this program specifically focuses on school administration. Students enrolled in this kind of program need to take a total of 36 credits, including 4 for thesis writing and complete a Master thesis to get the degree.

Purposes of this program include:

1. Developing students' knowledge and capabilities to do educational research;
2. Providing opportunities for teachers to develop professionally in order to increase the quality of educational research;
3. Developing students' capabilities of becoming competent teachers, administrators, and researchers.

This program emphasizes both theory and practice. It tries to introduce new technology of education. It also attempts to establish the ideas of school improvement and to increase teachers' professional growth. The program includes three categories of curricula, which are fundamental courses, professional courses, and dissertation writing or project design courses.

Based on the curriculum design above, the fundamental courses consist of: Research Methods of Education, Advanced Educational Statistics, Research on Elementary and Secondary Education, The Studies on Educational Administration, The Studies of Analysis of Educational Organizations, and Research on School Administration.

The professional courses focus on three areas: instructional leadership, management and administration, and politics and policy. The courses include Educational Administrational Leadership, Instructional Leadership, Curriculum Leadership, Instructional Supervision, Interpersonal Relations, Childhood Education, the Application of Computer on Education, Educational Supervision, Educational Evaluation, School Management, School Effectiveness, School Public Relations, School Business Management, School Building Maintenance and Purchasing, School Quality Management, Evaluation of School Education, Kindergarten Management, Educational Reform, Educational Politics, Educational Economics, Educational Finance, Educational Planning, Education Law, and Comparative Education etc.

Finally, dissertation writing or project design courses consist of independent studies and dissertation writing.

These programs are not restricted to principals. Teachers and administrators in schools and in the Bureaus of Education can also participate in the programs. Since the content covers a lot of areas; therefore, considerations of focus, coherence, sequence, and integration of the curricula need to be addressed in the future.

**Instructional strategies**

Although a wide variety of strategies are used in the process of professional development, lecture and discussion are still the most popular ones. Fortunately, this is changing. Other strategies include group interaction and discussion, role-playing, keynote speech, case study, school visiting, use of technology, and distance learning. Some teachers even utilize the method of problem-based learning.

One good example is the on-the-job Master program offered by the Institute of Education, National Chung Cheng University. Based on the author's observation and reflection, most of the professors have applied principles of adult learning in teaching and learning in order to stimulate students' interests of learning. The proportion of lecture has been significantly reduced. Assigned readings, small group and whole-class discussions and interactions based on the content of readings have been hugely increased.

In addition, the problems and challenges of the 'real world' of the principalship and school administration have been linked to discussions in the class. The design of group projects, group presentations, case studies, role-plays, school visitations, simulations, and experience sharing now occur in almost all courses. Some courses also utilize distance learning.

The major consideration for accommodating instructional strategies is to empower individual and group participants. Some of the participants have said that the courses and instruction offered by the program are both real world oriented and theory-based. They are helpful to their self-improvement and school innovation.

## Future Needs and Trends

The issue of professional development for principals becomes salient due to the great impact of principal leadership on school participants, processes, and outcomes (Lin, M., 2000a; 2000b), as well as the changes and challenges facing on principalship introduced by recent reform initiatives. In addition, because there are always new and complex changing forces inside and outside of the school, it is hard to specify and develop all the important knowledge, understanding of tasks, and skills during the preparation and pre-service training phases. Principals as professionals must engage in a journey of continuous improvement or 'Kaizen' (Ky'zen) (Imai, 1986). They must reflect on their practice based upon other frameworks, theories and practices and contribute to the development of other participants (especially school members) and the profession by sharing their most updated and best practice and insights.

Professional development can offer support for continuous improvement of knowledge, understanding of work, and skills of principals beyond the pre-service level. It also provides opportunities for principals to learn beyond what they might develop on their own in the practice of their work (Kelley & Peterson, 2000a). Furthermore, the traditional ideas of INSET (in-service education and training) should be transited to 'fully systematic professional learning' (West-Burnham & O'Sullivan, 1998, vii).

Due to the gap between the ideal practice and the status quo in the field of professional development, based upon recent efforts made by the researchers and practitioners and the findings of related literature, the author provides several recommendations as follows. These recommendations could represent both needs and future trends of further efforts in professional development for principals in Taiwan.

### Professional licensure for principals

It is imperative to introduce professional licensure for principals and to require periodic renewal. However, before that can be accomplished, we must define a common set of values and performance standards for principals. The Interstate School Leaders Licensure Consortium (ISLLC) in the US has established six Standards and 96 Performance Objectives for school leaders (Sharp, Walter, & Sharp, 1998; Shipman, Murphy, & Topps, 2001). Professional associations such as the National Association of Elementary School Principals (NAESP) and the National Policy Board for Educational Administration (NPBEA) in the USA have similarly developed standards and proficiencies for good principals (Kelley & Peterson, 2000b).

Concerning the process of establishing Taiwanese standards and performance objectives, understanding the ISLLC's Standards and Performance, the NAESP Proficiencies, and the NPBEA Standards could be useful first steps. We should consider the educational, social, economical, political, and cultural context of our nation and develop suitable standards. Moreover, because external stakeholders of the school, including parents, interested members of the business and representatives of the community, will play significant roles in education, they should be invited to participate in the process.

Linkages should be made between the principals' preparatory and developmental programs and the standards and objectives (Shipman et al., 2001). Also the task of principal evaluation needs to be considered. In short, the issues of school leader preparation, assessment, licensure, selection, induction, evaluation, and professional development should be considered as systemic reforms.

**Building multiple sources of professional development for principals**

People develop in different ways and at different time. For principals to develop to their full potential, it is important to establish various sources of professional development. As described above, many groups, associations, and institutes are now offering professional development for principals in Taiwan. In the future, the following approaches should be emphasized.

*Establishing a principals center or school leadership academy*

This Center or Academy could be located on the campus of universities or in a professional association. The management team could consist of scholars, researchers, principals, school administrators, teachers, and members of the schools. The Center can serve as a core for networking, problem solving, information changing, assessment, and training.

*Encouraging professional associations to provide longer training programs*

In addition to conference-type activities, professional associations such as AEA, ROC and SARA, ROC could offer several-week intensive programs on specific topics. If the programs could be designed in a coherent and sequential way, the quality and impact of the program would be dramatically enhanced.

*Establishing professional degree programs in universities*

For the purpose of professional development for principals, this kind of program could solely focus on training school principals. Considering their stages of career and school context (size), the content of the program could be designed closely related to relevant topics of the principalship.

*Extending principals' informal learning groups*

Study groups of principals could be extended throughout the nation. Experience sharing, reading journal articles and books, role-playing, and problem solving could be included in the development process.

**Refining curricula in professional development programs**

In defining the curriculum of professional development for principals, what kinds of issues are essential? The author contends that defining purposes and goals, adopting various models of development, authentically assessing principals' individual and organizational (or positional) needs, meeting diverse needs of principals at different career stages, taking contexts of schools into considerations, and linking professional development and school improvement initiatives and policies in a systemic way are some of the most important areas needed to be considered.

Although acquiring a common base of knowledge is important, the purposes and goals of professional development programs should go beyond merely helping principals master a body of knowledge. For the principalship, it is also important that both the key words of *education* and *leadership* are being considered. After reviewing related literature, Murphy (1992) contends that the purpose of training programs for principals is 'to provide leadership to communities so that children and young adults are well educated, in the deepest sense of the term' (p. 139). In light of this purpose, Murphy (1992) argues that the goals of training programs include:

1. [H]elp[ing] students articulate an explicit set of *values and beliefs* to guide their actions – to become moral agents …
2. Helping students become *educators* …
3. Facilitating the development of *inquiry skills*, or enhancing the thinking abilities of students …
4. [H]elp[ing] our students learn to work productively with people, to lead in the broadest sense of the term. (pp. 141–145)

Finally, Murphy (1992) further indicates that training programs should have purposes and goals first. '[T]he content in training programs should backward map from the goals. … ' (Murphy, 1992, p. 140)

Sergiovanni (1995) suggested that there are three models of teacher development. They are 'training models', 'professional development models', and 'renewal models' (pp. 209–211). The author argues that all these can be used to design the programs of professional development for principals. Different models have different assumptions, roles of principals, and practices. It is important that the professional development for principals to adopt all three models. However, the *emphases* can be different. For principals in the different stages of career, or for different types of professional development programs, the emphases could be different.

Part of the reasons that university-based preparation and professional development programs for principals are often viewed by principals as inappropriate is because the needs of principals are not taken into account (Bredeson, 1996). Walker and his colleagues (2000, p. 30) in Hong Kong further suggest the idea of combining personal and organizational or positional needs of principals into a 'Profile'.

The point they have made is similar to Peterson's (1977–1978) argument that, 'To begin to say what the principal should do, one needs to understand better what the principal in fact does' (p. 5). The more we understand the needs of principals, the more likely the program will succeed. However, the author must emphasize that the findings of needs assessment is only one of the resources for designing the curriculum. Research findings, related theories, policies, and past experience also need to be taken into account.

Principals have different requirements and needs at different career stages (Kelley & Peterson, 2000a). It is important to consider whether principals are in their early years, mid-career, or the later years of their careers. The curricula should adapt to the varying needs that principals tend to meet at these different stages.

Principals in different settings also achieve results in a variety of ways. Size of school and type of school influences the combination of values, knowledge, and skills needed by principals to be outstanding in their role. For example, 'In larger districts, or highly diverse districts training could focus on the needs of the communities served by the school itself' (Kelley & Peterson, 2000a, p. 6). Middle level principals are in need of knowledge and skills in building a respectful, cooperative, collegial school climate. They also need to understand, implement, and assess newly proposed approaches to teaching and learning, and remain up-to-date legally, financially, organizationally, and technologically (Brown, Anfara, Hartman, Mahar & Mills, 2001).

Professional development for principals should consider the context of the school, community, and region. Furthermore, the nature of the values held by the school community and the preferences of strategies of teaching and learning, to name a few should be taken into account in programs of professional development for principals.

In a world characterized by fast change, school administrators in general, and principals in particular, must both attend to changes initiated by *past* and *current* reforms. Guskey (2000) argues that if educators have no guidelines on how new initiatives fit with those espoused in the past, they will see the new ideas as passing fads. Therefore, the topics, contents, and potential measures of past and current reform initiatives and policies should be addressed in programs of professional development for principals.

**Improving instructional strategies**

Several articles have provided the delivery recommendations or instructional strategies for pre-service and professional development programs (Chang, C., 2001; Daresh, 1997; Evans & Mohr, 1999; Goldring & Rallis, 1993; Hallinger & Murphy, 1991; Leithwood, Begley & Cousins, 1994; Murphy, 1992; Tanner, Keedy & Galis, 1995). The major recommendations are as follows.

1. Consider the principles of adult learning;
2. Adopt problem-based learning and case studies;
3. Introduce the daily work of practicing principals into the processes of discussion;
4. Provide opportunities for direct observation, coaching, and feedback;
5. Offer opportunities for clinical learning;
6. Encourage thinking and reflecting publicly with colleagues;
7. Design mentoring mechanisms;
8. Adopt collaborative learning strategies;
9. Use action research;
10. Incorporate the technology of Internet and WWW into the delivery system.

Daresh (1997) has argued that the critical basis in this list is that there are some creditable (teaching) practices being tried. The effectiveness of the same instructional strategy may be various in different places. Nonetheless leaders and staff in Principals Centers should keep in mind that the effort of pursuing effective instructional strategies is a key task in design and delivery of programs.

**Constructing a community of professional practice among principals**

Normally, there are many students, teachers, and administrators in one school. But there is only one person who is called 'the principal'. Lin (1999) proposed the need for constructing a community of professional practice among school principals. Principals need more opportunities to gather together sharing what they do (Lemley, 1997). The ideas of building collegial learning, learning community, professional caring community in schools suggested by Sergiovanni (1994) could be extended to outside the school. The members of the community could consist of principals and administrators in different school settings.

## Professional Development for Principals: Making a Genuine Change

One thing that we must always keep in mind is 'Can we expect principals to make a genuine change in themselves and in their schools through professional development?' Can students learn better because of principals' participation in a two-week workshop? Traditionally, we care more whether 'principals liked the programs' or 'if they acquired the intended knowledge and skills.' However, we now need to expect more from professional development for principals.

For meaningful and effective evaluations of professional development, Guskey (2000) adapted Kirkpatrick's (1959) model into a five-level evaluation model. The five critical levels of professional development evaluation include 'participants' reactions, participants' learning, organization support and change, participants' use of new knowledge and skills, and student learning outcomes' (Guskey, 2000, p. 82). These critical levels should be considered when designing professional development programs for principals.

In other words, for planning, implementing, and evaluating an excellent program, we need to ask several important questions.

- Were the topics addressed relevant to principals' professional responsibilities?
- Were they satisfied with the program?
- Did principals obtain the expected attitudes, values, knowledge and skills?
- Are the goals of the program in alignment with the missions of the school?
- What was the impact on the school?
- Did principals effectively utilize the new skills?
- What was the impact of the professional development program and activities on students?

With regard to changes in the thinking and behavior of the principals themselves, the author (2000b) has reviewed related literature (e.g., Evans & Mohr, 1999; Guskey, 2000). Based on this review, effective programs should target the following changes in principals.

1. The ability to question their own daily practices;
2. The ability to reflect;
3. The willingness, habit, and ability to share with principals of other schools and colleagues within their school;
4. The ability of problem solving;
5. Increased self-efficacy and confidence to assume challenges;

6. The willingness, ability, and the understanding of possible measures for further learning;
7. Understanding where they could get help;
8. The encouragement to get help; and
9. The ability to apply ethical and critical judgment.

Finally, professional development programs must have an impact on teaching and learning and administrative processes. Eventually, they should have an impact on students' learning processes and outcomes.

## Conclusions

This chapter describes the status quo of professional development for principals in Taiwan. The chapter focuses on the source, content and instructional strategies of professional development programs. The author also outlined the future needs and trends in this area. These include introducing professional licensure for principals and the regulation of licensure renewal, building multiple sources of professional development for principals, refining curricula of professional development programs, improving instructional strategies, and constructing a community of professional practice among principals.

Several actions are essential in building multiple sources of professional development for principals. These include

- establishing a principals center or school leadership academy,
- encouraging professional associations to provide longer training programs,
- establishing professional degree programs in universities, and
- extending principals' informal learning groups.

With regard to the task of refining curricula of professional development for principals, the following aspects are important. These consist of defining purposes and goals, adopting various models of development, authentically assessing principals' individual and organizational (or positional) needs, meeting diverse needs of principals at different career stages, taking contexts of schools into considerations, and linking professional development and school improvement initiatives and policies in a systemic way.

Finally, it is important for all professional development programs to make a genuine change in principals themselves. That is to say, critical changes need to happen that influence the thinking and behavior of principals as well as practice in their schools and the capabilities of students.

## References

Beaumont, J. J. (1997). Issues in urban school district leadership: Professional development. *Urban Education*, 31(5), 564–581.

Bredeson, P. V. (1996). New directions in the preparation of educational leaders. In Leithwood, K., Chapman, J., Corson, D., Hallinger, P., & Hart, A. (Eds.). *International handbook of educational leadership and administration (251–277)*. Dordrecht, Netherlands: Kluwer.

Brown, K. M., Anfara, Jr., V. A., Hartman, K. J., Mahar, R. J., & Mills, R. (2001). *Professional development of middle level principals: Pushing the reform forward.* Paper presented at the American Educational Research Association Annual Meeting, Seattle, Washington.

Chang, D. J. (2000). *The research of educational administration* (3rd. ed.).Taipei, TW: Wu-Nan Publishing Company (in Chinese).

Chang, C. (2001). *Changing school leaders before changing schools: An in-service development program for school administrators.* Paper presented at the International Conference on School Leader Preparation, Licensure/Certification, Selection, Evaluation, and Professional Development. Taipei, Taipei Provincial Teachers College.

Chin, M. C. (1997). *Educational administration: The theoretical part.* Taipei, TW: Wu-Nan Publishing Company (in Chinese).

Daresh, J. C. (1997). Improving principal preparation: A review of common strategies. *NASSP Bulletin*, 81(585), 3–8.

Department for Education and Employment (D*f*EE) (2000a). *Leadership Programme for serving headteachers*. Department for Education and Employment.

Department for Education and Employment (D*f*EE) (2000b). *Professional development: Supporting for teaching and learning*. Department for Education and Employment.

Dunn, M. B. (1999). The NASSP assessment center: It's still the best. *NASSP Bulletin*, 83(603), 117–120.

Evans, P. M. & Mohr, N. (1999). Professional development for principals: Seven core beliefs. *Phi Delta Kappan*, 80(7), 530–532.

Goldring, E. B. & Rallis, S. F. (1993). *Principals of dynamic schools: Taking charge of change*. Newbury Park, CA: Corwin Press, Inc.

Guskey, T. R. (2000). *Evaluating professional development*. Thousand Oaks, CA: Corwin.

Hallinger, P. & Murphy, J. (1991). Developing leaders for tomorrow's schools. *Phi Delta Kappan*, 72(7), 514–520.

Hsieh, W. C. (1993). *School administration*. Taipei, TW: Wu-Nan Publishing Company (in Chinese).

Huang, C. C. (1990). The preparation and in-service training for secondary principals. In the Association of Teachers Education, ROC (Ed.), *The policies and problems of teachers education (pp. 51–82)*. Shih-Tai Book Story (in Chinese).

Imai, M. (1986). *Kaizen: The key to Japan's competitive success*. New York, NY: McGraw Hill Publishing Company.

Institute of Education, National Chung Cheng University (2001). *The introduction of on-the-job master program.* [WWW document] URL http://www.ccunix.ccu.edu.tw/~deptedu/ (in Chinese).

Kelley, C. & Peterson, K. D. (2000a). *Principal inservice programs: A portrait of diversity and promise*. Document prepared for the National Center on Education and the Economy/ Carnegie.

Kelley, C. & Peterson, K. D. (2000b). *The work of principals and their preparation: Addressing critical needs for the 21st century*. Document prepared for the National Center on Education and the Economy/Carnegie.

Ko, P. S. & Lin, T. Y. (1996). *The study of the preparation system for elementary and secondary principals and heads*. The Educational Reform Council of the Executive Yuan (in Chinese).

Leithwood, K., Begley, P. T. & Cousins, J. B. (1994). *Developing expert leadership for future schools*. London: The Falmer Press.

Lemley, R. (1997). Thoughts on a richer view of principals' development. *NASSP Bulletin*, 81(585), 33–37.

Lin, M. D. (1999). Constructing the professional community of the practice of school administration. *Educational Administration*, 1, 69–73 (in Chinese).

Lin, M. D. (2000a). The impact of principal leadership: An analysis of the research findings, 1970–1999. *Proceedings of the National Science Council, Part C: Humanities and Social Sciences*, 10(2), 233–254 (in Chinese).
Lin, M. D. (2000b). The curriculum design and instructional strategies of professional development for principals. *Educational Resources and Research*, 37, 10–20 (in Chinese).
Lin, W. L. (1999). *New trends of professional development for principals*. Paper presented at the Fifth Forum on Educational Administration. Association of Educational Administration of ROC (in Chinese).
Ministry of Education (MOE) R.O.C. (2001). [WWW document] URL http://www.moe.gov.tw (in Chinese).
Murphy, J. (1992). *The landscape of leadership preparation: Reframing the education of school administration*. Newbury Park, CA: Corwin Press, Inc.
National College for School Leadership (NCSL; 2001). *The leadership and management programme for new headteachers (HEADLAMP)*. [WWW document] URL http://www.nsclonline.gov.uk/
Peterson, K. D. (1977–1978). The principal's tasks. *Administrator's Notebook*, 26(8), 2–5.
Ruohotie, P. (1996). Professional growth and development. In Leithwood, K., Chapman, J., Corson, D., Hallinger, P., & Hart, A. (Eds.). *International handbook of educational leadership and administration (419–445)*. Dordrecht, Netherlands: Kluwer.
Sergiovanni, T. J. (1994). *Building community in schools*. San Francisco, CA: Jossey-Bass Publishers.
Sergiovanni, T. J. (1995). *The principalship: A reflective practice perspective* (3rd ed.). Boston, MA: Allyn and Bacon.
Sharp, W. L., Walter, J. K., & Sharp, H. M. (1998). *Case studies for school leaders: Implementation the ISLLC Standards*. Lancaster, PA: Technomic Publishing Company.
Shipman, N., Murphy, J., & Topps, B. (2001). *Linking the ISSLC standards to professional development and relicensure*. Paper presented at the International Conference on School Leader Preparation, Licensure/Certification, Selection, Evaluation, and Professional Development. Taipei: Taipei Provincial Teachers College.
Taipei Municipal Teachers College (2001). *The curriculum design of on-the-job master program*. [WWW document] URL http://www.tmtc.edu.tw/~primary/ (in Chinese).
Tanner, C. K., Keedy, J. L., & Galis, S. A. (1995). Problem-based learning: Relating the 'real world' to principalship preparation. *The Clearing House*, 68(3), 154–159.
Walker, A., Begley, P., & Dimmock, C. (Eds.; 2000). *School leadership in Hong Kong: A profile for a new century*. Hong Kong Centre for the Development of Educational Leadership, the Chinese University of Hong Kong.
West-Burnham, J., & O'Sullivan, F. (1998). *Leadership & professional development in schools: How to promote techniques for effective professional learning*. London: Pitman Publishing.
Wu, C. S. (1999). *School administration* (4th ed.). Taipei, TW: Wu-Nan Publishing Company (in Chinese).

# 13

# Principal Training in the People's Republic of China: Retrospect and Prospect

Feng Daming

*Associate Professor, Department of Educational Administration, East China Normal University, Shanghai, Peoples Republic of China*

The challenge of economic development faced by the People's Republic of China is as significant as for any other nation in the world. As the world's most populated country, China must raise the level of general education if it is to achieve its goals of economic and social development. As with other nations, the role of the principal is viewed by China's educational policymakers as key to success in educational development. This is reflected in an increased interest in and emphasis on training for the nation's principals.

This chapter will review the progress of education in China over the previous 50 years with special emphasis on the role and training of school principals. It will examine how the goals and process of education have changed in response to changing internal and external conditions. The role of the principal and the training provided to principals is described and the author speculates on current problems and future trends.

## Principal Training in China: 1949 to the Present

On October 1, 1949, Mao Zedong (formerly known as Mao Tse-tung), the Chairman of the Chinese Communist Party, was in Beijing to proclaim the birth of the People's Republic of China. From that moment, the unique history of the restructuring of China directed by Marxism and Mao Zedong thought began. Education formed a key feature of social development within this political restructuring of Chinese society. Generally, the history of education in the People's Republic of China can be divided into two periods of rapid development separated by a period of inactivity.

### The first period: 1949–1957

Meisner wrote in his *Mao's China*:

> What makes contemporary China unique and a matter of special historical interest is that the history of the People's Republic represents what seems to

> be a radical departure from the common pattern of the history of a 'post-revolutionary' society. … … This is not to suggest that revolutions necessarily are fruitless or undesirable. By destroying the political institutions and the ruling classes of regimes, social revolutions open the way for societies to 205 follow new directions and allow at least the possibility to create new and better social worlds. (Meisner, 1977, p. 56)

To keep going on the successful path of restructuring China, the new nation needed trained cadres who were loyal to the new ruling party. Schools were viewed very much by the new government as institutions of socialization Thus, teachers and principals of elementary and secondary schools occupied key roles in the institutionalization of the new educational system.

In February 1955, the first training program for school principals in the nation was implemented by the Ministry of Education. The stated purpose of that program were to:

- develop principals' ideological and political aspects allowing them to perform the task of school administration in the new educational setting, and
- train a certain number of promising personnel for school administration to suit the planned educational development of the nation.

Meanwhile, the State Council and the Ministry of Education asked local governments to develop local training programs for the principals to run alongside the centrally administered program. Local governments were to have completed the training in three to five years.

Shortly after the new training program was announced by the State Council and the Ministry of Education, a national training institution for principals was founded. This was named the *Central Institute of Educational Administration* and was operated under the auspices of the Ministry of Education. The main curriculum offered by the Institute were philosophy, pedagogy, and psychology (He Lefan et al., 1997). These emphases reflected the primarily political and social developmental goals of the government for the nation's educational system.

In subsequent years, many additional training institutions were established by local governments. These generally followed the example of the *Central Institute of Educational Administration*. This resulted in a great increase in the number of trained principals in the ensuing years. This movement to train the country's principals was intended to promote the development of school education in the nation.

**The period of inactivity: 1958–1979**

In 1958 and 1959, the nation's policy for principal training changed when an 'Ultra Left' ideology came to dominate the politics of the central government. A movement named *Fighting Against the Right-leaning Ideas* spread across China and affected all areas of society, including education. From the perspective of this new political movement, psychology was a pseudo-science unsuitable for introduction to school principals and teachers. Moreover, the content of the Pedagogical Curriculum was also criticized. Under these circumstances, many local training institutions for principals stopped offering courses. Eventually, the *Central Institute of Educational Administration* was closed down in 1959 due to these political pressures (Wu Zhihong & Feng, 1998).

This was the beginning of almost 20 years of inactivity in the area of principal training. In 1966, the Cultural Revolution was formally launched by late Chinese leader Mao Zedong. The training of principals was stopped by the government and all training institutions for principals were completely closed down. This situation continued for the next 10 years. Although the Cultural Revolution ended formally in late 1976, training institutions for principals didn't reopen until 1979.

**The third period: 1979–1989**

After the Cultural Revolution, the new central government began to focus on economic reconstruction. The nation set up new objectives: The Four Modernization Goals. These focused on achieving the comprehensive modernization of agriculture, industry, national defense, and science and technology. With the establishment of these new priorities, the government came to view education as the foundation for future growth.

Beginning in 1979, training institutions for principals were re-established in rapid succession. In 1982, the Ministry of Education issued a document entitled, *The Opinion on Strengthening the Training Work for Administrative Personnel of Elementary and Secondary Schools*. This document clarified the philosophy, purposes, requirements, content, and means of training school principals in China. Local governments were given the responsibility of offering training courses for all principals of elementary and secondary schools within a five year period. The publication of this document marked a new era for principal training in China (Wu Zhihong, Feng & Zhou Jiafang, 2000).

## The Development of Principal Training in China in the 1990s

**The first professional training program**

Although principal training in China began anew in 1979, the newly reestablished programs were considered unsatisfactory. Some local training programs for principals could not meet the various demands from principals. Many school principals still could not be adequately prepared for the fast change of educational reform. Under these new circumstances, the State Education Commission (SEC, renamed the Ministry of Education in 1998) issued in December 1989 a document entitled *On Strengthening the Training for Principals of Elementary and Secondary Schools Nation-wide*.

This document stated, 'Generally speaking, the performance of the principals of elementary and secondary schools in the whole nation, both in political and in professional aspects, cannot meet the demands of educational development and further reform' (SEC, 1989). This statement reflects a similar trend as found elsewhere around the world (Caldwell & Feng, 2001; Hallinger & Feng, 2001). The new policy asked local educational officials to try to give all school principals the chance to be trained once more in the next three to five years.

Furthermore, the document put forward a professional training program for principals of elementary and secondary schools nation-wide. As compared with former training programs, this new one emphasized professional educational qualifications and new requirements:

- After training, the principals, on the whole, should have the professional capacity as a school leader and should be able to meet new demands brought about through educational development and educational reform (An Wenzhu, 1997, p. 225).

- Local governments should build sufficient training institutions for principals with appropriately qualified staff and allocate special funds for principal training (SEC, 1989).
- The SEC would develop a general instructional outlines and syllabi for principal training (subsequently published in June 1991; see SEC, 1991). Under the syllabi, local training institutions would need to offer many more courses for principals than before. The textbooks and the other training materials corresponding to the outline and the syllabi would be published after they were issued.
- Local training institutions should make special efforts in their preparation of trainers of the principals in order to meet the requirements of the outline and the syllabi (SEC, 1989).
- A principal could receive a professional certificate after he passed the final test of the professional training program (SEC, 1989).
- Finally, the document suggested that no one get the position of principal without the professional certificate.

**The renewal training program**

The professional training program had a positive impact for China's principal training. However, professional development for principals was still far from perfect. With a view of taking principal training a step further, the SEC issued another important document in December 1995. This was entitled *The Training Direction for Principals of Elementary and Secondary Schools during the Ninth Five-Year Plan.*

The Chinese central government has set out a *National Five-Year Plan* every five years. The ninth plan was established to run from 1996 to the year 2000. In this government plan, the SEC designed a framework for a renewal training program for school principals. The program was highlighted by three innovative characteristics.

*Diversity*

The renewal training program was a professional training program with diversity in content and approach. Though similar in nature to the former professional training program mentioned above, this program didn't ask the local governments to train all school principals at the same pace and in same pattern. It proposed three types of principal training:

- *induction training* for school principal candidates;
- *continuing training* for a principal who had a professional certificate for the principal position; and
- *research training* for those principals who not only had a professional certificate but also had excellent performance in their positions.

The induction training offered basic knowledge and skills that were considered necessary for effective school leadership. The continuing training offered the trainees a broad curriculum in education and educational administration. The research training usually offered trainees selected topics for discussion and study. It also offered support lectures around the chosen topics.

*Localization*

The local governments were authorized to establish their own regulations for principal training. Meanwhile, local training institutions were allowed to develop courses with local features for principal training. Some characteristics of these localized courses were:

- The courses were expected to reflect the general trends of educational reform within the local educational surroundings.
- In large cities, some of the training content was revised to suit the needs of principals in metropolitan schools while smaller towns developed special training content suited for principals in their locales.
- The SEC asked the local governments to evaluate the effectiveness of the principal training after the training had finished (SEC, 1995).

By way of example, a local course featured in one of these training programs included the following:

- Educational philosophy of Xiake (Xu Xiake, an ancient Chinese geographer who was born and grew up in Jiangying, Jiangsu)
- Culture and Competence-oriented Education
- 'One, Two, Three' Reading Program and Educational Modernization
- The Theory and Practice on Leisure Education
- The Social Function of An Elementary School in Rural Areas
- Sub-qualified School Analysis
- Administrative Art for a Principal in a Rural School
- Healthy Township and Basic Education
- Analysis and Study of the 'Hot Issues' in Education (Yan Jun & Xiao Jie, 1999).

*University participation*

In order to raise the standard and promote quality assurance in principal training, the State Education Commission began to encourage universities to participate in the renewal training program and to take responsibility for some of the instructional and consultative responsibilities. Some universities began to prepare series of lectures, field study programs, case study programs and school leader workshops in response to the call of the SEC.

Generally, the universities participated in principal training in five ways:

- Principal centers were established in some top universities involved in teacher education. In 1990, for example, the *National Training Center for High School Principals* was operated in East China Normal University (ECNU). ECNU took responsibility for organizing the specific training activities as well as providing staff.
- The universities developed the advanced research training programs for local governments. Some selected professors from the universities participated in local training program design and local feature course appraisal directly.
- Professors in universities gave presentations about the frontiers of education for school principals. In 1998, for example, a lecturer group for principal/teacher training was established in Shanghai. Some professors from universities in Shanghai were the members of the lecturer group.
- Some selected professors participated in the process of government decision-making on policies and guidelines for principal training. In Shanghai, for

instance, several professors were appointed as the members of Shanghai Core Group of Professional Development for School Leaders, Shanghai Municipal Education Commission. These members contributed to the process of government decision-making for principal training.

- The local government sent some young promising principals to pursue Master Degrees in Educational Administration.

This university participation was designed to raise the level of China's principal training. Another outcome of university participation was bringing about change in research in principal training. It was only after university participation began that professional associations, academic journals and publications on principal training began to appear in China. Thus, it is fair to observe that this phase saw the beginning of the professionalization of principal training in China through the collaboration of universities and local education authorities.

## Achievements and Problems

Training for the principalship in China has improved in scope and quality since implementation of these professional training programs in the 1990s. Four featured achievements include the following.

First, policies regarding principal training have been set up by both the central and local governments. In December 1999, the Ministry of Education (formerly, the SEC) issued a document entitled, *Training Regulations for School Principals*. This policy document announced that one had to obtain the principal-qualified certificate before gaining the position of principal. Moreover, principals had to be re-trained every five years if they wanted to keep their positions (Ministry of Education, 1999).

In response to this training policy from the Ministry of Education, local governments developed systems to control and coordinate local principal training. For example, in the city of Wuxi, Jiangsu Province, a local government policy indicates: 'A principal will be forced to leave his position if he doesn't take part in the training program for principals within five years, or if he doesn't pass the computer test and foreign language test' (Wuxi Municipal Education Commission, 1998).

Second, more than one million school principals have taken part in these professional training and renewal training programs. A survey covering Beijing, Shanghai, Liaoning, Henan, Gwangdong, and Niningxia showed that the effect of principal training in those cities and provinces was quite positive. Most of the 1499 principals investigated concluded that the training was useful for the promotion of school leadership as well as school quality.

Third, the academic and professional level of principal training has increased with the participation of universities in the development and delivery of training programs. Especially, as mentioned above, the participation of university professors in designing and appraising local training programs has provided a more systematic framework and procedural input for local governments and training institutions. For example, the universities helped the Shanghai government develop a supervisory procedure for principal training in 1998.

Fourth, many educational researchers have seen principal training as a valuable research field. The number of research projects on principal training has significantly increased since 1990 (Wu Zhihong, Feng, & Zhou Jiafang, 2000). Some research projects on principal training were sponsored by the SEC and the Ministry of Education and have had far-reaching effects. For example, findings of research projects 'The Report on the Training Regularity and Supervisory System for School Principals' (He Lefan et al., 1990) and 'The Report on the Quality Appraisal of Principal Training' (He Lefan et al., 1997) had notable impact on central government policies on principal training during the 1990s.

This type of impact could hardly have been imagined before 1990. This change suggests that principal training had become viewed as among the most important fields in Chinese pedagogical education. Notably, two nation-wide research bodies concerned with principal training were founded during 1990s. They are *China's Research Association of Principal Work Study* and the *National Research Association for Principal Training*.

**Current problems**

There is no doubt that principal training in China has been improving since the beginning of 1990s. However, new problems in principal training have been created at the same time. The first problem is that many serving principals lack motivation to participate in principal training. Traditionally, the majority of principals are promoted directly from the faculty. After they are appointed, their salary still depends on their professional status as teachers, not on their new administrative duties or role.

In China, there are four professional status levels for a teacher: none-titled teacher, second-class teacher, first-class teacher and senior teacher. For a principal, whether excellent or not in leadership, salary will only increase if one moves from first-class teacher to senior teacher. The principal's salary does not depend upon excellence in school leadership. Hence, principals are usually more concerned about their specialized subjects than school leadership.

In addition, principals are busy with endless routine administrative tasks. Their working day is described as 'six out, six in;' that is they have to leave home at six a.m. and return home at six p.m. it is often difficult for them to find quality time for training.

Consequently, principal training is generally conducted as part-time learning. Training activities are usually held on weekends or holidays. Therefore, some principals have little interest in participating in training. They come late, leave early, are often absent, and drop out of the training quickly when they feel bored.

Secondly, the training capability of institutions in large cities is quite different from that of similar institutions in towns and rural areas. It is difficult for many towns and rural areas to carry out professional training programs at the desired level of quality because those institutions have insufficient numbers of qualified trainers. Hence, a professional certificate from some training institutions may be devalued.

Thirdly, most professional training programs continue to emphasize knowledge, and neglect administrative skills and leadership competencies. Consequently, despite obtaining a professional certificate, principals may still lack necessary skills and competencies for their position.

Finally, most training institutions lack modern technology and advanced methods in principal training (Feng, 1996). Many principal trainers are still used to adopting the 'chalk and talk' lecture model in principal training. These factors continue to limit the potential effectiveness and impact of principal training in China.

## Trends in Principal Training from the Year 2000 Forward

Based on these developments, several trends in China's principal training in the early years of the 21st century can be observed. The Trainers' Training has been placed on the government agenda of principal training. The Ministry of Education noted the problem of unqualified principal trainers a couple of years ago.

Qian Yicheng, the Director of the Personnel Department, Ministry of Education pointed out this problem in an official speech at a national conference on principal training in 1998. He concluded that the training staff in many training institutions was unable to keep up with the changing needs and rising expectations of the public. Especially, there were few trainers who could act in the proper role in these newly developed training programs.

Secondly, the younger training staff often had insufficient experience in principal training. Thus, they were ill prepared to assume the training tasks after a number of experienced trainers retired. Thirdly, even for experienced trainers, the knowledge base on principal training needed to change and be updated. Finally, the Director noted that the training approach used by trainers needed to improve (Qian Yicheng, 1998).

Since that time, the trainers' training has been placed on the agenda of some local governments. At the moment, the trainers' training appears in two forms: direct training and indirect training. In late 1999, SMEC and ECNU jointly organized an institute on principal/teacher training for the heads and deputy heads of 20 district training institutions in Shanghai. The participants were told that each would develop a draft of the blueprint on principal/teacher training and principal/teacher training management for his or her district during the Tenth Five-Year Plan (from the year 2000 to 2005). To assist the participants to accomplish the task, the institute organizer offered a series of lectures and presentations delivered by professors both from home universities and overseas institutions (Feng, 1999). This is an example of the direct training for principal trainers.

In addition, conferences and seminars on principal training have been held regularly since late 1990s. The principal trainers, professors who have the interest in principal training, and the government officials in charge of principal training get the opportunity to share their ideas and experiences to one another at such conferences and seminars. Therefore, these conferences and seminars are usually seen as providing, indirectly, training for principal trainers.

### The connection between the training participation and trainees income

A pilot program to establish an independent system for principal status has been carried out in Shanghai. According to this system, all serving principals in Shanghai are divided into four grades and twelve levels of pay. This system is an attempt to begin to link pay to level of qualification.

Every principal has the right to apply for the grade and level they judge appropriate. However, a special committee evaluates the file of each applicant and decides on the appropriate professional status. The committee uses a newly developed evaluation system based on 32 evaluation indicators. Evidence of the principals' performance is gathered by four methods: observation, interview, data-based review, and the principals' task report.

To ensure that all principals are treated fairly, the evaluation process ignores the school's historical achievements and teacher status (e.g., first class teacher or senior teacher). It mainly focuses on the current performance of the school and evidence of improvement in school reform under the leadership of the principal. To encourage unique and creative ideas of school leadership, a principal will get extra points if his school operates with some innovative features. Allocation of the principals to a particular grade and level determines their income as mentioned above.

This system provides not only a performance-related pay mechanism, but also a connection with participation in training. Moreover, since training participation is one of the evaluation indicators this approach has also acted as a stimulus for funding of the training system. Furthermore, principals will benefit from the training as a means of further enhancing their income if they think they can develop their school management skills and foster school change through the training activities. This pilot program developed by Shanghai Municipal Government has been encouraged by central government. Gradually, this independent system for principals may spread over whole nation.

### Administrative skills and leadership competencies emphasized in principal training

Administrative skills and leadership competencies are not new requirements for principals. In fact, the SEC document, *The Prerequisites and Requirements for the Principal Position*, issued in June 1991 expressed clearly that the principal should have the following essential skills and competencies (Personnel Department of SEC, 1991, p. 200):

- The ability to map out a school development plan and school routine task framework in the light of state guidelines, laws and policies.
- The ability to implement ideological and political education as well as moral education;
- The ability to facilitate students' overall growth.
- The ability to observe in classrooms, comment on classroom activities and guide teaching and learning, instructional studies, and other activities after class;
- The ability to guide teachers to promote their professional development and improve their competency in instruction.
- The ability to lead groups;
- The ability to coordinate the internal and external relationship of the school and to use the initiative of the community and parents to make the school operate effectively.
- The ability to study the evolving situation and new problems;
- The ability to implement experiments in teaching and learning, to pool together teacher experiences and to improve the quality of teaching and learning continuously.
- The ability to plan and write effective reports on the school's development;
- The ability to speak Pu Tong Hua (standard spoken Chinese) and to deliver good oral presentations.

Unfortunately, these currently represent goals, not the reality. Today, principal training does not yet reflect these characteristics. In 1996, this author argued that administrative skills and leadership competencies should be important parts in principal training. Furthermore, the author put forward proposals for competence-based training in China's principal training (Feng, 1996).

Nowadays, many trainers and training officials have recognized that one of the weak points in China's training programs was the neglect of skill improvement for school principals. However, things are changing. The current trend sees more and more training institutions attaching importance to administrative skills as they develop their training programs.

### IT-supported training program for principals

The internet was introduced into China around 1995. Many Chinese people had never heard the term 'IT' before 1998. However, IT related enterprises and companies have mushroomed everywhere, even in Tibet, since 2000. For example, between January 1 and December 31, 2000 the number of registered internet users in China increased from 8.9 to 22.5 million people. The Ministry of Information Industry revealed that an average increase rate of growth in information-related industries in the nation was 25% in the past two years and 32.9% in the first half of 2001 (Hui Xiaoyong, 2001). These figures reflect the scope and speed of change in China's participation in the information revolution. Even so, given China's huge population, the trend for further growth is enormous.

These technological changes create a positive context for IT applied in principal training. Some researchers have tried to establish the possibility and feasibility to develop and sue new approaches for principal training under China's cyber-context (Feng & Hallinger, 2001; Hallinger & Feng, 2001). The author of this chapter believes that the IT supported approach for principal training will increase dramatically in the near future.

## Conclusion

More than 45 years have passed since the first training program for school principals in the People's Republic of China was developed. The main objective during the first period in principal training was to train school leaders to be government cadres rather than professional personnel. Training focused on learning political ideology (Wu Zhihong & Feng, 1998) because late Chinese top leader Mao Zedong contended: 'After the enemies with guns have been wiped out, there will still be enemies without guns; they are bound to struggle desperately against us, and we must never regard these enemies lightly. If we do not now raise and understand the problem in this way, we shall commit the gravest mistakes' (Mao Zedong, 1949).

From 1958 to 1979, the nation's training institutions nation-wide did little work in the area of principal training because of the political situation. During the third period, the principal training in China began to operate regularly and has experienced notable improvement. The professional training and renewal training programs have given birth to a wholly new phase in the history of principal training in China. The rule of the professional certificate now implies that a school principal is more a professional than a government cadre.

By the end of 1997, more than one million principals had been trained. This is bound to make a great contribution to the cause of education in China. Looking to the future, it is possible to anticipate three key trends:

- increased emphasis on the training of principal trainers,
- greater focus on the development of administrative skills and leadership competencies among principals, and
- the application of information technology to principal training in the early years of the 21st century.

## References

An, Wenzhu. (1997). *School administration studies*. Beijing: Popular Science Press.

Caldwell, B. J. & Feng, Daming. (2001). The school leadership in the 21st century. *Instruction & Management*, 23(7), 1–5.

Feng, Daming. (1996). On competence-based principal training. *Journal of Shanghai Institute of Education*, 54(4), 89–95.

Feng, Daming. (1998). The supervisory system for school principals: An international comparison. *Secondary School Education*, 190(6), 46–50.

Feng, Daming. (1999). The training for training managers: Preparing for the new century. *Modern Educational Administration*, 22(4), 48.

Feng, Daming. (2001a). Confronting the Internet: Challenges and responses in school management for moral education. *Instruction & Management*, 124(3), 50–53.

Feng, Daming. (2001b). Trends of educational administration in the world in early 21st century: Dialogue with Professor Brian J. Caldwell, University of Melbourne, *Global Education*, 30(4), 1–6.

Feng, Daming & Hallinger, P. (2001). Professional development model of school leaders with the support of IT: Dialogue with American Professor Philip Hallinger. *Global Education*, 30(7), 1–3.

Hallinger, P. & Feng, Daming. (2001). The new characteristics of the principal training in the 21st century, *Instruction & Management*, 23(8), 3–5.

He, Lefan et al. (1990). *The report on the training regularity and the supervisory system for school principals* .

He, Lefan et al. (1997). *The report on the quality appraisal of principal training*. Beijing: Beijing Educational Publishing House.

Hui, Xiaoyong. (2001, July 30). China's development of information industry has come into an express way, *Wenhui Daily*.

Mao, Zedong. (1949). Report to the second plenary session of the Seventh Central Committee of the Communist Party of China.

Meisner, M. (1977). *Mao's China: A history of the People's Republic*. London & New York: The Free Press.

Ministry of Education. (1999). *Training regulations for school principals*.

Qian, Yicheng. (1998). High quality development of school principals for the 21st century. *Management of Primary and Secondary Schools* (the quarterly photocopy data). China Renda Social Sciences Information Center, No. 2, 20–23.

State Education Commission. (1989). *On strengthening the training for principals of elementary and secondary schools nation-wide*.

State Education Commission. (1995). *The training direction for principals of elementary and secondary schools during the ninth five year plan.*

Shanghai Municipal Education Commission (SMEC) and Shanghai Institute of Education. (1997, Ed.) *The foundation of the excellent schooling: Reviewing on the training practice for school principals in Shanghai.* Shanghai: Shanghai Institute of Education.

Shanghai Municipal Government. (1999). *On establishing the system of professional status for school principals (Draft).*

Teacher Affairs Office of SMEC. (1998). *The professional development for the K-12 teachers and administrators: Theory and action studies.*

The Personnel Department of SEC. (1991). *The syllabus for principal training institutions in the whole nation.* Tianjing: Tianjing Educational Publishing House.

Wuxi Municipal Education Commission. (1998). *On professional development for principals.* Wuxi: Wuxi Municipal Education Commission.

Wu, Xiujuan et al. (1998). *The successful principals: Practice and studies.*, Shenyang: Liaoning People's Publishing House.

Wu, Zhihong & Feng, Daming. (1998). *Management in elementary and secondary schools: The international comparison*, Shanghai: Shanghai Educational Publishing House (SEPH).

Wu, Zhihong, Feng, Daming & Zhou, Jiafang. (Eds.) (2000). *Educational administration: A new framework.* Shanghai: ECNU Press.

Yan Jun & Xiao Jie. (1999). The local course study of Jiangying's principal training, *School Teacher Training*, 11(2), 18–20.

# 14

# Supporting School Leaders in an Era of Accountability: The National College for School Leadership in England

Dr. Harry Tomlinson

*Professor, Department of Educational Administration, Leeds Metropolitan University, Leeds, England*

In January 1999 there were 8.31 million pupils in just over 26,000 maintained and independent schools in England. The average class size in maintained primary schools was 23.5 (normally for pupils aged 5–11) and in secondary schools 21.9 (normally for pupils aged 11–16 or 11–18). This gives some indication of size of context in which the preparation, licensure or certification, selection, evaluation and professional development of school leaders takes place in England.

Gunter, Smith and Tomlinson (1999) show how headship has historically been constructed and the new construction of headship today in England and Wales. This modernised construction identifies what is being presented to educational professionals as leadership through the centralised training programmes discussed below. In the same book headteachers from England and Wales explore, as the title 'Living Headship: Voices, Values and Vision' suggests, how they as successful headteachers interpret their leadership priorities and role.

## National Professional Qualification for Headship (NPQH)

The National Professional Qualification for Headship (NPQH) will become a mandatory qualification for all Headteachers seeking their first headship in England and Wales. Even in its initial period of implementation, however, the NPQH has changed significantly from the model presented in the first three years (1997–2000). The latest round of changes to the programme and processes came after a thorough consultation process.

The number of applicants has risen significantly from under 2000 in the year 2000 to over 5000 a year in 2001. This may suggest that the restructuring and redesign of the programme is viewed as better meeting the needs of those aspiring to headship. The first major modification has been that candidates are now much more responsible for managing their own learning. The second major alteration is that the assessment process has been greatly simplified, while sustaining the essential rigour. The third major change has involved the extensive use of information and communications technology (ICT) to support on-line learning.

The NPQH continues to be fully funded by the government. However, funding now comes directly to Regional Centres through the National College for School Leadership (NCSL). In the past, funding came from the Department for Education and Employment but flowed indirectly through the local education authorities (LEAs). They are now excluded from the process.

**Application process for the NPQH**

The application process begins with a preliminary consideration of candidates' developmental needs. In the first phase, the goal is to determine whether applicants are capable of completing the NPQH successfully. This process includes an initial self-assessment, with the possibility of an online diagnosis of strengths, the filling in of the application form itself, and a reference from the Headteacher. The application process requires – for information purposes only – career details, and qualifications.

Applicants have to present significant details of their Continuing Professional Development over the last three years. They must show how it has prepared them for a leadership role. They must explain why they want to be a Headteacher and they must provide examples of their achievements and expertise in the five key areas of headship. These are identified in the National Standards for Headteachers (www.dfee.gov.uk/headship/): Strategic Direction and Development of the School; Teaching and Learning; Leading and Managing Staff; Efficient and Effective Deployment of Staff and Resources; and, Accountability.

Judgments about eligibility are made against these seven elements for which there are clearly specified criteria to ensure consistency of judgment across England. At the end of the process, those selected will be assigned to one of three developmental routes through the NPQH. The total score determines whether candidates start on the Access Route, the Development Route or the Accelerated Development Route.

An increasing proportion of candidates apply online. It is anticipated that this will increase substantially and that all applications will be online within two years. Now that the National College for School Leadership (NCSL) is responsible for managing the delivery of the NPQH, Regional Centres rather than local education authorities (LEAs) determine eligibility. It is clear that the criteria for allowing candidates to start the NPQH have been loosened because of the addition of the new Access Stage. Candidates have the opportunity to start the NPQH who would not have done so under the old model because of the learning opportunities the new modes of training and development provide.

**Access stage**

In the Yorkshire and Humber Region in 2001, 60% of the candidates were designated for the Access route. They either selected this route themselves or were guided by those who evaluated their preparedness for starting the NPQH. Some 5% of those who applied were judged ineligible.

As the diagram below shows, there are 4 modules in the Access Stage. Each comprised of four units which Access route candidates will work through in six or twelve months, depending on their experience and skills. They have two half-day tutorials, one at the beginning and one at the end of the programme. A Candidate Handbook guides them through the learning and developmental process. There is also an ICT Handbook that will be explained more fully below.

Candidates have two days of face-to-face training: *Working with Staff* and *Working with the Community*. These days concentrate on skills development. There is a complementary new emphasis on online learning for the 16 study units. The units consist of about 25 pages of content each. These are available in hard copy, but also online so they can be continuously updated. There are activities associated with each unit.

**Candidate handbook and ICT handbook**

What is entirely new is the delivery mode. Candidates study and explore the content of the study units and the outcomes of the activities suggested in them in online discussion groups. For many units this has become the primary mode of learning.

This means that candidates can work at times that are convenient for them rather than being tied to a fixed workshop schedule as in the past. Candidates are also responsible for structuring their individual learning programmes. This requires very new skills of tutors for which there has been additional training.

The characteristics of highly effective online tutors are not yet clear, though theory is being developed from practice. Developing these new skills for online tutoring has been particularly exciting and challenging.

When the new model NPQH started sets of four tutors each had individual responsibility for 12 candidates. In addition they each managed the online learning of the whole group of 48 for one of the modules shown below. This was based on an assumption that groups of 48 were appropriate for online learning.

This process only started in January 2001. However, it has already been determined that there will be just one learning group for each of the three routes, rather than the previous group size of 48 candidates. This will result in one group of 144. Four lead tutors manage the learning for the whole group for each of the modules. The other eight tutors supporting the personal learning of their own groups of 12.

The candidates work through the units in a structured and sequential way with the tutors focusing discussion on the activities associated with each unit. Each candidate sets up an online Learning Journal accessible only to them self. The Summary of Learning, which for the first cohort was accessible only to the candidate and the tutor, is now shared with the tutor group of 12. The Learning Journal is intended to encourage candidates to:

- Reflect on what they do and how they do it,
- Identify their learning,
- Record their learning,
- Review their learning.

This learning process is designed to encourage reflective practice on the part of school leaders. During the Access Stage, candidates work in mixed-phase tutor groups with other candidates from primary, special and secondary schools.

The ICT Handbook presents the world of www.think.com and the Virtual Heads Community, the name for the NPQH community. Virtual Heads incorporates all the NPQH communities and discussion groups. The main focus of work in Virtual Heads is on the activities in the regional groups. However candidates are also encouraged to participate in national conferences with national experts in 'hot seats' on topics aligned with the study modules and units.

They also participate in online discussions with serving Headteachers to gain insights into good practice. Within think.com there are other educational communities as well. *Talking Heads* is a community of Headteachers, *Go Aheads*, the community of those who have completed the Leadership Programme in Serving Heads (see below). The NPQH ICT Handbook explains fully how the candidates can learn on line.

At the introductory tutorial candidates receive a *Virtual Heads* CD-ROM and receive their username and password and sufficient training to enable them to prepare to use the new learning. The ICT Handbook is presented at the Introductory Tutorial and provides full guidance for use of the site, not only on the technical process but also on associated ethical issues. The handbook offers a recommended code of practice and detailed guidance on how learning and development can be achieved through the virtual community.

The technicalities of this process have been presented in some detail since this cutting-edge development is being initiated even as this chapter is being prepared. The changes for Cohort Two starting in September 2001 have not been implemented as the chapter is finalised. Using online learning as the predominant means of training all future Headteachers has required an ambitious investment in time and resources. It is essential to ensure not only effective delivery of the technology, but more importantly to plan for effective online learning as a means of preparation for headship.

*Talking Heads* has been trialled now for well over a year (July 2001), and all Headteachers are encouraged to join. This has many of the characteristics of *Virtual Heads*, but provides learning through professional dialogue rather than through a structured learning and development programme. Inevitably many of the themes for the 'hot seats', brainstorms, conversations and debates that Headteachers have raised are precisely those which candidates aspiring to headship have as issues.

The titles of the units (Table 1) illustrate emphases in the preparation of Headteachers in England and Wales. They are based on the five key areas of headship in the National Standards for Headteachers.

Unit 3.4 on personal effectiveness will be used to illustrate the approach. The personal effectiveness unit focuses on the differences between leadership and management, and quickly recognises the importance of emotional intelligence. The Covey (1999) Time Management Matrix is used to clarify distinction between important and not important and urgent and not urgent.

The framework of the Management Charter Initiative in the UK is used to propose structures for managing time. *Managing Information* concerns managing the work schedule, office space, telephone, and effective use of e-mail. The section on *Communication* includes consideration of body language and listening skills. *Taking Control* focuses on managing pressures positively, particularly recognising the importance of assertiveness. The importance of self-development within the framework of personal and professional development is explored. There are a number of activities associated with these elements

Table 1. NPQH Curriculum Units.

| |
|---|
| Module 1 – Strategic Direction and Development of the School |
| 1.1 Determining the curriculum |
| 1.2 Vision into action |
| 1.3 School development planning |
| 1.4 Accountability for improvement |
| Module 2 – Teaching and Learning |
| 2.1 Analysis of data for school improvement |
| 2.2 Target Setting for school improvement |
| 2.3 Equal opportunities |
| 2.4 Monitoring, evaluation and review |
| Module 3 – Leading and Managing Staff |
| 3.1 Working with stakeholders |
| 3.2 Leading and managing teams |
| 3.3 Managing performance |
| 3.4 Personal effectiveness |
| Module 4 – Efficient and Effective Deployment of Staff and Resources |
| 4.1 Managing a budget |
| 4.2 Planning and implementing the curriculum |
| 4.3 Recruitment and selection of staff |
| 4.4 Health, welfare and safety |

to encourage learning. This brief presentation of the content shows that it is dealing with central current issues. It is enabling those seeking headship to practice skills that will enable them to be more effective people and Headteachers.

**Development stage**

Candidates will enter either directly at the Development stage or after having completed the Access Stage. In the Yorkshire and Humber Region some 35% of the candidates are using the extended Development route with training and development. This stage requires a maximum of eight days out of school. Schools are funded for a replacement teacher to cover the work of teachers in primary or special schools. Some 5% of candidates are using the route that requires no training before school-based assessment because they are very close to headship. The Development Stage is structured as follows.

*Induction day*

This will incorporate self-assessment tasks, meeting the tutor and an introduction to the ICT process. Candidates will receive printed copies of the Candidate's Handbook, an Introduction to the ICT, and all the Development Stage Units. The tutor group of 12 will be from the same phase of school, primary special or secondary. Access candidates are in mixed phase tutor groups but Development candidates are from the same phase. This is a response to the consultations in 1999 that recognised the necessity for candidates to gain a broad understanding of the issues at an Access level, but that, at the more strategic level, it was essential to focus on the phase in which the candidate would be applying for headship, primary, secondary or special.

*Contracting visit*
After the induction day candidates will complete further self-assessment tasks to map their achievements and expertise against the National Standards in order to identify their individual training and development needs. This is the Candidate's Professional Development Record (CPDR). The tutor will then visit the school to agree with the candidate and the Headteacher the personal training and development plan. Together they will draw up a contract outlining the planned work on school improvement, and the elements of the on-line training programme that will be undertaken as the focus of the developmental activities.

Candidates who are very close to headship will not undertake the training activities and school visits, but proceed straight to School-Based Assessment described below. As stated above this is some 5% of all candidates in the Yorkshire and Humber region.

**Training and development activities**
The Development Phase modules have the same titles as the Access modules but some of the Units have different titles because they are dealing at a more strategic and conceptual level, building directly on the Access Stage modules. This has meant very careful design because candidates are starting from different bases. The tutor will provide advice and support and monitor progress. The on-line training is very similar to that provided at the Access Stage but at a higher level.

Unit 3.4 at the Development Stage focuses more strategically than Unit 3.4 at the Access Stage. This Unit *Continuing Professional Development* explores how teachers can meet the challenge of change. It refers to a series of government papers on continuing professional development, the importance of staff development and on the role of teachers as researchers.

Candidates learn about DfEE support for teaching and learning, performance management and the wide range of national standards at different career stages. There is additional information about the Ofsted Inspection Framework, the National College for School Leadership, and the General Teaching Council (see below). The next section focuses on the school as a learning community (Senge, 2000). *Continuing Professional Development* (Development Unit) is clearly more strategic than *Personal Effectiveness* (the complementary Access Unit).

There will be four days of face-to-face skills based training for Development candidates which will focus on problem-solving activities, *Governance and Headship*, *Leadership and Developing Potential, Leadership and Professional Accountability,* and *The Public Face of Headship.*

There are a number of other interesting further refinements. Candidates will have the opportunity to make visits to schools to learn about different approaches to leadership with clear guidance on how to learn from the visits.

**School-based assessment**
One of the issues arising from consultation on the first model of the NPQH was that the assessment tasks were immensely time-consuming. Moreover, these tasks did not necessarily demonstrate that candidates met the National Standards for Headteachers and were ready for headship. Towards the end of the period when the earlier model was in place, school-based assessments were being developed. These proved much

more rigorous and less time consuming. All of the work done in preparation for the assessment was directly related to the candidate's current activities aimed at improving practice in the school.

The school-based assessment has four distinct activities:

- Review of the completion of the training and development plan
- Review and discussion of the learning points from the reflective journal
- Assessment of written and oral evidence of the school improvement work
- Assessment of written and oral evidence of capability against the National Standards for Headteachers.

A tutor who has not been involved in working with the candidate will carry out this assessment.

**Final stage**

Candidates must have been successful in the school-based assessment before they progress to the Final Stage that comprises two elements:

*Two-day residential conference*

This will be hosted by the National College for School Leadership, though until the college is actually built this is managed by the Regional Centres. The residential will focus on: school leadership and vision; future schools; national priorities and international perspectives; and, personal effectiveness. This provides an opportunity for candidates to meet aspiring Headteachers from many different contexts and to extent their professional networks working on an extensive case study together.

*Final assessment*

This one-day assessment centre provides the opportunity to demonstrate overall readiness for headship. There are a number of exercises that reflect issues that Headteachers meet regularly. There is an in-depth personal interview with an assessor. This part of the NPQH has not changed because it has been evaluated as being particularly effective in judging preparedness for headship.

**Conclusion**

The NPQH is underpinned by a model of leadership effectiveness developed by Hay Management Consultants for the DfEE. This model posits that leaders influence superior school performance through four interactive dimensions: individual characteristics of the principal, job requirements, leadership style, context of the school. The NPQH framework, therefore, assumes that training should focus on increasing principals' capacities to work effectively on these dimensions.

The refined NPQH framework is grounded in Kolb's learning cycle, designed to facilitate adult learning, and uses modern learning technology as the key foundation for its three distinctive routes for preparation and development. It honours the aspirations of those seeking headship as well as the societal demand to guarantee high quality training, development and assessment of future leaders of schools in England and Wales.

The model of adult learning recognises that adults are more self-directed than children, and that they use previous knowledge and experience to shape their learning.

Adults learn for specific purposes. They are problem-centred and want to apply what they have learned as they go through transitional phases in their career management.

The NPQH has been designed to help candidates achieve self-direction and to capitalise on their prior experience; the mode of learning is collaborative and fosters participation. In developing critical, reflective thinking, promoting learning for action and encouraging problem setting and problem solving the new NPQH appears to be successful. We await evaluations of the new NPQH in practice.

## The Selection of Headteachers

The current emphasis on preparation and development represents a clear departure from past practice in England and Wales. It has signaled the higher priority afforded towards the Headship by the government and society at-large. However, preparation and development alone will not ensure that world-class leaders lead our schools. The issue of selection is also paramount. This section of the chapter reviews recent measures in England and Wales taken to upgrade the selection of high quality school leaders.

### The role of governing bodies

The Governing Body of each school selects their Headteacher. From September 1999 onwards, government maintained schools have been restructured with new names, new constitutions, new categories of governors, and new rules for school governance. The governors, when they select the Headteacher at their school, have new rights and responsibilities (School Standards and Framework Act, 1998).

Schools are now organized into four new categories: community, foundation, voluntary aided and voluntary-controlled. The composition of the governing bodies has been significantly altered. New regulations specify the Instruments of Government and Governing Body Composition, Meetings and Proceedings of Governing Bodies and the Committees of Governing Bodies in great technical detail. This complex structure and detailed guidance provides a form of control by regulation.

The Regulations describe the roles and responsibilities of governing bodies as well as those of Headteachers. Governing bodies carry out their functions with the aim of taking a largely strategic role in running the school. This includes setting up a strategic framework for the school, setting its aims and objectives, setting policies and targets for achieving the objectives, reviewing progress, and reviewing the strategic framework in the light of progress. The Headteacher is responsible for the internal organisation, management and control of the school, and for advising on and implementing the governing body's strategic framework.

If the governing body delegates any function to the Headteacher it has the power to give reasonable directions in relation to that function, and can oblige the Headteacher to comply with these directions. The relationship between the Headteacher and governing body has parallels to that of a Chief Executive and a Board of Directors. Given that the governing body is responsible for evaluating the performance of the Headteacher, the quality of the relationship between the governing body and the Headteacher is crucial.

For Secondary Schools with more than 600 pupils the governing body is comprise of six elected parents, five local education authority (LEA) representatives, two elected teachers, one elected member from the other staff, five co-opted members, and the Headteacher who may elect not to be a governor.

For larger Primary Schools with more than 100 pupils, the governing body normally is comprised of: five elected parents, four LEA representatives, two elected teachers, one elected member from the other staff, four co-opted members, and the Headteacher who may elect not to be a governor. The chair of the governing body that is elected by the governing body is particularly important and influential.

In making a Headteacher appointment the governors must take account of the responsibilities of the post, the social, economic and cultural background of the pupils attending the school, and whether the post is difficult to fill in determining the salary of the Headteacher. There are some flexibilities and enormous complexities for the governors in determining how much they will pay the Headteacher. There is a salary range that relates to the size of school. A Headteacher of a primary school from approximately 150 to 300 pupils could be paid in the range £33,811 to £44,322 in 2000. The Headteacher in England, despite the apparent constraints discussed above, has enormous power and influence in the school.

### HEADLAMP

HEADLAMP (Headteachers' Leadership and Management Programme) is a more flexible programme for Headteacher's starting their first headship. The programme begins with a needs assessment. This can take various forms, but is viewed as an essential prerequisite for effective personal development. Most training and development activities are suitable for HEADLAMP.

Training needs, priorities, and the consequential professional development are planned over the two years for which the funding is accessible. It is significant that HEADLAMP does not specify any particular development programme. Its goal is to provide for headteachers a strong financial commitment to continued professional development after completion of the two year NPQH programme. HEADLAMP focuses solely on the further professional development of Headteachers to meet their individual needs. As such it is a unique and important facet in the scheme of leadership development in England and Wales.

## The Evaluation of the Headteacher's and School's Performance: Implications for Headteacher Professional Development

There are a number of individuals who contribute to the evaluation of the performance of Headteachers and schools. Headteachers are increasingly accountable for the performance of their schools. Local Education Authorities (LEAs) have traditionally monitored the performance of their schools and the effectiveness of Headteachers. It was partly because they were judged not to be sufficiently effective in this role that the Office for Standards of Education (Ofsted) was established to evaluate the performance of schools.

More recently, following the Green Paper Reforms, several new roles have been created to support the Headteacher and governors in evaluating and improving the performance of schools. These new roles, those of External Adviser, External Assessor and Performance Management Consultant will be defined below, as will the new roles of LEAs and Ofsted. The central role is arguably still that of the governing body that must interpret all the information available and both support and challenge the Headteacher in her/his leadership of the school.

The *School Teacher's Pay and Conditions Document 2000* (DfEE, 2000) states that the Headteacher's performance shall be reviewed by the governing body taking account of performance objectives agreed upon in the previous year. The governing body and the Headteacher will now seek to agree on performance objectives relating to school leadership and management as well as pupil progress. If there is no agreement, the governing body shall set the performance objectives.

There have been a number of significant developments in the last two years that build on the Ofsted evaluations of school performance that have been in place since the early 1990s. The government consultation Green Paper *Teachers: Meeting the Challenge of Change* (December, 1998) and the associated *Technical Consultation Document on Pay and Performance Management* (January, 1999) are about transforming the teaching profession. The objectives stated in the Green Paper were to:

- Promote excellent school leadership by rewarding leading professionals properly,
- Recruit, retain and motivate high quality classroom teachers, by paying them more, and
- Provide better training and support to all teachers, and to deploy teaching resources in a more flexible way. (1999, p. 4)

Implementing performance management, including a systematic assessment of the performance of all teachers linked to a suitable level of reward, demands new skills of Headteachers. A culture that provides greater incentives and rewards for good performance and for better career progression involves a profound and necessary change. The development and focus of new additional awards for school performance based on collective achievements, excellence in outcomes and strong improvement, illustrates the seriousness of the government's intent to improve schooling. The school's leadership team must be redesigned to support these moves towards accountability and reward for collective performance.

The chapter on leadership in the Green Paper reflects the new emphases in practice. When the present government was elected in 1997, it stated that it had three priorities: education, education and education. The absolute centrality of the Headteacher in delivering this policy by which the government insists on being judged is crucial. The direct quotation illustrates recognition of Heads' leadership competencies: 'the best heads are as good at leadership as the best leaders in any other sector including business' (Forde, Lees, & Hobby, p. 2).

**External adviser**

The role of the External Adviser is to use all available evidence to advise the appointed governors, normally three from within the governing body, on setting the Headteacher's objectives and reviewing the Headteacher's performance. The advisers have received DfEE-approved training to ensure that there is consistent advice across the 25,000 schools of England and Wales. There were some difficulties in making appointments and managing this process. Therefore, the date by which Headteachers were to have their objectives agreed upon was delayed from December 31, 2000 to April 6, 2001.

This process represents a much more rigorous approach to evaluating Headteacher performance. It is also tied in with pay rises for Headteachers that are paid from the school budget with the associated ethical issues. The role of External Adviser involves:

- Providing high quality and focused advice to governing bodies on setting performance objectives,
- Coaching and supporting governors,
- Supporting an effective process of performance monitoring and review,
- Helping governors to prepare for and conduct the review,
- Attending the review,
- Advising the appointed governors on the assessment of the overall performance of the Headteacher,
- Advising the governors, if requested, on how to assess the performance of the Headteacher in relation to whether a pay increase is merited,
- If requested, drafting the review statement,
- If the appointed governors are confident they may wish the adviser to simply give advice.

The implication is, very clearly, that governors are encouraged to trust the adviser who will have a breadth of understanding and experience across a number of schools. This will assist them in understanding the relative strengths of their Headteachers.

**Performance management consultant**

The DfEE-funded schools are encouraged to use performance management consultants (PMCs), with, in the first year, additional funding which could be accessed for this purpose only. This group of professionals provided one day's training for Headteachers in all 25,000 schools in England aimed at helping schools implement performance management.

This effort is aimed at enhancing the professional development of teachers to raise standards in schools. Team leaders will monitor the teaching and other responsibilities of all teachers in an annual cycle. This will focus on setting objectives, monitoring performance and reviewing performance. It is the Headteacher's responsibility to establish this new statutory system in school.

This informal process complements the more formal processes described above. The PMC's main role is to help the Headteacher implement performance management, a responsibility that represents one of their priorities by government policy.

**Office for standards in education (Ofsted)**

Inspection is to help schools to improve by building on their strengths and tackling their weaknesses. Inspection is also aimed at strengthening accountability by giving the parents and the community an independent report on the performance of every school. Schools are inspected at least every six years consistent with the principle that intervention should be in inverse proportion to success. The teams work to national standards set by Ofsted in their framework documents such as the *Handbook for Inspecting Primary and Nursery Schools, with guidance on self evaluation* (1999).

Inspection focuses on the school. Inevitably it is also an inspection of the performance of the Headteacher in leading and managing the school. At the end of the inspection the registered inspector provides feedback to the Headteacher and senior management. Feedback is intended to explain significant evidence and judgments, including issues identified as priorities for school improvement.

In a separate session the inspector will give feedback to members of the governing body who may also invite the representatives of the LEA to attend. These feedback sessions provide a basis for staff and governors to begin to consider post-inspection action plans. A report and summary have to be prepared by the registered inspector within six weeks of the end of the inspection. The school will be shown the report and have at least one week to comment on matters of factual accuracy.

The Ofsted report of a school is a most rigorous and public evaluation of the performance of the Headteacher who is held accountable for the performance of the school (Ofsted, 1999). The most significant section explores how well the school is led and managed (pp. 92–106 in the primary handbook). The guidance does not always distinguish particularly clearly between the Headteacher, the school and the governing body. This is because the Headteacher, in the UK, is seen as centrally responsible for the performance of the school, its staff and pupils (Hall & Southworth, 1997).

In determining their judgment, inspectors should consider the extent to which, through the Headteacher:

- Leadership ensures clear direction to the work and development of the school, and promotes high standards;
- The school has explicit aims and values, including a clear commitment to good relationships and equality of opportunity for all, which are reflected in all its work;
- There is rigorous monitoring, evaluation and development of teaching;
- There is effective appraisal and performance management;
- The school identifies appropriate priorities and targets, takes the necessary action, and reviews progress towards them;
- There is a shared commitment to improvement and the capacity to succeed. (Ofsted, 1999, p. 92)

The Ofsted process clearly provides a rigorous evaluation of the performance of the Headteacher. The extent to which this departs from traditional practice in England and Wales is dramatic and notable.

## Leadership Programme for Serving Headteachers (LPSH)

The LPSH started in November 1998. It is a four-day leadership development programme followed up by a further day normally one year later to evaluate progress in achieving the targets set for personal development and school improvement a year earlier. This is the only directly funded and DfEE-managed Headteacher development programme after the NPQH and HEADLAMP. This is also being transferred to the National College of School Leadership (NCSL).

Before the four-day programme, participants complete diagnostic questionnaires. These are based on personal characteristics related to the *Headteacher Models of Excellence, Leadership Styles*, and the *Context for School Improvement*. The data are designed to provide diagnostic information about the Headteacher and the school. The Headteacher and other members of the school community, staff and governors complete the questionnaires to provide a 360-degree feedback. Headteachers also bring to the training days their own analyses of appropriate school performance and evaluation data to provide a sharper focus for their training and development.

The exercise in gaining 360-degree feedback provides a comprehensive indication of how successful an individual is in the totality of her or his relationships at work. It focuses on the skills and competencies that those working within organizations believe will improve organizational performance. Those managed have more to contribute to an analysis of the performance of an individual than has been previously recognised. They know well the attitudes, beliefs, underpinning values and behaviours of those for whom they are providing feedback (Farrell, 1999). A *Model of Organisational Effectiveness* and a *Model of Leadership Effectiveness* (see above) presented at the start of the programme provide a contextual framework for the programme.

**Models of excellence**

The Models of Excellence refers to how highly effective Headteachers raise standards. Personal characteristics concern the deep-seated qualities the Headteacher brings to the role. They relate to self-image and values, traits, how the Headteacher habitually approaches situations, and at the deepest level the motivation that drives performance. These when combined with the knowledge and skills presented in the National Standards lead to excellent results. The fifteen characteristics can be grouped into five clusters based on the model below (Table 2).

The Headteachers receive an analysis of their own 15 characteristics mapped against the characteristics of highly effective Headteachers who were involved in the study that led to the programme. This profile has been further refined as the programme has been

Table 2. Clusters and Characteristics in the Framework.

| Clusters and Characteristics |
|---|
| **Personal Values and Passionate Conviction** |
| Respect for Others |
| Challenge and Support |
| Personal Conviction |
| **Creating the Vision** |
| Strategic Thinking |
| Drive for Improvement |
| **Planning for Delivery and Monitoring Improvement** |
| Analytical Thinking |
| Initiative |
| Transformational Leadership |
| Team working |
| Understanding Others |
| Developing Potential |
| **Getting People on Board** |
| Impact and Influence |
| Holding People Accountable |
| **Gathering Information and Gaining Understanding** |
| Understanding the Environment |
| Information Seeking |

implemented. Within the programme documentation there are definitions of each of these characteristics and how they are demonstrated at four distinct levels.

**Leadership styles**

The diagnostic instrument on leadership styles produces a similar analysis of relative strengths based on six leadership styles. Effective leadership for Headteachers is using the appropriate style in the particular context. Ideally the Headteacher has access to all six styles at a high level. Evidence is presented through the diagnostic instrument about the Headteacher's interpretation of their range of styles and the judgment of others within the school about their capacity to access these styles when appropriate: coercive, authoritative, affiliative, democratic, pacesetting, coaching.

**Context for school improvement**

The Headteacher's range of characteristics compared with those of highly effective Headteachers and the impact of these on the range of leadership styles they can access impacts on the climate in the school. The third instrument measures the context for school improvement, the climate in the school. These six dimensions are those that have a significant impact on school performance. The climate that the Headteacher creates impacts significantly on school performance. What staff perceive the climate to be is their reality.

- **Flexibility** – freedom to act, no unnecessary bureaucracy, new ideas are accepted;
- **Responsibility** – authority is delegated, no need to check everything with the manager, people feel fully accountable;
- **Standards** – the management emphasis on improving performance, setting challenging attainable goals, and mediocrity is not tolerated;
- **Rewards** – recognition and reward for good work, recognition related to quality of performance, individuals know how well they perform;
- **Clarity** – everyone knows what is expected; they understand how expectations relate to the wider objectives;
- **Team Commitment** – pride in the school, staff provide extra effort when needed, and trust all are working together.

The programme combines challenge and support in a setting that provides space for critical self-reflection. The opportunity to work on neutral territory, alongside other Headteachers from different types of schools and parts of the country in a confidential process, enables the sharing of expertise.

The first two days concentrate on feedback and understanding motivation. The second two days focus on relating all this information to the school data, a critical incidents exercise and on action planning as a basis for the setting of clear and challenging personal and school improvement targets.

Two additional complementary elements have not been as successful and are being restructured. The online discussion group of the 14 participants after the four-day event has rarely been sustained. This is now to be replaced by one national conference for all those who have participated in LPSH within the *Talking Heads* community discussed above. *Partners in Leadership* which was intended to provide Headteachers with a business partner has had considerable success where it has been possible to find a partner. Unfortunately this has only been possible for about half the Headteachers. Consequently there has been a problem associated with raised and unfulfilled expectations.

Hay McBer has compared the leadership skills of highly effective Headteachers to those of senior executives in business. Forde, Hobby and Lees (2000) explored the inputs of leadership, the characteristics and styles the individual brings to the role – their drive, problem solving and influencing skills, and the outputs of leadership – the measures of success – the motivation, engagement and effort inspired in those who are led. Heads exert strong and versatile leadership, adapted to the needs of their people. The role of Headteacher is stretching in comparison with that of the business leader. Heads think of leadership in terms of developing people whereas business leaders, comparatively better, create a sense of mission, drive up standards and communicate their vision.

The programme has only been delivered since November 1998 but already over 5000 Headteachers have participated in the LPSH. The evaluations have been significantly more positive than for the early NPQH. This might be attributable to because the preparatory work that could be completed before the programme was launched. The programme is half-funded directly by central government. Local education authorities have funded the other half from grants not specifically allocated for this purpose but also from central government.

## The National College for School Leadership and General Teaching Council

### National College for School Leadership (NCSL)

The Prime Minister formally launched the National College for School Leadership in November 2000 (www.ncsl.gov.uk). It heralds a new era of professional development opportunities for school leaders and will provide a national and international focus for school leadership training, development and research. It is intended to ensure that Headteachers have access to the support, recognition and inspiration they need by providing a framework for high quality training and development for school leaders at every stage of their careers.

The NCSL will award Associate, Fellowship and Companion status to those benefiting from its programmes. Heather Du Quesnay is the first Director and her senior management team is in place. It seems very clear that the NCSL will retain the NPQH, HEADLAMP and LPSH. If these are judged to be appropriate, the NCSL will use them as elements in a wider framework of professional development for school leaders, particularly after the first two years of headship.

Heather Du Quesnay, interviewed recently (Hellawell, 2000), suggests that there is a recognised need for a degree of coherence in planning support for Headteacher professional development. In addition, she has a complementary and equal awareness of the dangers of imposing one model of leadership. There will be a focus on the new leadership teams, established recently in a change to the structuring of teacher pay. This suggests that the NCSL could be supporting the full range of 100,000 school leaders that includes the nearly 26,000 Headteachers.

The College will have a limited role as far as research is concerned. It will, however, need to find means of contributing to what government perceives as the isolation of the research community; that is the community talks within its membership rather with practising professionals in the schools. The College will want to encourage Headteachers to become researchers and to learn from the study of and reflection on their own and other people's practice.

The NCSL is the emergent force for future School Leader Preparation, Licensure/Certification and Professional Development in the UK. Du Quesnay sees the College as part of a network of government agencies, and believes it is very important to work with Ofsted to support Headteachers in developing self-evaluation as a basis for future self-regulation.

### The General Teaching Council

The *General Teaching Council (England)* was also established in the year 2000 to advise on

- standards of teaching,
- standards of conduct for teachers,
- the role of the teaching profession,
- training, career development and performance management of teachers,
- promotion of teaching recruitment to the profession,
- medical fitness to teach.

The career-long entitlement for teachers to guaranteed professional development opportunities complements the opportunities provided by the NCSL for school leaders. The learning process for teachers will focus on the personal qualities that make a good teacher, to balance the current focus on competences and skills. The role of Headteachers in shaping this independent professional body will be important, and the new Council has a significant number of Headteachers. If teachers are to have a single professional voice, and the opportunity to lead and shape future education policy, this will require an appropriate lead from Headteachers.

## Conclusion

In the period 1997–2001 the Department for Education and Employment (DfEE), and the Secretary of State, David Blunkett, have overseen a total transformation in policy in England and to a lesser extent in Wales of School Leader Preparation, Licensure/Certification, Selection, Evaluation and Professional Development.

- The National Professional Qualification for Headship has been significantly changed.
- HEADLAMP has been redesigned.
- The selection of Headteachers by new governing bodies is increasingly rigorous.
- The evaluation of Headteachers and schools has become much more sharply focussed as a result of using External Advisers, External Assessors, and Performance Management Consultants with sharper accountabilities for Local Education Authorities.

This has all built on the work of the Office for Standards in Education (Ofsted) that inspects schools. The Leadership Programme for Serving Headteachers is the government programme for experienced serving Headteachers. However the National College for School Leadership is now exploring significant extensions of this professional development. All this is within a context, demonstrated by the new General Teaching Council for England, in which government policy is to enhance the status of teaching as a profession.

## References

DfEE. (1998). *Teachers: Meeting the challenge of change.* London: DfEE.

DfEE. (1999). *Technical consultation document on pay and performance management.* London: DfEE.

DfEE. (1999). *The education (school government)(England) regulations* 1999. No 2163. London: HMSO.

DfEE. (2000). *School teachers pay and conditions document 2000.* London: HMSO.

Farrell, C. (1999). *Best practice guidelines for the use of 360 degree feedback,* London: The Feedback Project UK.

Forde, R., Hobby, R. & Lees, A. (2000). *The lessons of leadership.* London: Hay McBer.

Gunter, H., Smith, P. & Tomlinson, H. (1999). Introduction: Constructing headship: Today and yesterday. In H. Tomlinson, H. Gunter & P. Smith (Eds.), *Living headship: Voices, values and vision.* London: Paul Chapman.

Hall, V. & Southworth, G. (1997). Headship. *School Leadership & Management,* 17(2), 151–170.

Hellawell, D. (2000). Learning to lead. *Managing Schools Today,* 9(10), 34–36.

House of Commons. (1999). *The role of school governors.* The House of Commons Education and Employment Committee Inquiry Report, London.

NCSL. (2001). *NPQH tutor handbook.* National College for School Leadership.

Ofsted. (1999). *Handbook for inspecting primary and nursery schools, with guidance on self evaluation.* London: The Stationery Office.

Senge, P. (2000). *Schools that learn: A fifth discipline fieldbook for educators, parents and everyone who cares about education.* London: Nicholas Brealey.

# Section III

## Global Trends, Conclusions and Implications

# 15

# Examining the Impact of Professional Preparation on Beginning School Administrators

Dr. Ronald H. Heck

*Professor and Chair, Department of Educational Administration and Policy, University of Hawaii, Manoa, Hawaii, USA*

In the United States, universities traditionally have been responsible for the preparation of administrators before their initial appointments in schools. Periodically, there have been calls for the reform of these initial preparation programs. Critics claim they are not based on the realities of the school as a workplace (e.g., American Association of School Administrators, 1960; Barth, 1997; Bridges, 1977; Bridges & Hallinger, 1995; Halpin, 1970; McCarthy, 1999; Parkay & Hall, 1986). Preparation programs have also been described as collections of random events, which lead to random socialization (Bridges, 1977; Hart, 1991).

Securing effective school leaders for the future involves many steps including recruitment, selection, preparation, and ongoing development (Hart & Wending, 1996; Pounder & Young, 1996). Recently, a number of external forces have contributed to calls for change in the programs that prepare and support the ongoing development of school leaders in the United States (McCarthy, 1999). These forces include:

- demands for greater accountability in K-12 education,
- upgraded standards in the form of government initiatives for the preparation and continuing professional development of educational professionals such as teachers and administrators, and
- alternative types of school reforms (e.g., site-based management, charter schools) that focus on reducing bureaucracy and identifying the school as one locus of educational change.

Moreover, practitioner-oriented (e.g., National Association of Secondary School Principals), professor-oriented (e.g., University Council for Educational Administration), and foundation-supported (e.g., Kellogg, Danforth) organizations have also been concerned with improving the preparation and ongoing support of school administrators

(McCarthy, 1999). All groups anticipate significant shortages of administrators over the next several years (Educational Research Service, 1998).

Several literature reviews have appeared recently on the selection, preparation, and ongoing development of school leaders (e.g., Hart & Wending, 1996; McCarthy, 1999; Milos, 1988; Pounder & Young, 1996). These reviews have identified three weaknesses relevant to the purpose of this chapter. First, there is meager research relating program innovations to either administrator success or to the use of new knowledge on the job. Second, there is little systematic information about what happens to graduates of these programs; that is, how they move through the system, become socialized into their roles, and obtain their first principalship.

Finally, while some argue that the complexity of schools makes the evaluation of school administrators on the basis of outcomes unrealistic, as Hart (1992) suggests, today's accountability-focused policy context increases the need for models that tie the evaluation of administrators more closely to valued outcomes (e.g., achieving goals, improving student results). Therefore, a remaining question is whether students from redesigned preparation programs have different characteristics, function differently in the role, and ultimately move ahead more quickly in their careers (McCarthy, 1999).

The purpose of this chapter is to develop a theoretical framework for understanding the impact of initial professional preparation on the performance and career paths of beginning school administrators. More specifically, the goals are to:

1. discuss professional and organizational socialization as a theoretical lens for understanding the transition from administrative trainee to school administrator, and
2. provide empirical evidence from a longitudinal study undertaken in Hawaii to examine how administrative socialization affects the performance and evolving professional practice of new assistant principals and principals.

As Duke (1987) argues, becoming a school leader is an ongoing process of socialization. The process encompasses personal experience in school settings, formal job orientations, university courses, interactions with mentors and supervisors, and a variety of other occasions through which individuals learn the norms and expectations of their profession and of the organization in which they work. Providing empirical data about how new administrators experience their initial preparation, their probationary years as vice principalship, and how they then move from the vice principalship to their first principalship should be useful in developing meaningful evaluation and support for school administrators at different points of their early career socialization.

## Challenges to the Preparation of School Administrators

McCarthy's (1999) review of the literature on school leader preparation revealed that from the 1970s through mid-1990s there was considerable stability in the structure of university units, degrees offered, and curriculum comprising preparation programs at the university level. As Cooper and Boyd (1988) described this, the typical model of leadership preparation was 'state controlled, closed to non-teachers, credit-driven, and certification bound' (p. 251). Often, the emphasis consisted of candidates gaining a broad knowledge of the social sciences, as opposed to a deeper understanding of the specific educational needs of schools (e.g., the curriculum, instruction and learning) and strategies for implementing appropriate changes. Moreover, linkages between the planned preparation experiences and what administrators actually do in schools were largely nonexistent (Barth, 1997; Bridges, 1977; Bridges & Hallinger, 1995; McCarthy, 1999). Thus, administrative preparation was criticized as not being rigorous enough, consisting of outdated content, and having course work lacking cohesion and grounding in principles of cognition or leadership (McCarthy, 1999; Murphy, 1993).

Such criticisms have produced a series of changes in recent years in some university administrative preparation programs (McCarthy, 1999). Redesigned preparation programs have attempted to incorporate policy changes that have occurred during the 1990s involving decentralization and the shifting of greater authority to the school site. These programs focus on developing integrated and sequenced formal course work taken through the university. They may emphasize the school's curriculum and instruction, the social context of education, school culture, organizational change, and values (Murphy, 1993), as opposed to the more traditional administrative curriculum consisting of personnel, finance, and facilities management. Redesigned programs also require students to spend considerable time in schools serving as administrative interns, where they can receive mentoring from practicing principals.

Many of these changes also required an emphasis on providing leadership skills including team building, goal setting, collaborative decision making, self-assessment, and conflict resolution (Crews & Weakley, 1995). There was also an increased emphasis on building the skills aimed at increasing student learning in the school (Cambron-McCabe, 1993). Required in this new approach is 'learning-in-action' or learning as students reflect on their own actions (Silver, 1987), through a variety of case studies, problem-based learning, and internships (Bridges & Hallinger, 1995). These programmatic changes require closer working relationships between university leadership preparation programs, school districts, and school sites. Some programs have implemented innovative degree programs in cooperation with school districts to locate, nominate, interview, and select potential candidates (Murphy, 1993; Ogawa & Pounder, 1993). These changes have also led to efforts to increase the quality of assessment regarding students' progress through the program, as well as their socialization into the field during and after their preparation program has been completed.

## The Socialization of New Administrators

One lens through which to understand preparation, induction, and early career mobility in school administration is socialization theory (Hart, 1991; Hart & Wending, 1996). Previous research has examined various parts of the administrative induction and socialization process (e.g., Bridges, 1965, 1977; Duke & Iwanicki, 1992; Gaertner, 1978–79; Greenfield, 1977; Leithwood et al., 1992; Heck, 1995; Ortiz, 1978; Marshall, 1981; Marshall & Greenfield, 1985; Parkay, Currie, & Rhodes, 1992; Wolcott, 1973). Socialization refers to the processes through which individuals acquire the knowledge, skills, norms, values, and operating procedures needed to perform an organizational role effectively (Hart, 1991; Milos, 1988; Parkay et al., 1992). In *The Man in the Principal's Office*, Wolcott provided one of the early examinations of the administrative socialization process (see also Bridges, 1965). Wolcott suggested that socialization consists of the:

> ... processes by which the schools manage to maintain stable cultural systems in spite of the constant change of personnel assigned to their relatively few statuses. The underlying thesis here is that schools, like other cultural systems, are perpetuated through the processes of socializing new members into the statuses that must be occupied. (p. 192)

Theorists further define *professional* socialization as the process through which one becomes a member of a profession and over time identifies with that profession. With respect to school administration in the United States, these experiences include the administrative candidate's formal preparation (e.g., university course work, fieldwork or internship) and the various responsibilities and requirements met during this initial phase of induction. While significant field experience has been identified as an important need in preparing administrators, the clinical aspects of most university programs have been described as weak. Moreover, their impact on subsequent performance in administrative roles is largely unknown (McCarthy, 1999). Besides this initial induction to the profession, professional socialization also includes the longitudinal process by which the recently trained beginning administrator develops into a competent educational leader (Parkay et al., 1992). New principals must learn to master a complex set of professional challenges after they assume the principalship.

In contrast to professional preparation, *organizational* socialization refers to the process through which one learns the particular knowledge, skills, norms, and behavior (i.e., 'learns the ropes') of a particular organizational role in a specific work setting (Hart, 1991; Van Maanen & Schein, 1979). Environmental and organizational factors exert a powerful influence in shaping the norms, values, and behavior of new members (Hart, 1991). Some of these organizational influences include responsibilities of the role, expectations of others, and the ongoing relationships that develop with teachers, parents, and the school principal. Greenfield (1977) suggested that how the new administrator was socialized to the principal's role depended on both the organizational context and the personal skills and training they brought to the job.

From early work on the socialization of school administrators (e.g., Wolcott, 1973; Greenfield, 1977), it became apparent that the norms and values of the work context often conflict with the formal preparation (e.g., university course work) an individual has received. In those cases, organizational socialization is hypothesized to be more pervasive with respect to the performance and ongoing development of the individual (Hart, 1991). The need to fit into the immediate work environment can make organizational socialization more crucial to job success than the professional preparation experiences that precede it (Guy, 1985), even if those experiences have been carefully planned. This aspect of administrative socialization theory, however, has not been extensively applied to educational settings and the socialization of new administrators (e.g., see Duke, 1987; Duke et al., 1984; Leithwood et al., 1992).

Descriptive studies on new school administrators identified a variety of problems they face during their first years on the job as they struggle to take over the leadership of the school (e.g., Daresh, 1987; Parkay & Hall, 1992). The initial socialization new administrators experience in making the transition from teachers to administrators was reported to be intense, short, and informal, rather than planned (Duke et al., 1984; Greenfield, 1977; Khleif, 1975). Duke (1987) suggested new administrators are often very frustrated by their relationships with peers and supervisors. Moreover, new administrators reported experiencing stress resulting from time constraints, loneliness, and a perceived lack of skills to manage the demands of the job (Duke et al., 1984). A number of studies also identified differences in socialization experiences for males and female administrators (e.g., Marshall, 1981; Ortiz, 1982; Ortiz & Marshall, 1988; Valverde, 1980).

Those new to the school administrator role need to adjust to the divergent views of faculty, to develop flexibility with respect to alternative points of view, and to develop a broader, school-wide perspective (Alvy, 1984). Daresh (1996) summarized a variety of these problems and concerns and concluded they could be classified into three types: issues of role clarification (i.e., understanding who they were as principals and how they were to make use of their new authority), limitations with technical expertise (i.e., how to do the things they were supposed to do according to formal job descriptions), and difficulties with socialization to the profession (learning how to be successful on a day-to-day basis) and to the school system (learning how to do things in a particular organizational setting).

**Socialization and the career path to the principalship**

Most previous work on beginning school administrators focused primarily on describing problems they face in learning the role. Less is known about the interplay between their professional and organizational socialization and how this socialization may affect the evaluation of their performance and their subsequent career paths (Hart & Wending, 1996; Milos, 1988; Parkay et al., 1992). Early studies on socialization suggested that school principals learned most of their role behavior in the context of the school, through which they begin to develop a 'world view' of the principal's role (Greenfield, 1977; Wolcott, 1973). Often, their initial beliefs about what schools should be like were re-shaped by their actual experiences within their work setting (Marshall, 1992).

In contrast, Leithwood and colleagues (1992) found that both professional socialization (e.g., formal training) and organizational socialization (e.g., relationships with superordinates and peers) patterns were helpful in contributing to administrators' abilities to provide instructional leadership within their schools. Similarly, Kraus (1996) found several aspects of initial preparation programs were related to individuals' perceptions of their successful mobility from trainee to school principal. These features were 'cohort' preparation program designs (i.e., where students completed core courses, attended reflection seminars, and over time developed a sense of 'family'), an internship experience, long-term relationships with mentors (which helps in learning the role and securing sponsorship), and an emphasis on reflective practice.

Parkay and colleagues (1992) provided perhaps the most extensive in-depth view of beginning principals from a socialization perspective. They examined high school principals' developing leadership competency during the first three years of their initial appointment. Parkay and colleagues found that professional socialization to role of the principal (i.e., learning the principal's world view) unfolds in several stages. Importantly, principals entered the position at different levels of professional socialization. More specifically, the stages ranged from learning day-to-day survival (e.g., frustration, powerlessness, personal inadequacy), gaining control and stability in the role, to eventually becoming more expert (i.e., exercising educational leadership and becoming professionally actualized). Principals operating at higher levels of professional socialization had internalized what it means to be a principal (e.g., developing a code of ethics, an administrative philosophy, and a cooperative vision for the school) and, therefore, were likely to see beyond the boundaries of their school settings. This type of expertise developed with experience, reflection, and further education.

While professional socialization is obviously important in how the individual learns to view the principal's role, organizational socialization impacts the evaluation and reward structures for compliance with social norms and expectations (Hart & Wending, 1996; Marshall, 1992). Learning respected values, attitudes, and beliefs in the school context helps new administrators gain acceptance of those in similar and superordinate leadership roles and serve as primary criteria for later success (Duke, 1987; Greenfield, 1986). Organizational socialization often instills a strong value on the maintenance of regularities within the school. Experienced members of the organization find ways to ensure that newcomers do not disrupt ongoing activities, embarrass or disparage others, or change established cultural solutions and practices worked out previously (Van Maanen & Schein, 1979).

The process of organizational socialization suggests that sponsorship and subsequent career mobility are related to the new administrator receiving successful performance evaluations at each step (e.g., trainee, assistant principal). For example, Marshall found that assistant principals develop orientations to the principal's role and career mobility in response to the range of opportunities they experience during their time in the assistant principal's position. As supervisors of new assistant principals, principals have major control over the organizational socialization and subsequent promotion process, providing resources for training experiences in the school as well as access to information and opportunities for visibility (Marshall, 1992; Milos, 1988).

## Overview of the Study

In the remainder of the chapter, results from a longitudinal study are presented that examine aspects of professional and organizational socialization on the performance of new assistant principals and their eventual transition to their first school principalship. The first part concerns the impact of professional and organizational socialization on the leadership performance of probationary assistant principals. The second part examines how professional and organizational socialization affected the evolving professional practice and career paths of a subset of the assistant principals who attained their first principalship within five years of their initial training.

### Examining the effects of socialization on administrative performance

While a number of studies have described the socialization process, trying to determine how professional and organizational socialization impact the evaluation of the performance of new administrators represents a needed extension of this previous work.

The intersection of performance evaluation and school administration is a relatively new domain of inquiry. There has been little consensus about how evaluation models for school administrators should be constructed and what variables should be included in evaluating their effectiveness (e.g., Glasman & Heck, 1992; Hart; 1992). Part of the problem is the absence of theoretical work linking important aspects of socialization and the responsibilities and expectations associated with the role to effective performance. The quality of evaluation systems and the corresponding assessments of principals' performance at various points in their careers have, therefore, been suspect.

There are a number of different ways that have been proposed for the evaluation of school administrators (e.g., see Glasman & Heck, 1992 for further discussion). For example, outcomes-based evaluation focuses on the results achieved (e.g., improving the school over time, meeting a set of achievement standards). Yet, this method of evaluation is largely untested for school administrators. In contrast, role-based evaluation of school administrators is anchored in formally articulated responsibilities and tasks (e.g., job descriptions) as well as in contextual expectations (Hart, 1992). The tasks imply responsibilities that go with the role of school administrator. The expectations imply meeting the needs (e.g., vision, goals, values) of the community, staff, and students in each unique school context.

To sum up many of these activities, researchers have concluded that schools are successfully managed today primarily through the human and social dimensions (e.g., Hallinger & Heck, 1999; Leithwood, 1994). The key evaluation criterion becomes the extent to which the administrator meets the school community's expectations. The tasks can include analyzing the needs of the school and community, organizing personnel and resources, and implementing strategies that correspond to chosen goals (Glasman & Heck, in press).

**Specifying the model**

Of course, no single evaluation model is likely to capture the entire richness of the school administrator's role. Any of six basic school leadership conceptualizations identified by Leithwood and Duke (1999) could constitute the focus of role-based evaluation, but each might have slightly different purposes and criteria for judging performance. For this study, a number of performance indicators were chosen which are representative of administrators' responsibilities. These include responsibility to oversee the school's governance, its culture, and its instructional processes. In a number of previous studies using various leadership models, these broad responsibilities have been found to be associated with overall school effectiveness (Hallinger & Heck, 1996; Leithwood, 1994).

One hundred and fifty beginning assistant principals (i.e., first- to third-year) and their supervising principals (N = 150) participated in the study. This represented the total number of probationary assistant principals in the state (i.e., consisting of approximately 250 public schools with 186,000 students). Principals rated the quality of their assistant principal's performance of 21 tasks comprising:

- *school governance* (e.g., understands and takes action on school problems, includes staff in decision making, responds to parent concerns and requests, involves parents in the school, establishes cooperative relationships with other organizations and agencies in the community),

- *school culture* (e.g., assists in communicating the school's vision and purpose to staff, develops effective two-way communication, uses mediation skills to resolve conflicts, provides support to staff), and
- *school instruction* (e.g., helps coordinate the school's curriculum improvement efforts, helps develop goals for school performance, works with teachers to improve classroom management, works with teachers to provide staff development, makes classroom visitations and provides strategies to improve instruction).

Assistant principals also responded to a questionnaire designed to gather information about their professional and organizational socialization. These data included:
- the length and type of fieldwork they experienced during initial preparation,
- the quality of their university course work, and
- the contextual conditions in the school where they completed their probationary administrative assignment (see Heck, 1995, for further description of the variables).

From this instrument, *professional socialization* was defined as the experiences assistant principals had during their initial preparation including:
- the type of preparation program they were in (i.e., teachers pulled out of the classroom and given on-the-job training only as opposed to a cohort-type program consisting of formal university course work, reflective seminars conducted by practitioners, year-long internship with a mentor principal),
- the extent to which they had opportunities to develop leadership skills during their preparation (e.g., working with staff on school improvement, acting as a facilitator with groups, shadowing the principal, receiving mentoring), and
- the extent to which they learned operational skills (e.g., handling discipline, personnel issues, budget preparation).

Organizational socialization was defined as the experiences that took place during the probationary vice principal period within the context of the work setting. Variables comprising this construct included the extent to which individuals developed a support network (i.e., consisting administrators, staff, and peers), worked in a community that supported the school through involvement, had a supportive relationship with the principal, was satisfied with the job as vice principal, worked in a school with cohesive social relations and safe environment (climate), and received an adequate orientation to the position during the first few weeks on the job.

**Summarizing the findings**

The goals of the analysis were to estimate the relative strength of the variables in the proposed theoretical model in explaining administrative performance and to assess how much variance in performance could be accounted for by the socialization constructs and individual demographic controls, as opposed to sources outside the model (i.e., random error, supervisor errors of measurements, differences in individual skills, and other variables not included). The model as tested is summarized in Figure 1. The estimates represent the simultaneous contribution of the demographic variables and socialization constructs in explaining performance.

Organizational socialization exerted the strongest direct (and total) effect on administrative performance (.55). In contrast, professional socialization was only weakly related to performance directly (.10). Because professional socialization was directly related to organizational socialization (.38), however, a substantial indirect effect of professional socialization on performance (through organizational socialization) was also observed (.21, not shown). Therefore, the combined direct and indirect effects of professional socialization on administrative performance were moderate (.31, not shown in Figure 1).

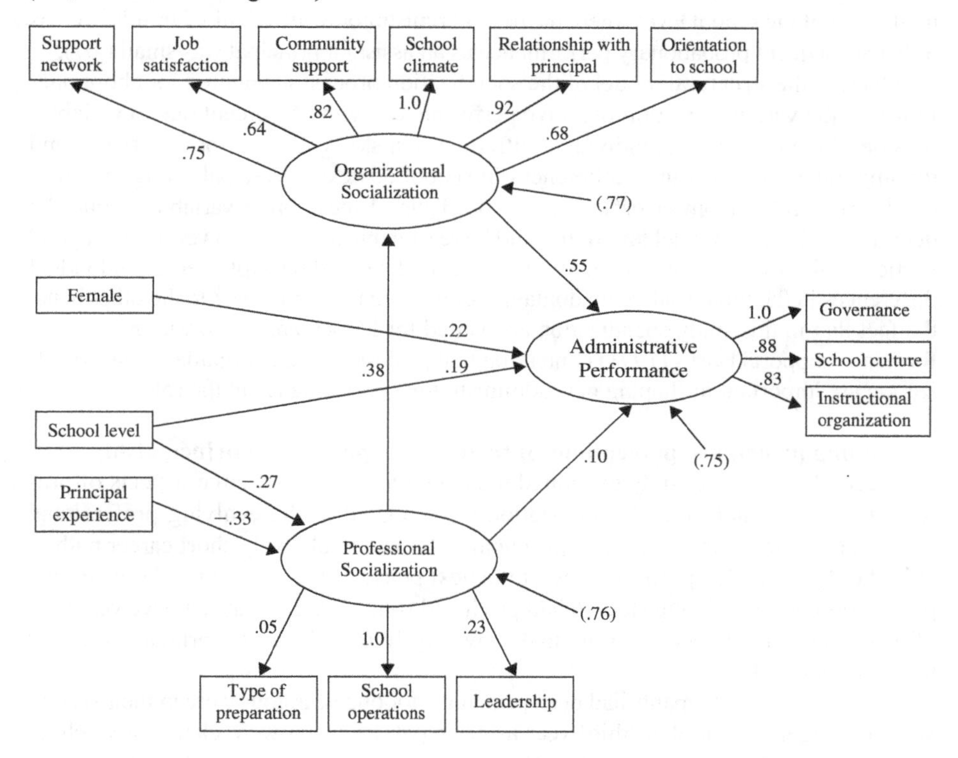


Figure 1. Standardized LISREL parameter estimates.

Regarding background variables, female assistant principals were rated somewhat higher in performance than males (.22). Further inspection suggested that part of this finding may be related to the increased time that women in the study spent in the classroom before becoming administrators, compared with men (e.g., see Ortiz, 1982 for further discussion). Interestingly, the years of experience of the supervising (mentor) principal during the respondents' initial training was negatively related to their perceptions about the quality of their initial professional socialization experiences (−.33). More specifically, less-experienced principals (who likely view the preparation needs of new administrators somewhat differently from their more experienced colleagues) were perceived as providing more extensive opportunities to engage in a variety of day-to-day operations and to learn leadership skills during their initial preparation.

While assistant principals initially prepared in secondary schools received higher performance ratings (.19), the quality of their initial professional socialization experiences tended to neutralize this direct effect. To explain this in more detail, the direct effect favoring secondary-trained assistant principals (.19) was negated by an indirect effect (−.12, not shown) concerning the quality of their initial preparation experiences. Follow-up analyses suggested this was largely because those trained in the secondary schools concentrated more on learning operational skills (e.g., discipline) rather than developing leadership skills (e.g., group facilitative skills) during their initial preparation. Thus, the total effect of the school level where the new administrator received initial training on his or her subsequent probationary performance as an assistant principal was small (.07).

Overall, the structural model of the socialization process accounted for about one-fourth of the variance in administrative performance, with 75 percent due to variables outside of the model (e.g., individual differences in skills, other prior experience and training, other non-measured differences in school contexts, principal rating measurement error, and random errors). The actual effects of these other variables cannot be determined from the model tested. It could be reasonably assumed, however, that a good portion of the variance unaccounted for by the model could be explained by individual differences in the three leadership domains comprising performance. On the other hand, the fact that in this study socialization accounted for 25 percent of the variance in performance supports Hart's (1991) contention that professional and organizational socialization are important in shaping new administrators' performance in the role.

**Examining principals' perceptions of their career path to the principalship**

The second part of the study examined the extent to which particular aspects of professional and organizational socialization were related to the evolving professional practice of a subgroup of assistant principals who had a relatively short career path to the school principalship. At the time of the next phase of the study, the administrative preparation program being studied had produced some 180 graduates in five years, 36 of which had already become principals.[1] Twenty-four of these new principals agreed to be interviewed.

Twelve of the participants had been principals for one year, four were in their second year, and eight were in their third year as principal. Eighteen were elementary school principals and six were secondary school principals. Eight were male and 16 were female. Previous to this appointment, all had been vice principals in the same educational system. Fourteen had been vice principals under three years, while 10 had been vice principals for three or four years. Eleven had been elementary school vice principals and 13 had been secondary school vice principals.

The administrative preparation program was designed in a cohort-type format with integrated university course work, reflective seminars conducted by principals and university faculty jointly, and a yearlong paid internship in a school. The intent of the analysis was to determine whether key preparation program goals would be present in new principals' discussions about their initial preparation, their probationary period as vice principals, and their readiness to become principals.

[1]Principals in Hawaii must serve a minimum probationary period of two years as assistant principals before they are eligible to become principals. At the time of the study, therefore, the most experienced program graduate was in her third year as a principal.

Important preparation program goals included:

- developing leadership skills aimed at establishing a shared school mission,
- using collaboration in school decision-making,
- implementing school improvement that leads to increased student achievement, and
- developing a professional support network through fostering relationships with other beginning administrators and mentor principals.

The study was constructivist in nature, in that it examined how the new principals made sense of their professional and organization socialization processes. Data analysis occurred in three parts. First, interviews were transcribed and patterns and experiences that were unique to each individual were identified around the three phases (i.e., preparation, vice principalship, principalship) of the study. Second, emergent themes were identified in each of the three phases that resulted from the patterns observed in the data. Finally, the data were combined across the separate phases of the study. Three themes emerged that appeared salient to a successful, relatively fast, career path to the principalship. These themes are discussed and illustrated (i.e., with a few representative comments due to space limitations) in the following section.

**Clear understanding of the principal's role**

First, even though they were relatively new, the principals had developed a clear understanding of their role and responsibilities, in contrast to many first-year principals identified in previous studies who struggle with role clarification (Daresh, 1996). As Hart and Wending (1996) argue, when new school administrators begin to take charge of the school, role innovation takes place, where missions and goals may be redefined and the individual may reject some of the norms governing past practice at the school. Many of the new principals reported that the principal's role was to be the 'overall leader' and 'facilitator' at the school level. Principals commented on their commitments to collaboration and empowerment of others within the school setting. As one beginning principal stated about her first year,

> I think one of the greatest challenges that we've had this year dealt with the attitudes I've mentioned about parents and the community, and the role of the school. … The challenge is being able to work with those groups, so it's not only working with your staff and students in your school … but how you deal with parents who have their own agendas. And with SCBM (shared decision making) now, it has become – we talk about it a lot as principals – because it's beyond operational. You know, how do we set the parameters for SCBM within our schools? So it's the idea of dealing with all kinds of people and their personal agendas.

Efforts for some new principals focused on developing a common understanding of the school's mission and vision and implementing school improvement processes, as well as combining the knowledge and strengths of different role groups within the school. As one beginning principal summarized her initial change efforts:

> The first year was rough for me in the sense of staff understanding that we were going to shake things up a bit, and I always believe as an administrator that we

> have to do things right away. If I just laid back and allowed the school to function … you know, as it was left by the other principal, I would have a harder time trying to initiate change, so I initiated change slowly, but quickly … you know, in the very beginning, we are going to make change.

Another principal also commented on facilitating change:

> When you get thrown into this position, it's what you make it, you know, and the boundaries are only what you limit yourself to. So somebody asked me, 'You've been here only two and a half years, and you did all this stuff? When did you start all this?' The first year we came on board (new principal and vice principal), you set the vision and make sense of the curriculum, and you start working.

As a group, the new principals suggested that planned professional socialization experiences during their formal preparation helped them understand the responsibilities of the principal's role clearly and led them to seek the principalship relatively quickly. More specifically, 20 of the respondents mentioned that the preparation program's yearlong internship helped them to understand the principal's role. As one principal summarized this initial preparation:

> I know their [the preparation program's] purpose was to focus on the role of a principal, but it did a great job for both the role of a VP and a principal. The school improvement process I find I use almost on a daily basis … working with people, understanding human nature. The Cohort (preparation) program focuses on the process and the contextual factors that play into what happens in the school improvement process.

Another suggested that

> [During training] my mentor spent a lot of time with me, asked critical questions of me. That internship dialogue with my mentor helped develop the shaping of my role. I was afraid of her at first, but we had an open talk and things worked out. She's still my mentor today. I admire her. She instilled the professional ethics, how to work, support each other, give personal time to each other. … I honestly remember changing. It was the beginning of learning to deal with others. Before you were king or queen of your classroom. I learned what was expected of administrators.

From new principals' perspectives, their initial professional socialization was supported by a broadening set of experiences gained in the vice principalship that helped them prepare for the principalship. Their comments highlight the effects of the organizational socialization that take place over time within the system and extend beyond initial formal preparation. As one principal noted, 'I was like a sponge, trying to absorb everything. My vice principalship was built on strengths such as grant writing, vocational education, and special education, and I was given the opportunity to be in the role of head administrator.'

Another principal discussing his transition from trainee to vice principal suggested,

> 'I had to implement school improvement efforts and operations and show the school that I had grown from intern to VP. I had multiple experiences at high-powered schools focused on school improvement planning and school-wide change, and so I learned solid skills.' A third new principal identified previous experiences with increased responsibility, 'I was twice acting principal of a school and felt ready for the responsibility.'

**Support networks**

A second theme was that these new principals had established and frequently made use of a strong support network. This helped in resolving school problems and served to orient them further to the responsibilities of the role. A number of new principals also maintained friendships with the other students in their preparation program. As one suggested, 'We get together once a quarter. Whenever we get bummed out or feel like killing someone, we call each other.' Another new principal commented about support networks:

> I don't think there's a particular principal I'd consider a mentor. I look at it as a network of administrators. We basically call each other when we need help and direction, opinions, and ideas. [Administration today] is a lot of sharing, a lot of teaming and collaboration, which we learned in Cohort [preparation program]. That was one of the main things I think I learned in the program. But it's different because not too many administrators have experienced this new way of administration, but I think it has changed how we should do things, and I think we all are trying to adapt to the changing ways of running the school.

A high percentage of new principals also commented on the close relationship that developed with their principal during the few years while they were a vice principal – for example, 'My principal was a second mother to me' or 'The principal pushed me to higher levels.' The relationship was viewed as providing the variety of activities that prepared them for the principalship. This was important in having a widening set of experiences beyond student discipline. All but one new principal also mentioned developing positive working relationships with members of their school community, suggesting the importance for individuals to possess interpersonal skills.

Maintaining support relationships with superiors also led to sponsorship that new principals suggested was important when potential principal positions became available. For example, three of the principals were encouraged to apply for their initial principalship by district supporters, and three were encouraged to apply by their supervising principal. For others in the group, support came from previous training mentors. A number of the new principals also mentioned that they lived in parts of the state where principals were needed (e.g., due to turnover from retirements and principals seeking more desirable positions through seniority).

**Evolving professional practice**

A third theme that emerged across the set of interviews was the new principals' sense of evolving professional practice in gaining a more complete understanding of their school and career situations. This suggests that professional socialization may unfold as a series of stages from survival to greater expertise as the new principal gains experience (Parkay et al., 1992). Although a majority of the new principals interviewed indicated that they experienced first-year difficulties in running their schools, many found ways to cope with their situations. One expressed initial difficulty in coping with the new demands of the principal's role:

> I was going to quit. It was November. I was asking what did I get myself into? Those first couple of months no one helps you. You're by yourself, you're on your own. And you have to make your own decision based on what you already know or what you can find out … and those decisions sometimes may be right or wrong, but you have to make them.

Another reflected on making the jump from training and the assistant principal's position to the principal's role:

> It wasn't that tough to do what I was doing up until becoming a principal. And then, whoa, this job is hard, man. It's a tough job. There is no Cohort program, no vice principalship that prepares one for the principalship. I had no idea it was going to be that hard. You know, when the buck stops with you, it's like you don't realize how many things get thrown at the principal. Even as a VP, you don't. You hear about it, you might participate in the discussion, but when you're the one that has to, you know, on a daily basis, be the final arbitrator, I didn't realize how hard it was going to be.

A third comment suggests the difficulties some new principals have in gaining more control over the nature of the principal's work:

> You think [to yourself], I can come in and I think I know, but you're scared because you're not sure what kind of a leader you're going to be. Even the parameters of leadership, you don't know about. You know, it's really funny that we get this idea that you're going … to be a curriculum leader, instructional leader, all this kind of stuff, but you end up being a fireman. So for the first two years, I felt like a fireman, because there were so many things going on. Everyday some fire hits you from a different direction.

In contrast, another comment speaks to this new principal's perception of her growth during her short time in the role:

> It's funny how you look at the things you don't do, or you haven't accomplished, rather than what you have done. I've done a lot of reflection. I think that the single biggest way I describe my view is of myself evolving. I'm certainly not where I was a year ago, and I know I'm not where I want to be … and I think it's just going to continue. When you evolve, it's two steps forward and one step back, and I think it's just going to continue like that.

As these glimpses suggest, the socialization framework is a useful lens for understanding new administrators' evolving professional practice and, to some extent, their subsequent career paths. Almost all of the new principals appeared to focus on the overlapping commonalities among their growing understanding of the principal's role from their planned induction and organizational experience, their use of organizational support networks, and their developing awareness of their own evolving professional practice. The new principals expressed this as being able to see the 'big picture' and being able to view situations comprehensively, as opposed to focusing on individual factors. These developmental experiences led them (and their superiors) to feel they were ready to take on the responsibilities of the principal's role. This readiness was expressed in comments such as 'I was ready for the principalship because of background – having lots of leadership positions,' 'in my heart I was ready,' and 'I was twice acting principal of a school and felt ready for the responsibility.'

## Implications and Recommendations

Recent calls for changes in the preparation of school administrators have led to substantial changes about how programs are conceived and delivered. Despite these changes, the overall impact to date on the majority of educational leadership programs has been modest (McCarthy, 1999). As a result, evidence about whether changes in preparation actually lead to improved practice and changes in school outcomes has been slow to accumulate. The purpose of this chapter was to develop a theoretical framework for investigating how features of redesigned preparation programs may affect the practice of school administration. As researchers have suggested, few school systems have presently given adequate attention to the socialization and evaluation of their new administrators (Duke, 1987; Glasman, 1990; Parkay et al., 1992).

The data presented in the chapter add to a growing number of studies that have examined how professional preparation and organizational context affect what administrators do (e.g., Greenfield, 1977; Hart, 1991; Leithwood et al., 1992). Perhaps more importantly, these studies begin to assess how well they do those activities. This research responds to long-standing calls for more empirical studies of principal preparation and development. The field has been long on rhetoric and short on empirical data (e.g., Hallinger, 1992; Leithwood & Steinbach, 1992).

The first part of the study confirmed a model of how organizational and professional socialization affect administrative performance. The model estimates provided empirical support for the view that organizational socialization (i.e., those conditions in the organization in which individuals work) may be more important in shaping new administrators' performance as assistant principals than their initial professional socialization. More specifically, the combined (or total) effect of organizational socialization on performance was somewhat stronger than the total effect of professional socialization. Importantly, however, the quality of professional socialization experiences received during training also significantly influenced performance observed later in time, although this influence was almost entirely indirect.

The new administrator's organizational context mediated the effect of professional socialization on the outcomes studied. The second part of the study corroborated these findings by describing the relevance of professional socialization (e.g., internalizing initial preparation program goals such as collaborative leadership, focusing on school improvement, developing a support network, learning to reflect on professional practice) and organizational socialization (e.g., having increasing responsibilities as a vice principal, securing organizational sponsorship) to new principals' relatively fast mobility to their first principalship.

In truth, new administrators inherit a variety of differing school contexts and needs. Some schools contexts are more difficult settings in which to work, and this has a measurable influence on how well the new administrator fulfills role-related responsibilities and expectations. A focus on socialization implies more context-grounded approaches to administrator preparation (Hart & Wending, 1996). Possible responses to the unique features of the school that the new administrator can choose range from a custodial path (i.e., maintaining existing practices and values) to a more innovative path – perhaps reshaping the school's mission and goals, along with its norms, values, and operational practices (Hart & Wending, 1996).

The school administrator's role is best learned in the field by doing and under the guidance of experienced, exemplary mentor principals. These experiences can be integrated with intentional, structured activities (e.g., seminars, readings) associated with formal preparation (e.g., Barth, 1997). Much remains to be discovered, however, about the interaction of formal preparation, organizational socialization, and effective administrative practice. Such knowledge may inform policy about administrative preparation, evaluation, and support.

Several implications and recommendations follow from the previous discussion of socialization and administrative practice. First, variation in administrators' performance early in their career is, at least in part, related to their socialization process, especially because much of this socialization takes place within the context of the school. Attention should be directed toward identifying those experiences and skills that are related to success within a diversity of school conditions. As Schon (1983) argues, the essence of professional work is the ability to draw on a complex body of knowledge to assess a unique situation and apply that knowledge to take the action that is warranted by the set of facts and desired outcomes. This can include getting others' to participate in reshaping the school's goals, educational strategies, and assessment practices.

Second, the results imply that defensible appraisal of administrative performance during the early years (e.g., preparation, probationary period) must take into consideration the individual's professional socialization experiences, as well as the contextual conditions in the schools in which she or he works. The school environment exerts an influence on what principals do and how they are evaluated. Evaluation for new school administrators should reflect the interplay between preparation and socialization on the one hand, and school needs and expectations on the other. This is important to consider because initial experiences as trainees and assistant principals can have an effect on new administrators' subsequent professional socialization and career paths.

Third, redesigned, integrated administrative preparation programs should maximize formal learning opportunities and in-context growth once students occupy administrative positions to allow new administrators to develop behavioral options for varying school contexts and needs. Many of the school processes needed to improve outcomes require collaboration, community building, data for decision making, recognition of nonproductive patterns of behavior, and a focus on change. Carefully designing preparation programs that maximize formal learning and in-context growth over time, however, will require a longer-term view of the development of school leaders extending from initial induction through the assistant principalship and principalship.

## References

Alvy, (1984). *The problems of new principals*. Unpublished doctoral dissertation. Missoula: University of Montana.

American Association of School Administrators (1960). *Professional administrators for America's schools*. Thirty-eighth yearbook, American Association of School Administrators, National Education Association, Washington, DC.

Barth, R. (1997). *The principal learner: A work in progress*. The International Network of Principals' Centers, Harvard Graduate School of Education, Cambridge, MA.

Bridges, E. (1965). Bureaucratic role and socialization: The influence of experience on the elementary school principal. *Educational Administration Quarterly*, 1(2), 19–28.

Bridges, E. (1977). The nature of leadership. In L. Cunningham, W. Hack, and R. Nystrand (Eds.), *Educational administration: The developing decades* (pp. 202–230). Berkeley, CA: McCutchan.

Bridges, E. & Hallinger, P. (1995). *Implementing problem-based leadership development*. Eugene, OR: ERIC Clearinghouse for Educational Management.

Cambron-McCabe, N. (1993). Leadership for democratic authority. In J. Murphy (Ed.), *Preparing tomorrow's school leaders: Alternative designs* (pp. 157–176). University Park, PA: University Council for Educational Administration.

Cooper, B. & Boyd, W. (1988). The evolution of training for school administrators. In D. Griffiths, R. Stout & P. Forsyth (Eds.), *Leaders for America's schools* (pp. 251–272). Berkeley, McCutchan.

Crews, A. & Weakley, S. (1995). *Hungry for leadership: Educational leadership programs in the Southern Regional Education Board (SREB) states*. Atlanta: SREB.

Daresh, J. (1987). *The highest hurdles for the first year principal*. Paper presented at the annual meeting of the American Educational Research Association, Washington, DC, April.

Daresh, J. (1996). *Lessons for educational leadership from career preparation in law, medicine, and training for the priesthood*. Paper presented at the annual meeting of the American Educational Research Association, New York, April 8–12.

Duke, D. (1987). Why principals consider quitting. *Phi Delta Kappan*, 68, 308–312.

Duke, D., Issacson, N., Sagor, R. & Schmuck, P. (1984). *Transition to leadership: An investigation of the first year of the principalship*. Portland, OR: Lewis and Clark College, Educational Administration Program.

Duke & Iwanicki, (1992). Principal assessment and the notion of 'fit.' *Peabody Journal of Education*, 68(1), 25–36.

Educational Research Service. (1998). *Is there a shortage of qualified candidates for openings in the principalship? An exploratory study*. Arlington, VA: Author.

Gaertner, K. (1978–79). The structure of careers in public school administration. *Administrator's Notebook*, 27(6), 1–4.

Glasman, N. (1990). *Skills in educational administration*. Paper presented at the annual meeting of the University Council for Educational Administration, Arizona, October.

Glasman, N. & Heck, R. (1992). The changing leadership role of the principal: Implications for assessment. *Peabody Journal of Education*, 68(1), 5–24.

Glasman, N. & Heck, R. (in press). Principal evaluation in the United States. In D. Stufflebeam & T. Kellaghan (Eds.), *International handbook of evaluation research in education*. Norwell, MA: Kluwer Academic Publishers.

Greenfield, W. (1977). Administrative candidacy: A process of new role learning. Parts 1–2. *Journal of Educational Administration*, 15, 30–48, 170–193.

Greenfield, W. (1986). *Moral imagination, interpersonal competence, and the work of school administrators*. Paper presented at the annual meeting of the American Educational Research Association, San Francisco.

Guy, M. E. (1985). *Professionals in organizations: Debunking the myth*. New York: Praeger.

Halpin, A. (1970). Administrative theory: The fumbled touch. In A. Kroll (Ed.), *Issues in American Education* (pp. 100–125). NY: Oxford University Press.

Hallinger, P. (1992). School leadership development: An introduction. *Education and Urban Society*, 24(3), 300–315.

Hallinger, P. & Heck, R. (1996). Reassessing the principal's role in school effectiveness. A review of empirical research, 1980–1995. *Educational Administration Quarterly*, 32(1), 5–44.

Hallinger, P. & Heck, R. (1999). Can leadership enhance school effectiveness? In T. Bush et al. (Eds.). *Educational management: Redefining theory, policy, and practice* (pp. 178–190). London: Paul Chapman Publishing.

Hart, A. (1991). Leader succession and socialization: A synthesis. *Review of Educational Research*, 61(4), 451–474.

Hart, A. (1992). The social and organizational influence of principals: Evaluating principals in context. *Peabody Journal of Education*, 68(1), 37–57.

Hart, A. & Wending, D. (1996). Developing successful school leaders. In K. Leithwood, J. Chapman, D. Corson, P. Hallinger, & A. Hart (Eds.). *International Handbook of Educational Leadership and Administration* (pp. 309–336). Boston: Kluwer Academic Publishers.

Heck, R. (1995). Organizational and professional socialization. Its impact on the performance of new administrators. *Urban Review*, 27(1), 31–59.

Khleif, B. (1975). Professionalization of school superintendents: A sociocultural study of an elite program. *Human Organization*, 34(3), 301–308.

Kraus, C. (1996). *Administrative training: What really prepares administrators for the job*? Paper presented at the annual meeting of the American Educational Research Association, New York, April 8–12.

Leithwood, K. (1994). Leadership for school restructuring. *Educational Administration Quarterly*, 30(4), 498–518.

Leithwood, K. & Duke, D. (1999). A century's quest to understand school leadership. In J. Murphy and K. Seashore-Louis (Eds.), *Handbook for Research on Educational Administration 2nd Edition* (pp. 45–72). San Francisco: Jossey-Bass.

Leithwood, K., Steinbach, R. & Begley, P. (1992). The nature and contribution of socialization experiences to becoming a principal in Canada. In G. Hall & F. Parkay (Eds.), *Becoming a principal: The challenges of beginning leadership*. New York: Allan & Bacon.

Leithwood, K. & Steinbach, R. (1992). Improving the problem-solving expertise of school administrators. *Education and Urban Society*, 24(3), 317–345.

McCarthy, M. (1999). The evolution of educational leadership preparation programs. In Murphy & Seashore-Louis (Eds.), *Handbook of Research on Educational Administration 2nd Edition* (pp. 119–139). San Francisco: Jossey-Bass.

Marshall, C. (1981). Organizational policy and women's socialization in administration, *Urban Education*, 16, 205–231.

Marshall, C. (1992). *The assistant principal: Leadership choices and challenges*. Newbury Park, CA: Corwin Press.

Marshall, C. & Greenfield, D. (1985). The socialization of the assistant principal: Implications for school leadership. *Education and Urban Society*, 18, 3–6.

Marshall, C., Mitchell, B., Gross, R. & Scott, D. (1992). The assistant principal: A career position or a stepping stone to the principalship? *NASSP Bulletin*, 76(540), 80–88.

Milos, E. (1988). Administrator selection, career patterns, succession, and socialization. In N. Boyan (Ed.) *The Handbook of Research on Educational Administration* (pp. 53–76). New York: Longman.

Murphy, J. (1993). *Preparing tomorrow's leaders: Alternative designs*. University Park, PA: University Council for Educational Administration.

Ogawa, R. & Pounder, D. (1993). Structured improvisation: The University of Utah's Ed. D. program in educational administration. In J. Murphy (Ed.), *Preparing tomorrow's leaders: Alternative designs* (pp. 85–108). University Park, PA: University Council for Educational Administration.

Ortiz, F. (1978). Midcareer socialization of educational administrators. *Review of Educational Research*, 48, 121–132.

Ortiz, F. (1982). *Career patterns in education: Women, men, and minorities in public school administrator*. New York: Praeger.

Ortiz, F. & Marshall, C. (1988). Women in educational administration. In N. Boyan (Ed.), *The Handbook of Research on Educational Administration* (pp. 123–142). New York: Longman.

Parkay, F., Currie, G. & Rhodes, J. (1992). Professional socialization: A longitudinal study of first-time high school principals. *Educational Administration Quarterly*, 28(2), 43–75.

Parkay, F. & Hall, G. (1992). *Becoming a principal: The challenges of beginning leadership*. NY: Allyn & Bacon.

Pounder, D. & Young, P. (1996). Recruitment and selection of educational administrators: Priorities of today's schools. In K. Leithwood, J. Chapman, D. Corson, P. Hallinger & A. Hart (Eds.). *International Handbook of Educational Leadership and Administration* (pp. 279–308). Boston: Kluwer Academic Publishers.

Schon, D. (1983). *The reflective practitioner: How professionals think in action*. San Francisco: Jossey-Bass.

Silver, P. (1987). The center for advancing principalship excellence (APEX): An approach to professionalizing educational administration. In J. Murphy & P. Hallinger (Eds.), *Approaches to administrative training in education* (pp. 48–67). Albany, NY: SUNY Press.

Valverde, L. (1980). Promotion socialization: The informal process in large urban districts and its adverse effects on non-whites and women. *Journal of Educational Equity and Leadership*, 1, 36–46.

Van Maanen, J. & Schein, E. (1979). Toward a theory of organizational socialization. *Research in Organizational Behavior*, 1, 209–264.

Wolcott, H. (1973). *The man in the principal's office: An ethnography*. New York: Holt, Rinehart & Winston.

# 16

# Toward a Second Generation of School Leadership Standards

Dr. Kenneth Leithwood

*Professor and Associate Dean, Faculty of Education, University of Toronto/O.I.S.E., Toronto, Canada*

Rosanne Steinbach

*Research Associate, Centre for Leadership Development, Faculty of Education, University of Toronto/O.I.S.E., Toronto, Canada*

Standards have become a pervasive part of the educational landscape – standards for students, teachers, parents, and almost everyone else with a stake in schooling except those policy makers so enamored of them (for others). While school leaders have no cause to feel neglected by the standards movement, they don't have much to cheer about, either. Whatever other reasons may exist for sobriety on the matter, a growing awareness of just how indefensible is this first generation of standards must be viewed as central. On many different grounds, these standards should be considered on 'life support'. If something is not done soon, the plug will be pulled and the standards will vanish along with the purposes for which they were designed.

Now this sounds like exceptionally harsh treatment of the prodigious efforts expended by many talented people to develop first generation school leadership standards. So it should be clear that those efforts have managed to capture considerable ground; much has been learned through their development, and especially their implementation. For the most part, first generation leadership standards should be considered an impressive and, we expect, necessary step in the direction of something more defensible. The primary direction of our argument in this chapter is that we need to continue the work pioneered by the first generation standards developers, given the considerable opportunities we have had to learn from our experiences with what they produced.

While this chapter aims to set the groundwork for developing a more defensible second generation of leadership standards, it does not propose such standards. Rather, it briefly summarizes the nature of five prominent sets of first generation leadership standards and then argues for the adoption of seven standards that a second generation of standards should meet if they are to be a significant improvement.

Moving toward a more defensible set of second-generation standards is not the only direction that could be proposed in the face of the first generation's inadequacies. Perhaps the most obvious alternative would be to do away with leadership standards altogether. We do not argue for that alternative, however, for two reasons. First, a defensible set of standards really does have the potential to help move the practice of school leadership to a higher level of professionalism in much the same way that it has aided other professional groups. Second, arguing for no standards would entail swimming against the current, extremely powerful, tide of accountability – oriented school reform, an effort we believe would be turn out to be quite futile.

## First-Generation Standards for School Leaders

Five influential sets of school leadership standards are examined in this section to illustrate some of the key features of the first generation. Two of these sets are from the US, one is from the UK, one is from New Zealand and one is from Australia.

### United states: Interstate School Leaders Licensure Consortium (ISLLC)

After two years of work, 'model standards for school leaders' were adopted by the Council of Chief State School Officers on November 2, 1996. These standards were 'forged from research on productive educational leadership and the wisdom of colleagues' by 'people from 24 state education agencies and representatives from various professional associations'. They had two purposes: 'to stimulate vigorous thought and dialogue about quality educational leadership' and to 'provide raw material that will help stakeholders… enhance the quality of educational leadership throughout the nation's schools' (p. 3).

There are six ISLLC standards, each specified in more detail as component knowledge, dispositions, and performances. Each standard begins with the phrase: 'A school administrator is an educational leader who promotes the success of all students by. …' and each standard addresses a different aspect of leadership: shared vision; school culture and instructional programs conducive to student learning and staff professional growth; safe and efficient management of the organization; collaboration with families and communities; acting with integrity, fairness, and in an ethical manner; and understanding and responding to and influencing the larger political, social, economic, legal, and cultural context.

### Queensland Australia: Standards framework for leaders

The Standards Framework for Leaders was developed as the basis for professional development and training, recruitment and selection of leaders and the credentialing of Education Queensland leaders. It 'outlines the competencies required … to achieve the shared vision of "Excellence in Education" '. These competencies, considered broad enough to allow 'leaders to reflect on their performances and achieve desired outcomes in individual ways' (p. 2), are intended to guide both leadership evaluation and professional development. The framework was developed by Education Queensland with the claim that it is based on 'current theoretical knowledge' (p. 5). Depending on the leader's work context, there are three credential levels.

The standards framework consists of six key roles: leadership in education; management; people and partnerships; change; outcomes; and accountability. For each role, both 'best practices' and 'personal performance' competencies are specified. Best practice competencies, 'the knowledge, skills and behaviors of the leader as exemplified by collective site-based actions of the personnel at the work site' (p. 6), were identified from the workshop interaction of 70 educational leaders. In addition, seven hundred leaders analyzed and refined these data and reacted on three occasions to draft material.

Each best practice competency is made up of:

- collective site-based actions (i.e. performances/behaviors of individuals or groups);
- underlying knowledge and understanding;
- context 'indicators' (examples of behaviors relevant in particular contexts);
- evidence to help demonstrate competency (e.g. plans, minutes, etc).

Each key role is comprised of from three to six best practice competencies.

**United Kingdom: National standards for headteachers**

The UK standards for headteachers were developed by the Teacher Training Agency (TTA) for the explicit purpose of improving student achievement by improving the quality of leadership.

Emphasizing national priorities, the standards are intended to 'provide the basis for a more structured approach to appraisal' and professional development for aspiring and serving headteachers. The standards, to be regularly reviewed, were initially developed and subsequently revised in consultation with individuals in all levels of education and with agencies both inside and outside the profession.

These national standards define expertise in terms of the knowledge, understanding, and skills and attributes related to: core purpose of the headteacher; key outcomes of headship; professional knowledge and understanding; skills and attributes; and key areas of headship. Skills and attributes are, in turn, classified as leadership skills, decision-making skills, communication skills, self-management, and attributes (e.g., such 'dispositions' as self confidence or enthusiasm).

Key areas of leadership specified in the standards include: strategic direction and development of the school; teaching and learning; and leading and managing staff. Also included as key leadership areas are efficient and effective deployment of staff and resources, and accountability.

**New Zealand: Principal performance management**

The New Zealand standards, developed in close consultation with a number of principals and governing boards, are based on the premise that 'the leadership and management skills of the principal have a huge impact on whether a school is successful or not.' (p. 2). The relationship of the principal and the governing board is central to these standards since the board is responsible for principal performance. 'Clear direction and agreed priorities will ultimately lead to a stronger partnership between boards and principals and to improved learning outcomes for students.' (p. 2).

More specifically, the standards were developed to:

- 'help schools clarify the knowledge, skills and attitudes all principals are expected to demonstrate;
- improve the quality and outcome of principal performance management;

- provide a framework for identifying the professional development needs of principals;
- provide a means of linking performance management and decisions on remuneration' (p. 7).

The Professional Standards for Principals are grouped in six categories or professional dimensions including: professional leadership; strategic management; staff management; relationship management; financial and asset management; and statutory and reporting requirements. There are standards for primary school principals as well as revised standards for secondary and area school principals. Also, the standards include the suggestion that schools may want to add other standards to fit their particular context or develop specific 'indicators' or 'performance criteria' 'to help clarify what is being expected' (p. 7). Schools are urged to check their performance agreements to see if they accurately reflect the standards.

**United States: State of Connecticut standards**

Research on principal effectiveness and '... an explicit set of assumptions about the nature of future schools ... ' formed the initial basis for Connecticut's professional standards and related appraisal procedures (Leithwood & Duke, 1999, p. 303). This original formulation was then subject to widespread refinement and validation based on feedback from a large number of school leader focus groups in the state. The purpose for developing these standards was to improve the quality of school leadership by serving as a guide to the preparation, licensure, and selection of new school leaders, as well as the professional development and performance appraisal of incumbents.

The framework for these standards consists of seven components of a school design that, a synthesis of research evidence suggested, school leaders are able to influence and which, in turn, have positive effects on students. These features include: mission, vision and goals school culture, policies and procedures organization and resources, teaching faculty programs and instruction, and school-community relations. Within each component, assumptions are stated about schools of the future, and implications specified for effective school leader practices.

## Standards for a Second Generation of Leadership Standards

Based on a critical examination of the five first generation standards along with an extensive literature review, in this section we propose seven standards for evaluating the adequacy of school leadership standards. Several of these standards require very little justification, in our view, while a much more extended defense of several others is provided.

### 1. Standards should acknowledge persistent challenges to the concept and practice of leadership.

Existing standards appear to view the concept and practice of leadership as unproblematic. But nothing could be farther from the truth. On the matter of leadership as a concept, the leadership literature routinely grapples with such problems as:

- How leadership is to be defined, for example: as the exercise of interpersonal influence, the initiation and maintenance of structure (Yukl, 1994), or a unique set of relationships (e.g., Yukl, 1994; Brower, Schoorman & Tan, 2000);
- Whether, and in what ways, leadership is to be distinguished from management (e.g., Kotter, 1990; Bennis & Nanus, 1985);
- Whether leadership is best understood as a set of replicable behaviors with predictable effects, or as a process of problem solving leading to unique behaviors depending on the context (e.g., Lord & Maher, 1993; Leithwood & Steinbach, 1995);
- What are, and ought to be, the sources of leadership in an organization, for example, individuals, groups, and non-human features of the organization such as policies or culture, sometimes referred to as 'substitutes' for leadership (Ogawa & Bossert, 1995; Podsakoff, MacKenzie & Fetter, 1993).

As to the practice of leadership, an equally persistent set of challenges exists, for example:

- What models, if any, are sufficiently robust to act as guides to practice? As Leithwood and Duke (1999) point out, the educational leadership literature regularly includes analyses of instructional, transformational, strategic, moral, participative and contingent models or approaches to leadership, each with their own advocates and evidence;
- Are the contributions of leadership to the productivity of schools sufficient to warrant devoting significant resources to its development? Evidence on this matter is really quite contradictory. Recent quantitative evidence suggests that principals' leadership likely accounts for no more than 3 to 5 percent of the variation in student achievement (e.g., Hallinger & Heck, 1999; Leithwood & Jantzi, 1999). But there is a considerable body of qualitative, case study, evidence that argues that school leadership effects are actually quite large (e.g., Morris, 2000; Gronn, 2000).
- Do some leadership practices deserve to be used in most organizational contexts, as Bass (1997) has argued in relation to transformational practices, for example, or are most effective leadership practices context or culturally dependent (e.g., Hallinger, Bickman & Davis, in press; Hartog et al., 1999), perhaps even varying by individual follower (e.g., Mumford, Dansereau & Yammarino, 2000).

What might it mean for a second generation of leadership standards to acknowledge these persistent challenges? The least it should mean, in our view, is that these challenges be made explicit and a position adopted in relation to each. For example, a set of standards should note the debate about definitions of leadership and at least stipulate one for its own purposes; it might, as well, point to a model of leadership that comes closest to capturing the range of performances described by the standards. Of the five sets of standards described above, the Connecticut standards do both of these things in a quite limited way, while the remaining four do neither.

**2. Standards are claims about effective practice and should be justified with reference to the best available theory and evidence.**

Most first-generation school leadership standards fail on this score. That is, they do not systematically describe the evidence on which they are based, and of the five sets described above, all but the Connecticut standards fail to report research related to their validity or their effects. ISLLC developers, for example, simply claim that: '... we relied on the research on the linkages between educational leadership and productive schools, especially in terms of outcomes for children and youth' (1996, p. 5). Since no further mention of that research is made in the remainder of the document, readers are required to take the claim as an act of faith. Developers of the UK leadership standards claim, in part, that 'The standards reflect the considerable work undertaken on management standards by those outside the education profession' (1998, p. 3), with no further indication of what this might entail.

While simplicity, readability, and conciseness for purposes of dissemination no doubt influence a decision of this sort by ISLLC and UK standards developers, the decision then becomes an implicit standard of its own. This implicit standard says, in effect, that users and consumers of the standards should not be concerned about systematic evidence. Perhaps it also says to some readers that leadership is 'an evidence-free enterprise, that the research supporting good professional practice does not warrant the attention of you busy people, that you do not need to be critical consumers of such research'.

It is hard to imagine a more damaging message to be conveying about school leadership at this moment in time. With the immense growth in theory and research about education and leadership over the past 20 years, the time is ripe to take leadership practice to a significantly higher level of professionalism in much the same way that medical education and practice was able to leap forward after the publication of the Flexnor Report in the United States in the early 1900s (Barzansky & Gevetz, 1992). Indeed, included among the UK standards (1998) are two assertions that confirm their developers' agreement with this position:

- 'Headteachers should be able to make decisions based upon analysis, interpretation and understanding of relevant data and information' (p. 7); and
- 'Headteachers should have knowledge and understanding of the contribution that evidence from inspection and research can make to professional and school development' (p. 6).

This seems about right to us.

Failure to justify standards with systematic evidence risks both errors of commission and omission. Errors of commission arise when standards include qualities that have little to do with what is known about effective leadership. This not only wastes the scarce resources available for leadership development but also encourages leadership practices that may be dysfunctional for schools. Errors of omission are failures to include, among the standards, leadership qualities critical to effective practice in schools.

How well do the five sets of first-generation standards summarized in this chapter meet this standard? This is a complex question to answer comprehensively. Given the space restrictions of this chapter, we limit our response to errors of omission, and describe the results of a recent review of theory and research (Leithwood, Jantzi & Steinbach, in press). This review compared the five first generation leadership standards with evidence about productive school leadership responses to government accountability policies. Since this focus for school leadership is a limited one, gaps found in the standards underestimate by an unknown, but likely quite substantial, amount the actual failure of the standards to reflect current theory and evidence about effective school leadership.

While results of our review indicated that, as a whole, the five sets of first generation standards reflected a significant number of practices important for leaders in accountable contexts, there were a number of important errors of omission, as well. For example:

- Except for the Connecticut standards, there is neither a strong nor explicit focus on teacher leadership, although the Queensland standards assert that the effective school leader 'challenges successful individuals to meet higher levels of performance' (p. 44). Identified in the review of literature as an important contribution to greater accountability through shared decision making and distributed forms of leadership, this omission also seems inconsistent with the newer forms of management and school organization generally advocated by the standards.
- None of the five sets of first generation standards explicitly mention the importance of being able to balance the full range of duties our review suggested were expected of school leaders in more accountable contexts. Some of these duties are quite new for school leaders and often are the consequence of site-based management policies, along with a substantially increased workload (the New Zealand and UK standards did mention the importance of 'prioritizing').
- Attention devoted to teaching staff in the five sets of standards is largely concerned with performance appraisal or curriculum innovation. Yet it is clear from the literature review that managing teacher stress and morale is a huge challenge for school leaders in most 'high stakes' accountability contexts. The word 'morale' is never mentioned in the five sets of standards unless one assumes it to be encompassed in the term 'support'.
- The importance of endorsing new programs in order to aid implementation of accountability policies may be implied, at best, but is not explicitly mentioned in the first generation standards.
- Also missing (except for Connecticut) is explicit reference to the consequences of high-stakes testing. General practices like 'dealing with barriers to learning', 'expressing concern and respect', and 'identifying potential problems' may be interpreted as including some aspect of such consequences, but in the current policy context more explicit attention to practices for effectively managing these consequences is warranted.
- Marketing, perhaps the most noticeable new practice demanded of school leaders in contexts where schools are held accountable by having to compete for students, is absent from all five sets of standards with two minor exceptions: Queensland's standards assert that 'ideas and activities are marketed' by school leaders, and the ISLLC standards mention marketing as 'knowledge'! New Zealand emphasizes the unique characteristics of its population.
- An equally surprising omission in the two sets of U.S. standards is any reference to leaders working effectively with school councils, a dominant structural component of approaches to increasing accountability by decentralizing decision making and empowering parents.
- Finally, except for ISLLC, the outreach or entrepreneurial function demanded of leaders by most approaches to accountability that require schools to compete for students is not mentioned explicitly, although the Queensland and Connecticut

standards do mention 'accessing necessary resources'; otherwise the focus is on the effective use of already available resources.

Clearly, the first generation leadership standards that we examined are a less-than-satisfactory reflection of current theory and evidence about effective leadership practices.

**3. Standards should acknowledge those political, social and organizational features of the contexts in which leaders work that significantly influence the nature of effective leadership practices.**

The analysis described above suggests that the first generation standards referred to in this chapter are, at best, uneven in their acknowledgement of the broad political context of national and state reform shared by school leaders in many developed countries. What about recognition of the social and organizational contexts of schooling? In relation to social context, evidence indicates that the nature of the school community and student population – urban/suburban/rural, advantaged/disadvantaged – has an important bearing on what constitutes effective leadership practice (Hallinger, Bickman & Davis, 1996). In addition, study after study demonstrates that such organizational features as school size and level make an important difference to how leadership is exercised and with what effect. When context is defined in this much more focused manner, existing standards provide little or no insight about the effective practices of school leaders. As a consequence, their value is diminished considerably.

What could respond to this standard mean for second-generation leadership standards? A straightforward and quite useful response would be for such standards to speak to the different leadership practices warranted by these key elements of context. Using organizational size as an example, a standard concerned with the delivery of instructional support to staff for the improvement of teaching and learning might call on leaders of small schools to deliver such support directly themselves, whenever possible. For leaders of large schools, however, the same basic standard might well focus on providing incentives, training and opportunities for teacher leaders, department heads and the like to deliver such instructional support.

**4. Standards should specify effective leadership practices or performances only, not skill or knowledge.**

While most first-generation standards specify effective school leader practices, performance, or overt behaviors of school leaders, some also attempt to identify the knowledge and skills that school leaders should acquire in order to engage in those practices. This is an example of making claims that dramatically exceed what we can defensibly claim to know.

Under the best of circumstances, our present, research-based, knowledge identifies effective school leadership practices only. There is essentially no research-based evidence available about the knowledge that a leader must possess in order to engage in those practices. While virtually every leadership development program, save those using problem-based learning strategies, teaches specific bodies of knowledge, the choice of knowledge to teach is based on an assumed ('logical') relation between knowledge and practice.

Such an assumption is not uncommon, underlying, for example, undoubtedly the most ambitious effort in North America to date to specify the knowledge and skills required of principals. This was the National Policy Board for Educational Administration's *Principals for Our Changing Schools: Knowledge and Skill Base* (Thomson, 1993). Consisting of 21 domains of knowledge and skill, the preface to this document summarizes its methods as follows:

> The domains project began with the question 'What are the tasks and expectations and responsibilities of the elementary, middle, and high school principal today and in the near future?' The project then identified the knowledge and skills, and attributes required of principals to meet these challenges. (p. xviii)

In the case of each domain, teams of developers reviewed the literature in order to discover the knowledge and skills required to successfully carry out tasks associated with the domain.

Using the domain 'leadership' for example, three categories and 21 more specific sets of knowledge and skills are identified. The category 'shaping school culture and values' includes such specific sets of knowledge and skills as 'well developed educational platform', 'high expectations', and 'understanding school culture'.

However, within each of these sets, it is mostly leadership practices or performances that are described. This is because that is what the educational leadership literature has identified. To discover the actual knowledge or skill required by a principal to shape school culture, for example, would require research about the procedural and declarative knowledge structures used by principals expert at this task. This type of research, to the best of our knowledge, has never been conducted in the educational leadership field. Furthermore, recognizing the socially constructed nature of knowledge and the unique cognitive resources each school leader brings to their work suggests considerable variation in the knowledge needed by leaders to engage in effective practices, whatever they may be. Any given effective leadership practice could be the product of many different types of knowledge.

As for the inclusion of 'skills', most existing standards do not include them (the exception being the UK standards). This is a good thing because, for the most part, skill statements typically amount to nothing more than statements of a desired performance preceded by words such as 'must be able to', or 'is able to'. From the UK standards, for example, we read that 'headteacher should be able to use appropriate leadership styles in different situations (1998, p. 7). This is not so much a skill as a rhetorical sleight of hand.

**5. Dispositions should not be included in any standards.**

Two of the five sets of standards included in our analysis specified 'attributes' (e.g., enthusiasm, energy) or 'dispositions' (e.g., trusting people and their judgments) of effective leaders. The UK standards justify the inclusion of attributes only by asserting that '... it is the sum of [the professional knowledge, understanding, skills, and attributes] ... which defines the expertise demanded of the role ... ' (1998, p. 1).

Defined as beliefs, values, and commitments, the ISLLC dispositions are more directly linked to performance than are the UK's attributes.

The case for omitting dispositions from second-generation leadership standards is more complex than the case against knowledge and skills and will take more space to describe. We use the ISLLC standards to illustrate problems that can arise with the inclusion of such dispositions, since ISLLC developers have attempted an explicit justification. As part of the justification, ISLLC developers invoke Perkins' assertions that 'dispositions are the soul of intelligence, without which understanding and know-how do little good' (1995, p. 278). Dispositions, the ISLLC developers report, often occupied center stage in their work: 'in many fundamental ways they nourish and give meaning to performance' (CCSSO, 1996, p. 8).

*The general case against dispositions*

Assuming, for the moment, that Perkins (1995) is a defensible source of knowledge about dispositions, there are three reasons why it is hazardous to include dispositions in standards for school leaders that are to be used as guides for the preparation, licensure, and perhaps evaluation of administrators. First, we know almost nothing about how to systematically influence or change the attributes and dispositions of mature adults and so are likely to squander scarce leadership development resources if we feel compelled to try.

Perkins' (1995) treatment of how to develop dispositions, for example, is outlined in a six-page treatment under the title of 'The Metacurriculum' (pages 332–338). About a third of this text is consumed by an example of (presumably) a secondary school history lesson illustrating a general approach to the teaching of dispositions apparently, although it is just as readily an example of teaching 'historical reasoning' or 'critical thinking'. In these pages Perkins alludes to two other books (his own) that develop these ideas more fully. But there is not even an allusion to empirical evidence about the effects on children, never mind adults, of these general approaches to instruction.

A second reason it is hazardous to include dispositions in leadership standards is that, even if we did know how to change or develop the attributes and dispositions of mature adults, we do not have the knowledge base to determine which dispositions should be developed. One need not disagree with Perkins' argument concerning their importance to human behavior in order to reject the appropriateness of including dispositions as a central feature of leadership standards. It is only necessary to acknowledge the lack of evidence available to standards developers for linking together, perhaps even in a causal relationship, specific dispositions and leadership performances. Certainly, Perkins' general argument for the important contribution of dispositions to intelligent performance provides no assistance whatever in making such specific links.

This problem is especially severe with ISLLC since there is a significant discrepancy between what Perkins means by a disposition and the 'dispositions' which appear in ISLLC. Perkins identifies a small number of general dispositions, whereas ISLLC identifies in excess of 40. Indeed, ISLLC developers appear to have confused dispositions with what Perkins identifies as sources of dispositions (beliefs, values, and commitments). So one part of the problem with including attributes or dispositions is the same problem we associated with the inclusion of knowledge.

A third reason not to include dispositions in leadership standards, a reason rooted on the inadequate knowledge base we have been discussing, is that there are no grounds for responding to legal challenges brought by aspiring leaders who fail to receive administrative certification, or by incumbents who are dismissed, partly on the grounds that they do not possess suitable dispositions.

*Unique problems with ISLLC's source of authority*

ISLLC has acquired an additional set of problems, however, by invoking Perkins (1995) as their authoritative source of advice about dispositions. Clearly Perkins has an enviable and well-deserved reputation as a cognitive psychologist and has been willing to work out the applications of his research to schooling much more systematically and fully than most of his colleagues. Nevertheless, there are serious reasons to question the ISLLC developers' choice of his 1995 book as the only source of justification for dispositions that they cite.

In the process of making the case that some dispositions are more likely to be possessed by members of some professions than others, Perkins offers the example of 'academic professions, in general, put[ting] a high premium on matters of evidence, argument, and solid conclusions' (p. 281). In its central treatment of the contribution, nature, and sources of dispositions, however, the Perkins book demonstrates only modest signs of this disposition at work.

About the contribution of dispositions to intelligent behavior, Perkins' claim that 'dispositions shape our lives' (p. 275–6) and represent an important explanation of intelligent behavior, is initially supported with brief allusions to the work of two philosophers (Ennis & Paul), and two psychologists (Baron & Perkins et al). He then claims that:

> 'For all of these authors and more, the appeal of the concept of dispositions emerges from a fundamental point of logic: skills are not enough. However technically adroit a person may be at problem solving, decision making, reasoning, or building explanations, what does it matter unless the person invests himself or herself energetically in these and other kinds of thinking ...' (p. 276).

But this 'point of logic' would lead us, just as readily, to a defense of the concept of motivation and the long-standing and substantial bodies of research and theory concerning that concept (e.g., Bandura, 1990; Ford, 1992). Here Perkins seems to be demonstrating what he calls an unfortunate 'my-side bias' (see below), although in his introduction to the meaning of dispositions he does make reference to the 'motivational side of thinking' (p. 275).

Under the title 'A Taste of Empiricism', Perkins makes reference to three additional lines of evidence in support of the importance of dispositions as a concept in explaining intelligent behavior. The first line of evidence concerns research by Carol Dweck and her colleagues (Dweck & Bempechat, 1980; Dweck & Licht, 1980). This evidence, however, is more readily viewed as helping make the case for the importance of self-efficacy and its role in motivation as an explanation for intelligent behavior. Indeed one of the Dweck papers that Perkins cites is published as a chapter in a book entitled *Learning and motivation in the classroom*.

A second line of evidence cited by Perkins in the same section of his book includes 'several experiments' by Tishman, Jay and Perkins (Tishman, 1991; Tishman, Perkins & Jay, 1995) which demonstrate what they label the existence of 'my-side bias' (the tendency for some people to focus only on evidence which supports their own beliefs and to ignore evidence which does not). While Perkins argues that such a bias should be reconceptualized as a disposition, others would most certainly interpret this as an entirely unoriginal contribution to the long-standing body of research on cognitive bias and heuristics (e.g., Tversky & Kahneman, 1974).

The third line of evidence that Perkins invokes to support the concept of dispositions in explanations of intelligent behavior is Facione and Facione (1992, in press). This evidence, however, seems to be about (a) the beliefs of 'people involved in the critical thinking movement' who, as it turns out on close reading, are not terribly supportive of dispositions as a concept, and (b) factor analytic work aimed at developing a classification system for dispositions, not demonstrating their contribution to intelligent behavior.

As to the nature of dispositions, the Perkins' book defines psychological dispositions as 'a tendency to behave in a certain manner', or 'the proclivit[y] that lead us in one direction rather than another within the freedom of action that we have.' (p. 275). Throughout Perkins' discussion of the nature of dispositions, he cites examples of dispositions in order to give further meaning to the concept. These examples include the tendency to be suspicious, close-minded or open minded, uneasy with new ideas or welcoming of new ideas, and broad or narrow minded. Other examples used in the text are tendencies to think across multiple frames of reference, tendencies toward hasty, narrow, fuzzy and sprawling thinking, the tendency toward ego defensiveness, and (oddly, in our view) problems with limited short-term memory.

But these qualities are intended to be simply illustrative because Perkins has developed his own taxonomy of dispositions that, he claims, overlaps considerably with the product of the Faciones' factor analytic work. This taxonomy of dispositions consists of: broad and adventurous thinking; sustained intellectual curiosity; clarifying and seeking understanding; being planful and strategic; being intellectually careful seeking and evaluating reasons; and monitoring and guiding your own thinking (metacognitive self-management). In defense of this taxonomy, the Perkins' book says only that '... it seems well-justified by the existing literature' (p. 278). No specific sources of such justification are offered for those disposed to be suspicious or intellectually careful.

The Perkins' book offers no empirical evidence concerning the sources of dispositions. It does claim, however, that these sources are to be found in the 'complex psychological mechanisms of the mind' (p. 279). They depend on a person's repertoire of concepts, beliefs, and values about what is important and what is practical. They may, as well, be related to underlying cognitive styles, and while they are not just emotions and feeling, they are invested with such emotions and feeling.

Dispositions, according to Perkins, develop as a consequence of direct experience, the models provided by others and perhaps the norms or the standards of the occupation in which one works. While there may be considerably more evidence than Perkins cites in support of the sources of what he is calling dispositions, in its absence it would be just as reasonable to conclude that the causes of dispositions are entirely genetic.

For all these reasons, first-generation leadership standards that include dispositions are skating on very thin conceptual, empirical, and legal ice.

## 6. Standards should describe desired levels of performance not just categories of practice.

Most first-generation 'standards' actually are standards-free 'criteria'. So at best they identify areas for capacity development without indicating how much capacity is needed to do a good job on behalf of the school. For example:

- 'The principal assists staff in finding opportunities for collaboration' (Connecticut Standards, p. 318);
- 'The administrator: examines personal and professional values' (ISLLC Standards, p. 14);
- '[Headmasters] determine, organize, and implement the curriculum and its assessment; monitor and evaluate them in order to identify and act on areas for improvement' (UK Standards, p. 10).

As these examples illustrate, most first generation standards identify broad categories of practice that can be 'implemented' with considerable expertise, in an entirely unproductive fashion, or at some level in between. In this form, standards amount to nothing more than 'duties' of the sort found in a job description and their potential to improve practice seems quite limited.

As a distinct improvement, second-generation leadership standards should specify, for most of their criteria, at least minimum entry-level standards, standards expected for competent and mature performance, and standards that define high levels of expertise. Among other things, this would assist those involved in leadership development to clarify the size of their task with each of their client groups.

**7. Standards should reflect the distributed nature of school leadership.**
The current generation of leadership standards focuses almost exclusively on the capacities of individuals, whereas the collective or distributed leadership of the school is at least as critical to its effectiveness (Hunt & Dodge, 2000; Spillane, Halverson & Diamond, 2000). Furthermore, such distributed or shared leadership is more than the sum of the leadership provided by individuals (Chrispeels, Brown & Castillo, 2000). It includes the collective capacities identified in recent research about 'professional learning communities' (Louis & Marks, 1998) and 'organizational learning' in schools (Leithwood & Louis, 1999).

Second generation standards would do well to adopt a more distributed conception of the sources of school leadership. This might entail the development of *leadership* standards that take the school to be the unit of analysis, in addition to standards for individual administrative leaders, which is what the first generation quite explicitly are intended to be.

## Conclusion

Under the best of circumstances, a defensible set of standards should provide guidance for those involved in leadership preparation, selection and evaluation processes, thereby reducing the ambiguities and uncertainties associated with a central issue in these processes. Such standards also have the potential to foster a professional consensus about the meaning of effective school leadership, something that could produce more effective communication with policy makers and school-level stakeholders about what are reasonable expectations for the role. Defensible standards may put a 'floor' under the practice of school leadership, reducing the likelihood of performances that fly in the face of what we know about 'best practice'. And, perhaps as important as any of these purposes, defensible standards may serve as an explicit set of expectations against which leaders may judge and reflect on their own and their colleagues' practices.

The widespread use of first-generation standards suggests that at least some of these purposes are being achieved. So it is especially important that the influence of whatever standards are used nudge the preparation, practice and assessment of leadership in the most defensible direction. On this score, first-generation standards are in need of considerable improvement. A second generation, reflecting the standards for standards outlined in this chapter, would serve us better.

As we move toward a second generation of leadership standards, we should also be addressing the limitations of standards of any sort for the improvement of practice. Adapting and building on arguments and evidence from many sources (e.g., Apple, 1998; Darling-Hammond, 1996; Eisner, 1995; Levin, 1998; Wise, 1996), standards may restrict variation in educational leadership practice, thereby reducing an important source of organizational learning about the nature of effective school leadership. Leadership standards have the potential to minimize the importance attached to critical differences in leaders' contexts. As well, over time they may harden into static views of effective practice that no longer reflect the best evidence currently available. This is likely to result in resistance to changing leadership practices in the future, in response to newer and better evidence.

None of these issues will be resolved by simply developing a second generation of standards; what they call for are regular cycles (perhaps every five to seven years) of standards review and modification. In addition, if national or state standards are to serve local purposes well, they should be positioned as the foundation for local standards development, such development aimed at truly reflecting what counts as a productive form of leadership in real individual schools, not the inevitably hypothetical average school. This means that standards will lose the appeal that comes from being viewed as a one-size-fits-all solution to the improvement of school leadership. And that would be a very good thing, indeed.

## References

Apple, M. W. (1998). How the conservative restoration is justified: Leadership and subordination in educational policy. *International Journal of Leadership in Education*, 1(1), 3–17.

Bandura, A. (1990). Self-regulation of motivation through anticipatory and self-reactive mechanisms. *Nebraska Symposium on Motivation*, 38, 69–164.

Barzansky, B. & Gevitz, N. (Eds.) (1992). *Beyond Flexner: Medical education in the twentieth century.* New York: Greenwood.

Bass, B. (1997). Does the transactional/transformational leadership transcend organizational and national boundaries? *American Psychologist*, 52, 130–139.

Bennis, W. G. & Nanus, B. (1985). *Leaders: The strategies for taking charge.* New York: Harper & Row.

Brower, H. H., Schoorman, F. & Tan, H. (2000). A model of relational leadership: The integration of trust and leader-member exchange. *Leadership Quarterly*, 11(2), 227–250.

Chrispeels, J., Brown, J. & Castillo, S. (2000). School leadership teams: Factors that influence their development and effectiveness. *Advances in Research and Theories of School Management and Educational Policy*, 4, 39–73.

Darling-Hammond, L. (1996). What matters most: A competent teacher for every child. *Phi Delta Kappan*, 78(3), 193–200.

Dweck, C. S. & Bempechat, J. (1980). Childrens' theories of intelligence: Consequences for learning. In S. Paris, G. Olson, & H. Stevenson (Eds.), *Learning and motivation in the classroom* (pp. 239–256). Hillsdale, NJ: Erlbaum.

Dweck, C. S. & Licht, B. (1980). Learned helplessness and intellectual achievement. In J. Garbar & M. Seligman (Eds.), *Human helplessness*. New York: Academic Press.

Eisner, E.W. (1995). Standards for American schools: Help or hindrance? *Phi Delta Kappan*, 76(10), 758–764.

Facione, P. A. & Facione, N. C. (1992). *The California critical thinking dispositions inventory*. Millbrae, CA: The California Academic Press.

Facione, P. A. et al. (in press). The disposition toward critical thinking. *Journal of General Education*.

Ford, M. (1992). *Motivating humans: Goals, emotions, and personal agency beliefs*. Newbury Park, CA: Sage.

Gronn, P. (2000). Distributed properties: A new architecture for leadership. *Educational Management and Administration*, 28(3), 317–338.

Hallinger, P. & Heck, R. (1999). Next generation methods for the study of leadership and school improvement. In J. Murphy & K. Louis (Eds.), *Handbook of research on educational administration, second edition* (pp. 141–162). San Francisco: Jossey-Bass.

Hallinger, P., Bickman, L. & Davis, K. (1996). School context, principal leadership and student achievement. *Elementary School Journal*, 96(5), 498–518.

Hartog, D. N., House, R. J., Hanges, P. & Ruiz-Quintanilla, S. (1999). Culture specific and cross-culturally generalizable implicit leadership theories: Are attributes of charismatic/transformational leadership universally endorsed? *Leadership Quarterly*, 10(2), 219–256.

Hunt, J. G. & Dodge, G. (2000). Leadership déjà vu all over again. *Leadership Quarterly*, 11(4), 435–458.

Council of Chief State School Officers (1996). *Interstate School Leaders Licensure Consortium*. Washington, DC: Council of Chief State School Officers.

Kotter, J. P. (1990). *A force for change: How leadership differs from management*. New York: Free Press.

Leithwood, K. & Duke, D. (1993). Defining effective leadership for Connecticut's future schools. *Journal of Personnel Evaluation in Education*, 6, 301–333.

Leithwood, K. & Duke, D. (1999). A century's quest to understand school leadership, In J. Murphy & K. Louis (Eds.), Handbook of research on educational administration. San Francisco: Jossey-Bass.

Leithwood, K. & Jantzi, D. (1999). The effects of transformational leadership on organizational conditions and student engagement with school. *Journal of Educational Administration*, 38(2), 112–129.

Leithwood, K. & Louis, K. (Eds.) (1999). *Organizational learning in schools*. The Netherlands: Swets & Zeitlinger.

Leithwood, K. & Steinbach, R. (1995). *Expert problem solving*. Albany, NY: SUNY Press.

Leithwood, K., Jantzi, D. & Steinbach, R. (in press). School leadership and teachers' motivation to implement accountability policies. *Educational Administration Quarterly*.

Levin, H. M. (1998). Educational performance standards and the economy. *Educational Researcher, May*, 4–10.

Lord, R. G. & Maher, K. (1993). *Leadership and information processing*. London: Routledge.

Louis, K. S. & Marks, H. (1998). Does professional community affect the classroom? Teachers' work and student experiences in restructuring schools. *American Journal of Education*, 106(4).

Morris, A. (2000). Charismatic leadership and its after-effects in a Catholic school. *Educational Management and Administration*, 28(4), 405–418.
Mumford, M. D., Dansereau, F. & Yammarino, F. (2000). Followers, motivations, and levels of analysis: The case of individualized leadership. *Leadership Quarterly*, 11(3), 313–340.
Ogawa, R. & Bossert, S. (1995). Leadership as an organizational quality. *Educational Administration Quarterly*, 31(2), 224–243.
Perkins, D. (1995). *Outsmarting I.Q.: The emerging science of learnable intelligence.* New York: The Free Press.
Podsakoff, P. M., Mackenzie, S. & Fetter, R. (1993). Substitutes for leadership and the management of professionals. *Leadership Quarterly*, 4(1), 1–44.
Spillane, J., Halverson, R. & Diamond, J. (2000). *Toward a theory of leadership practice: A distributed perspective.* Paper presented at the annual meeting of the American Educational Research Association, New Orleans, April.
Teacher Training Agency (1998). *National standards for headteachers.* London: Teacher Training Agency.
Thomson, S. D. (Ed.). (1993). *Principals for our changing schools: The knowledge and skill base.* Fairfax, VA: National Policy Board for Educational Administration.
Tishman, S. (1991). *Metacognition and childrens' concepts of cognition.* Cambridge, MA: Harvard University Graduate School of Education, unpublished doctoral dissertation.
Tishman, S., Perkins, D. & Jay, E. (1995). *The thinking classroom.* Boston, MA: Allyn & Bacon.
Tversky, A. & Kahneman, D. (1974). Availability: A heuristic for judging frequency and probability. *Cognitive Psychology*, 5, 207–232.
Wise, A. E. (1996). Building a system of quality assurance for the teaching profession. *Phi Delta Kappan*, 78(3), 190–192.
Yukl, G. (1994). *Leadership in organizations: Third edition.* Englewood Cliffs, NJ: Prentice-Hall.

# 17

# School Leader Development: Current Trends from a Global Perspective

Dr. Stephan Gerhard Huber

*Researcher Associate, Research Centre for School Development and Management, University of Bamberg, Bamberg, Germany*

In view of the ever-increasing responsibilities of school leaders[1] for ensuring the quality of schools, school leadership has recently become one of the central concerns of educational policy makers. In many countries, the development of school leaders is high on the agenda of politicians of different political wings. At the turn of the century, there is broad international agreement about the need for school leaders to have the capacities needed to improve teaching, learning, and pupils' development and achievement.

At first sight, there may appear to be an international consensus about the important role of school leaders and their development. On looking more carefully, however, it is apparent that a number of countries have engaged in this issue more rigorously than others. While in some countries discussions of school leader development are mainly rhetoric, elsewhere concrete steps have been taken to provide significant development opportunities for school leaders. Hence, a comparison of school leadership development opportunities in different countries is instructive.

This chapter draws on data from an international study of school leadership development[2] (see Huber, 2002, 2003). The report surveys the development models for school leaders in the countries included in the study. It describes international patterns in school leadership development and makes comparisons and recommendations based on current trends.

[1]The term 'school leader' is in this chapter used instead of principal, headteacher, administrator, rektor or other terms describing the person who is in charge of an individual school.

[2]The comparative research project was conducted at The Research Centre for School Development and Management, University of Bamberg, Germany, in the years 1998–2001. The methods used comprised two surveys, extensive documentation analysis, and additional country-specific investigations.

## Approaches to School Leadership Development: An Overview of International Efforts

The following table summarizes school leadership development models in 15 countries. It is meant to provide an accessible overview of predominant approaches in use across Europe, Asia, Australasia, and North America.

Table 1. Overview of Current Approaches to Develop School Leaders.

**Europe**

**Denmark**
Optional offers made by municipalities, universities and private suppliers without any central framework or delivery system

**Sweden**
A national preparatory programme offered by universities through a basic course plus additional offers by the municipalities

**England and Wales**
A centrally organised programme delivered by regional training centres; combines assessment and training with a competency-based and standards-driven approach; the programme is embedded in a three-phase training model

**France**
A mandatory, centrally-designed, intensive, full-time, half-year preparation programme with internship attachment for candidates who have successfully passed a competitive selection process; completion guarantees a leadership position on probation (during which further participation in training is required)

**Netherlands**
A broad variety of different optional preparatory and continuous development programmes by different providers (e.g., universities, advisory boards, school leadership associations) in an education market characterised by 'diversity and choice'

**Germany**
Courses conducted by the state-run teacher training institute of the respective State, mostly after appointment; differs from State to State in terms of contents, methods, duration, structure, and extent of obligation

**Austria**
Mandatory centrally-designed, modularised courses post-appointment; delivered by the educational institute of each State; required for continued employment after four years

**Switzerland**
Quasi-mandatory, canton-based, modularised programmes offered post-appointment; delivered by the respective provider of the canton, most often the teacher training institute, wherein the aim is nationwide accreditation (national standards are currently being developed)

**South Tyrol, Italy**
A mandatory programme for serving school leaders to reach another salary level as becoming 'Diricente'; delivered by a government-selected provider that combines central, regional, and small group events with coaching attachment

**Asia**

**Singapore**
A mandatory, centrally-controlled, preparatory, nine-month, full-time programme provided through a university; comprised of seminar modules and school attachments

**Hong Kong, China**
A centrally-designed, mandatory, nine-day, content-based induction course immediately after taking over the leadership position

**Australia**

**New South Wales, Australia**
An optional, modularised, three-phase programme offered by the Department for Education; centrally-designed, yet conducted decentralized via regional groups; besides there are offers by independent providers

**New Zealand**
A variety of programmes with variation in contents, methods and quality; conducted by independent providers, but also by institutes linked to universities; no state guidelines, standards or conditions for licensure

**North America**

**Ontario, Canada**
Mandatory, preparatory, university-based, one-year, part-time programme delivered through several accredited universities following a framework given by the 'College of Teachers' (the self-regulatory body of the profession)

**Washington, New Jersey, California, USA**
Mandatory, intensive, preparatory, one-year, university programmes that include extensive internship attachments; programmes use a broad variety of instructional methods

Although this table merely provides an overview, a broad variety of school leader development approaches is recognizable. In spite of differences in cultural and institutional traditions, there are common tendencies and trends across these countries. While some of them may be viewed as differences in emphasis, others may be so large as to represent paradigm shifts. The largest differences are evident in those countries that have longer experience in school leader development and school leadership research. This chapter focuses on nine of these tendencies, trends, and shifts (for a fuller account see Huber, 2003). These include:

1. Central quality assurance and decentralized provision;
2. Preparatory training and development;
3. Comprehensiveness of programmes;
4. Multi-phase designs and modularisation;
5. From administration and maintenance to leadership for improvement;
6. Developing the leadership capacity of schools;
7. From acquisition to creation and development of knowledge;
8. From role-based training to personal, professional development;
9. New leadership conceptions and an orientation towards values.

**1. Central quality assurance and decentralized provision of programmes**

As shown in Table 1, provision of development opportunities for school leaders varies broadly across the countries. There are different degrees of centralization and decentralization with regard to how much choice prospective participants have over available providers and development programmes. Here, the interrelation between the qualification approach and the educational policy and school system background is of particular interest. The countries can be categorized in terms of these two dimensions (see Table 2).

Table 2. Centralization and Decentralization of School Systems and School Leader Development.

| | | **Approach to School Leader Development** | |
|---|---|---|---|
| | | **Predominantly Centralized or Using Standards or Guidelines** | **Entrepreneurial** |
| Level of Central Control over School Management | Predominantly Centralized | A France; South Tyrol; Austria; Germany; Hong Kong; Singapore | B |
| | Substantially Devolved | C Ontario, Canada; USA*; NSW, Australia; Sweden; England and Wales; Switzerland | D Denmark; Netherlands; USA*; New Zealand |

**Double listing is due to differences in the approaches of the different States*

In some centrally organised school systems (see Table 2, Cell A), there is a centrally regulated development programme. It has a standardized approach and its delivery is centrally organised. The programme is mandatory for all school leaders. In contrast, in some decentralized school systems (see Cell D), there are a variety of programmes offered by competing providers. The choice of which programme(s) to attend is up to the individual (aspiring) school leader. Here, the governments abstain from any regulation or control of professional development. Countries with a predominantly centralised school system and with an entrepreneurial approach to school leader development could – not too much surprisingly – not be found in the study.

Another existing variant, however, is represented by countries with decentralized school systems (see Cell C), whose programmes are designed according to central guidelines, but are not standardized in every detail. Their general approach seems particularly progressive and pioneering. Teachers who want to qualify for a leadership position can choose among various service suppliers with assurance that the programme is accepted and recognized by the state and/or employing bodies. In North American countries, responsibility for designing and conducting qualification programmes lies primarily with universities (e.g., Ontario, Canada as well as in the US examples included in the study). However, these universities are not completely independent when setting up their development programmes. They must take centrally developed goals and standards into account.

Most recently in the US, a cross-state 'catalogue' of standards has been set up by the Interstate School Leaders Licensure Consortium (ISLLC; cf. Murphy & Shipman). This has been approved by the Council of Chief State School Officers (CCSSO) and by

36 States. Washington and New Jersey, for example, grant state certificates for the participants after completing the development courses based on those standards. This is the case even though – due to the decentralized character of the American school system – selection and employment of school leaders remain the responsibility of local committees. In Ontario, the self-regulatory body of the teaching profession, the Ontario College of Teachers (OCT), has established guidelines for development of school leaders. Only universities accredited by this body may offer the development programme. In Europe as well, some other countries are moving towards assuring quality by centrally regulating qualifications for school leadership positions. For instance, in England and Wales, a central institution, the National College for School Leadership (NCSL) has been established (cf. Bolam; cf. Tomlinson). This institute is responsible for designing and conducting all national development opportunities for school leaders.

A fundamental level of quality assurance is undoubtedly important to participants, as is nationwide acceptance and recognition of programmes by employing bodies. A popular approach has been to set up a 'central institution' responsible for establishing guidelines, standards and content. Accreditation of programmes by the profession itself (e.g., Ontario) seems to have particularly high value in terms of the acceptance by the participants. Therefore, it seems advisable that recognition, approval and control be shared between the state and the profession. Here, another trend is that provision is then offered by several providers. This enables training and development to be more flexible and adaptable to participants' needs.

To sum it up, across the countries there is a developing trend in which responsibility for designing goals and programmes, and assuring quality lies with a central institution, whilst delivery is decentralized.

## 2. Preparatory training and development

Another shift observed in the international comparison concerns the target group and the timing of the qualification in the participants' career. In more than half of the countries included in the study, development opportunities are scheduled before taking over school leadership. These countries offer pre-service preparation instead of relying solely on in-service training. Moreover, the programmes differ as to whether they are optional or mandatory (see Table 3).

Table 3. Timing in Participants' Career and Nature of Participation.

| | Preparatory | Induction |
|---|---|---|
| Mandatory | A Ontario, Canada;<br>USA;<br>France;<br>Singapore | B Germany*;<br>Austria;<br>Switzerland*;<br>South Tyrol;<br>Hong Kong |
| Optional | C England and Wales;<br>Netherlands;<br>NSW, Australia;<br>New Zealand | D Denmark;<br>Sweden;<br>Germany*;<br>Switzerland* |

**Double listing due to differences in the approaches of the German 'Laender' or Swiss 'Kantone'*

In countries that have mandatory preparation (see Table 3, Cell A), taking part in the programme is an important selection criterion for future employment as a school leader. For example, France has a unique interrelation of selection, training, and appointment. Here, successful completion of the competitive 'Concours' makes it possible to participate in the state-financed training. The state training is a pre-condition for employment in a leadership position. Subsequently, retaining one's post as a school leader depends on having successfully completed the second phase of qualification, the 'Formation d'Accompagnement'. In Singapore, the government has mandated specific career regulations. It is only possible to obtain a leadership post after taking part in state-financed, full-time training. This is offered through a single institution. The situation of teachers aspiring to school leadership in North America is less certain. Preparation is a precondition for application. However, successful completion of a preparation programme and subsequent certification does not automatically guarantee employment in a leadership position.

In countries where preparation programmes are optional (see Cell C), there is a tendency among employing bodies towards expecting some preparation for the position. An alternative trend finds the provision of in-service training immediately after appointment and before taking over the leadership position. This is the case in Hong Kong or in some States in Germany.

What are the arguments in favour of preparatory qualification? First a preparatory training and development is supposed to respond best to the relevance of school leadership. On one hand, the key role of school leaders is increasingly accepted internationally. On the other hand, pressure has increased on policymakers to ensure that the occupants of these positions can fulfil system expectations. Second, adequate preparation may reduce the 'practice shock' experienced by new entrants to the role (Storath, 1995). Particularly if pre-service learning and reflection is combined with practical experiences at school, new school leaders get the chance to develop a new perspective when changing from 'teaching' to 'management'. Third, pre-service training offers the chance of assessing one's own interests and strengths. This may help leaders to make career decision more consciously. Fourth, international experiences indicate that the provision of pre-service preparation may stimulate the number of women applicants to educational leadership positions. Women may be more self-critical, and may also be less connected to influence networks that are related to employment decisions. Obviously, development opportunities are helpful in this case. Fifth, experience shows that participants who do not obtain a leadership position may still enrich the leadership resources of their schools. Sixth, the assumption that 'on-the-job-training' alone is the most effective and efficient one has not adequately been empirically validated. In this context, a cost-benefit analysis – in terms of educational economy – would have to be complex and long-term.

All of these arguments clearly favour orientation and preparation opportunities[3]. More and more countries are considering preparatory courses in addition to existing

[3]Even more extensive are approaches to make orientation elements for leadership part of initial teacher training in order to identify and foster potential for leadership at the earliest possible stage. This has been done recently by the Australian State of Victoria. In Sweden, there is a project that offers enrolment in a school management course during initial teacher training, and in Canada, too, long-term promotion is intended by a portfolio-system.

in-service programmes. This reflects a movement away from the concept that the school leader is nothing more than a teacher with a few extra responsibilities.

Effective school leadership requires a demanding set of attitudes, attributes, skills, knowledge and understanding. A thorough training and development starting with appropriate preparation prior to assuming the position has been recognised as undoubtedly vital. This may be regarded as a kind of paradigm shift in the view of school leadership and leadership development.

**3. Comprehensiveness of programmes**

The tendency to regard school leadership as a profession in its own has implications concerning the extent of training and development provision for school leaders. Several countries in the survey originally started with short courses of a very practical orientation. As these providers gained more experience, they extended the programmes so that the courses might 'add up' into a more comprehensive package. In some cases this also reflected an effort to ground the programmes in a stronger theoretical framework (e.g., Kolb's framework of life-long learning).

These development opportunities have become quite extensive. By way of illustration, some examples are given here from North America, Europe, Asia and Australia/New Zealand (see Fig. 1). It is important to mention that all of the programmes listed here are preparatory, which means that they all take place before appointment (except the offer from the Netherlands, which may also be attended after appointment). This suggests the increasing recognition of school leader professionalization.

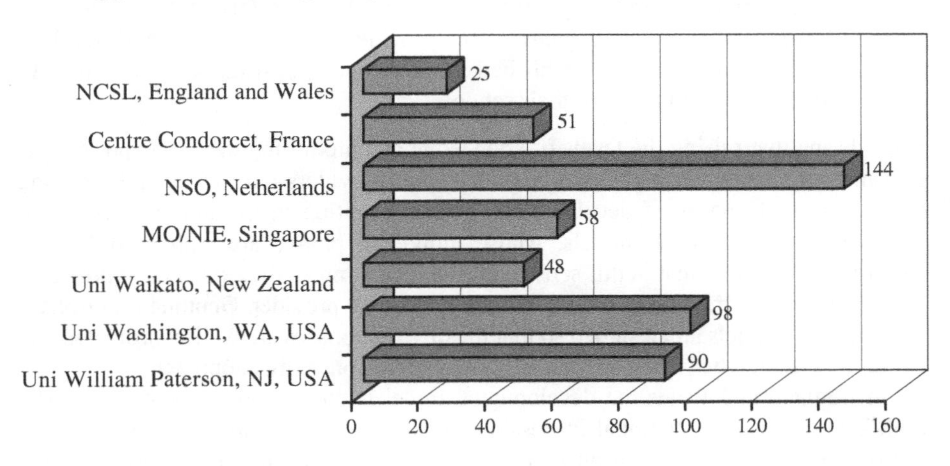


Figure 1. Length of school leader preparation programmes (contact time).

Whilst Figure 1 indicates only the number of course days, the real demands on the time of the participants is apparent when we consider that beyond 'contact time' there is other time committed to preparation. This includes individual study time for readings and writing assignments, but moreover time for internships or school-based projects, and the documentation of one's progress and reflection as by writing a 'learning journal'.

For example, at the University of Washington, preparation requires 39 credit hours (assuming 15-week semesters) and an additional 720 internship hours (i.e., 16 hours per week). The programme of the Nederlandse School voor Onderwijs-management is comprised of four semesters with around 350 working hours for each semester. This includes for each semester: 20 hours for seminars, 175 hours for training sessions, up to 20 hours for consultation sessions, further time for literature studies, and 140 hours for internships in the first three semesters, and time for a written assignment in the fourth semester. The University of Waikato offers a programme, comprised of 24 credit hours (assuming 12-week semesters). In addition there are 1.600 hours assumed by the provider for individual studies, participation in an email-forum and for conducting school-based projects.

In summary, there is a clear trend towards requiring an extensive set of quite time-consuming preparatory activities prior to assuming positions of leadership responsibility in schools across the countries included in this study.

## 4. Multi-phase designs and modularisation of programmes

The international comparison shows that school leader development is more and more regarded as a continuous process. This could be divided into several phases:

- *Orientation phase*: This provides the opportunity for teachers interested in leadership positions to reflect on the role of a school leader in respect to their own abilities and expectations.
- *Preparation phase*: This occurs prior to taking over a school leadership position or even before applying for it.
- *Induction phase*: After taking over a leadership position, development opportunities are provided to support the school leader in her/his new position.
- *Continuous development phase*: This provides various training and development opportunities for established school leaders, best tailored to their individual needs and those of their schools.

Considering that raising the levels of knowledge and modifying the behaviour of participants requires a serious commitment of time, providers are increasingly moving towards several phases of development. This is resulting in the implementation of multi-phase development models, whose individual phases are well co-ordinated. Multi-phase development in this sense does not merely mean the existence of pre-service and in-service training options offered by the same provider. Genuine multi-phase development models are designed so that the different phases are well-coordinated and match with each other. They are based on a coherent conceptual approach.

In England and Wales the development model is comprised of three phases (cf. Tomlinson). First, the National Professional Qualification for Headship (NPQH) is a preparatory programme for aspiring heads. Second, the Headteacher Leadership and Management Programme (HEADLAMP) addresses the needs of newly appointed school leaders. Third, the Leadership Programme for Serving Headteachers (LPSH), is a programme for school leaders who have served for more than six years. The overall conceptualization of this three-phase programme as well as the content design within each phase represent good examples of the multi-phase model. Other providers in different countries offer similar approaches. For example, in the US, the California School Leadership Academy offers a combination of programmes that fit various stages in the career cycle of participants; in New South Wales, the Department of Education has developed a 'five-phase development programme', also trying to meet different needs.

There is also a trend towards providing professional development through a series of modules. This takes two general forms. In the first form, the models are conceived of as a mandatory sequence of 'rounded' single programmes. In the second form, there is no specific sequence for completing the modules. Rather participation in the modules depends on the professional position and development needs of the individual participant. The modules may be 'collected' in a kind of personal portfolio. The individual school leader may well fall back upon them as support in crucial career phases.

Consequently, there is a tendency away from 'one-size for all' designs and towards programmes tailored for the individual participant. The basic idea is that an adequate qualification cannot be completed in one pass through a standardised training programme. Instead, there is an increasing trend towards development linked to the career cycle and to specific needs of the leader, both: personally and school context-related. Figure 2 shows some of the ideal type school leader development models in regard to phasing.

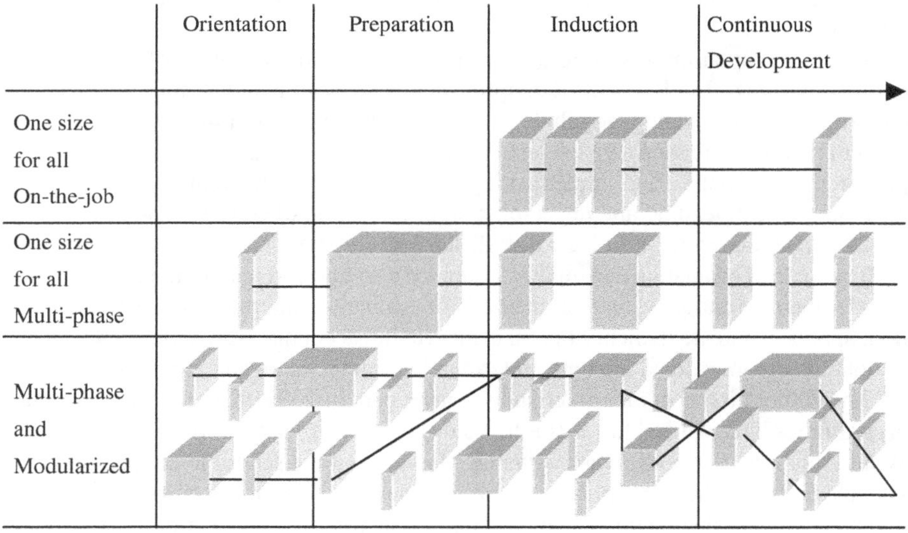

Figure 2. Phased models of school leadership development.

**5. From administration and maintenance to leadership for improvement**

Changes in the provision of development programmes also affect contents. In spite of the increased strain on school leaders due to task overload by additional administrative responsibilities – particularly in countries with more decentralised school systems – school leader development has not become dominated by administrative issues. On the contrary, its overall focus is no longer on administrative and legal topics, but has shifted to a focus on leadership and school improvement. The emphasis has clearly shifted towards the human dimensions of leading schools.

As communication and cooperation play an essential role in school leadership, this is mirrored in the choice of contents and methods. First, in the context of 'communication', one finds module titles such as leading conferences and meetings, leading a professional dialogue, problem- and conflict-solving, and creating structures of relation and communication. In the context of 'co-operation', it is about gaining the co-operation of all stakeholders, creating a shared vision, a shared school programme, shared leadership (in the sense of spreading responsibility), and team work.

It is no longer the primary aim of school leadership to make the school function within a fixed legal framework. Today schools and their leaders must respond to the challenges of social, cultural and economic change. Schools are more and more viewed systemically as learning organizations, each with their own specific conditions, rules, and cultures. Consequently, leading schools entails developing learning organizations (Senge, 1990; Fullan 1993, 1995).

This paradigm shift from managing and maintaining to leading and improving schools is mirrored in the themes of many development programmes reviewed in this study. For example, Danish and Canadian programmes place educational leadership explicitly in the context of school change. They view the school leader as a first class 'change agent'.

In many programmes, similar themes are evident within the areas of school development and staff development. Examples include: school as a learning organization, culture of an organization, psychology of organizations, school quality and development of quality, setting up a vision and implementing the vision, management of school programmes, initiating and implementing change, school improvement projects, project management, leading and developing staff, allocation of staff, teamwork and team development, in-service-training for staff, staff development and teachers' supervision.

The topic of evaluation and quality assurance also plays an important part within this broader theme: school evaluation, methods of evaluation, internal and external evaluation, appraisal and assessment of pupil achievement, accountability, action research and evaluation, organizational learning and evaluation, supervision and evaluation.

The international comparison shows quite conclusively that these development programmes have shifted towards a focus on the role of leadership for improvement. Within this role, the central task is the development of the school in cooperation with all stakeholders. The conception of school leadership as administration of the status quo has to a high degree given way internationally to a new conception of school as a learning organization and of its leadership as a driving force and safeguard of effective improvement processes.

## 6. Developing the leadership capacity of schools

One tendency suggested above that may be developing into a paradigm shift is the conceptualisation of school as a 'learning organization'. This conceptualization also shifts the focus away – somewhat – from the development of the individual school leader to the development of each individual school's leadership capacity. Hence, the school leadership development programme becomes a means of school development.

With this in mind, some providers explicitly have changed their programmes and have widened their target groups. They focus not only on (aspiring) school leaders, but also on teachers who want to enhance their leadership competencies even if they are not planning to apply for school leader positions. This is the case, for instance, in New Jersey and New South Wales. If school development is the explicit goal, programmes may target whole school leadership teams (e.g., the Danish programme Leadership in Development), and may include parent and community representatives. The California Leadership Academy has programmes that target established leadership teams from schools. While the trend towards team-based training is only apparent in a few programmes, an increasing number of providers state that they intend to focus on developing leadership teams. They express the belief that this approach is necessary in order to establish stronger leadership and change capacities within schools.

As an additional note, this new focus on developing team leadership capacity suggests a shift towards focusing on the individual school rather than the individual participant. This has interesting implications for programme content. When a programme focuses on a team, development activities must become even more contextualized: It is no longer context-free training, but context-specific applied development.

**7. From acquisition to creation and development of knowledge**

In many programmes, two considerations seem increasingly to be taken into account: First, when rapid social and economic change and changes in the educational system are coupled with a global increase in information production, it is insufficient for programmes to focus solely on enlarging the quantity of leaders' knowledge. The qualification must prepare for an unknown future environment. This suggests still another paradigm shift. It is a shift away from imparting a stable knowledge base and towards the development of procedural knowledge that can be applied. The notion of 'acquiring' knowledge is being replaced by the concepts of 'developing' or 'creating' knowledge and by information management. The participants will enhance their ability to learn, understand cognitive processes and achieve what is called 'conceptual literacy' (see Giroux, 1988). They have to be enabled to act in a complex, sometimes chaotic work environment (see Murphy, 1992).

Second, there is consensus that delivery methods must address the learning needs and competences of adult learners. Hence, fundamental andragogic principles must be taken into account. This reflects the belief that new knowledge is built on previous experiences and the knowledge of the adult learners. Adults bring personal and professional experiences, prior knowledge, and their own personal ways of seeing themselves to bear on the learning process to a greater degree than children (see Siebert, 1996). Themes that cannot be linked to previously existing cognitive systems are mostly forgotten. The reality and the experiences of the participants, their needs and problems, should therefore become the starting point of new learning. Consequently, methods of learning tend to favour a problem-centred rather than theme-centred approach. According to Gruber (2000), gaining experience for professional competences means learning in complex application-relevant and practice-relevant situations (see also Joyce & Showers, 1988, 1995). New competences are mostly gained in a process of practice and feedback. For this, sufficient theoretical foundations should be imparted as well in order to foster reflection on practice.

In many development programmes there is a clear tendency towards experience-oriented and application-oriented methods. Indeed, methods of learning and processing of information are apparent as programme themes as well, either implicitly or explicitly. There is a shift of emphasis in school leader development towards practice-with-reflection-oriented learning. This can be seen in the attempt to bring practical work experiences from the schools to bear during the programmes through cases, learning journals, and discussion groups.

Moreover, increasingly the participants are placed in a workshop surrounding, and confronted with modelled situations of school leadership work life and carefully constructed cases. They may be involved in teams in problem-based learning (PBL) where learning is cooperative, interactive, participative, and, to a certain degree, group- and self-organised. More consequently than the case studies and simulations often applied in development programmes, the PBL approach starts with real-life experiences and then looks for supportive knowledge as a tool. The slogan here is: 'First the problem, then the content.' (Bridges & Hallinger, 1995, p. 8). Here, the problem is seen as a stimulus for learning that then leads to the content required to solve it. Problem-based learning has become a consistent part of a number of programmes for school leaders internationally (e.g., at the University of Washington). It is meant to offer a greater practical relevance and thus addresses the theory vs. practice conflict. Within PBL, team learning is especially critical in order to achieve solutions to problems. Problem-solving is an interactive participative process.

Certainly, problem-based learning is an interesting attempt to get practice relevance by using concrete problems taken from real life. Yet in PBL, the problem remains constructed and imagined. This surely has advantages: However close to the complexity of school leadership reality the constructed problem may be, it always remains consciously designed and structured enough to enable exemplary learning experiences.

Going one step further means using genuine cases that are taken from real schools, either from the schools of the participants or from partnership schools. Within this approach, participants of the project group become external counsellors for the leaders of these schools. Through this interaction both parties benefit. This method is, for example, used by York University via an online conference system. Two experienced school leaders present a problem every seven to ten days, taken from their work life, to the group of which they are in charge.

Some development programmes take another step further, leave the workshop and turn to the authentic workplace, using it as a clinical faculty. It is argued that only the authentic working context can assure an adequate complexity and authenticity leading to learning processes required. For the participants of pre-service school leader development, internships at one school or several schools are organized parallel to the training. They can observe the school leader by shadowing her/him, can partially take over leadership tasks themselves, and can carry out projects independently. The school leaders at the internship schools then function as mentors or supervisors and will also benefit from this co-operation. In general, new partnership arrangements between universities, other providers of school leadership development, and schools are an important basis for learning opportunities like these. Thus, certainly the best possible practice relevance is created: Exemplary learning processes take place in the reality of school (cf. Littky & Schen).

As Huber and West (2002) show, the training provision can be conceptualised as being spread across two continua of course-based and experience-based learning opportunities. Hence, it is possible to distribute the programmes worldwide according to the relative emphasis given to these two strategies (see Figure 3).

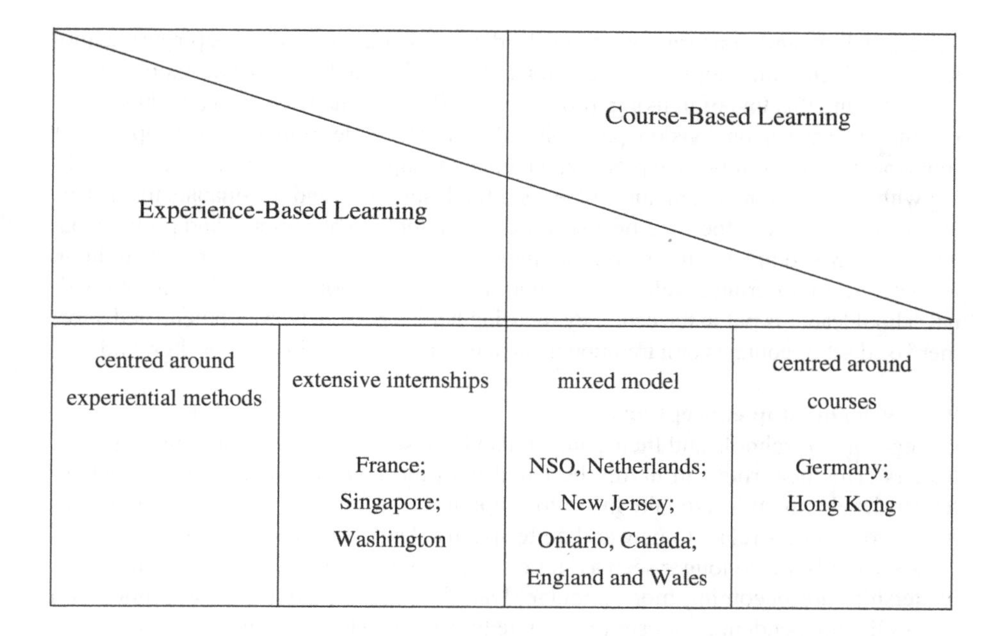


Figure 3. Emphasis of learning opportunities within school leader development programmes[4].

Project work and/or internships are included, for example, in the National Professional Qualification for Headship in England and Wales, in the Managementen Organisatieopleidingen of the Nederlandse School voor Onderwijsmanagement, in the Master programme in Educational Leadership at the William Paterson University of New Jersey, in the Principal's Qualification Programme in Ontario, and particularly extensive in the central programme in France, in the Diploma in Educational Administration in Singapore, and in the Danforth Educational Leadership Programme at the University of Washington. However, countries which still favour more or less an approach to leadership development which is centred around courses also indicate that certain modifications are under consideration.

Hence, it is obvious that in many countries there is a shift from solely course-based learning towards experience-based learning in development programmes. Increasingly, programmes are centred around experiential methods.

## 8. From role-based training to personal and professional development

Within this era of constant change, it is no longer sufficient to train participants for a fixed role. For some observers this suggests the need to focus on basic professional

[4]It has not been taken into account whether the offers are made to teachers aspiring to leadership or to school leaders newly appointed and in position. Besides, the different emphasis could be viewed in reference to the total amount or length of training available; since offering experiential learning opportunities inevitably means expanding the programme accordingly.

values, beliefs and concerns of school leaders. This demands that (aspiring) school leaders reflect upon their own conceptions of schools and the role of a school leader.

Following this line of thinking, programmes offered in many countries include components such as personal vision, personal and professional development, development of fundamental values, reflective practice, 'cognitive mapping' -strategies in terms of working with one's own mental pictures of one's school, and time- and self-management. This results in a shift from focusing on a specific role to looking at personal and professional needs within a complex setting. Programmatically there is a shift from 'providing training to someone' to 'offering development opportunities for someone'. Therefore, the individual school leader is put in the centre by focusing her/his personal development, and a former fixed set of contents or a traditional curriculum are pushed into the background.

**9. New leadership conceptions**

Changes in the schools and their context also have some impact on the role of school leaders. This new role can hardly be filled with old concepts of leadership. School leader development has to take this into account. Consequently, some of the development programmes relate to new and quite specific leadership conceptions.

As schools are no longer seen as static systems, conceptions like 'transformational leadership' are becoming more popular. Transformational leaders view school as a culturally independent organism that is able to develop. Hence, they exercise an active influence on the culture of the school. They are not only expected to manage structures and tasks, but to concentrate on people and their interpersonal relationships. They make an effort to win their cooperation and commitment. Leadership of this type is considered more suitable for the tasks of school development (see Leithwood, 1992).

If school is to become a learning organization, this implies the active empowerment and cooperative commitment of all stakeholders. Then, the previous division between the positions of teachers on one hand and learners on the other hand cannot be maintained. Nor can the division between leaders and followers. Leadership is no longer statically linked to the hierarchical status of an individual person, but empowers as many staff members as possible as partners in various parts. This is conceptualised by the notion of 'post-transformational leadership' (see Jackson & West, 1999).

Another concept, for example, is 'integral leadership'. It views school leaders primarily as leaders with genuinely educational tasks and emphasizes an integrating perspective, which overcomes the divide of management and leadership for the sake of the educational aims of schools (see Imants & de Jong, 1999).

## Final Remarks: Leadership, Values, and the School's Core Purpose

The comparison of school leader development in these 15 countries gives a dominant impression of global approaches and shifts. What can be clearly stated about school leader development from this international perspective is that there were many changes during the last years in nearly every country. It is apparent that a number of countries have acted more actively than others. In some countries, school leader development opportunities have improved particularly during the last 10 years. However, there is still some way to go.

One issue seems particularly interesting: Increasingly the programmes are organized around new conceptions of schooling. The old notion of the school as an unchanging, maintaining and very static organization is no longer suitable. Increasingly, schools are seen as learning, problem-solving, creative, self-renewing or self-managing organisations.

If change is on the agenda of schools and school leaders, it is crucial to have a vision which gives them a direction. Leaders (of any kind) need to know which are the goals and aims for real improvement (and not change for its own sake). Basically, what is needed is to have criteria to judge the overall leadership approach and the day-to-day decision making. This should be back-mapped against the core purpose of school.

In some countries, this notion was taken into account when designing school leader development. The schools' core purpose – namely teaching and learning – and the specific current and future aims of schools have increasingly left their traces in the concepts of various development programmes. The principle that a 'school has to be a model of what education aims at' (Rosenbusch, 1997) thus has consequences not only for defining the role of school leaders but also for the design of development programmes.

As a solid base for what education aims at, in some of the programmes an orientation towards a specific value-based attitude is intended. Thus, the understanding of leadership in this context includes moral and political dimensions. Leadership in a democratic society emphasizes values such as equality, justice, fairness, welfare and a careful use of power. In the compilations of topics, the role of values, ethics and morals, the question of power, and how to legitimate leadership in a democracy and for social justice are increasingly central themes. This holds true, for example, for the programme of Danmarks Paedagogiske Universitet and that of the University of Waikato, New Zealand (both doing without any state guidelines), but also in the standards- or guidelines-oriented programmes of the US examples, Canada, and some others.

This comparison indicates certain current trends and contributes to the discussion in the field, yet there is still much to be done. For example, there still is an obvious lack of analyses of the training and development needs of school leaders in the different stages of their careers. Moreover, the quality and the effectiveness of school leader development programmes have to be evaluated. Further internationally in-depth comparative studies to identify best practice have to be conducted. Very important is to establish networks, which could provide further co-operation and collaboration between those planning and providing school leadership development and those conducting research in different countries.

## References

Bridges, E. & Hallinger, P. (1995). *Implementing problem-based learning in leadership development.* Eugene, OR: ERIC Clearinghouse on Educational Management.

Fullan, M. (1993). *Change forces.* London: Falmer Press.

Fullan, M. (1995). Schools as learning organizations: Distant dreams. *Theory into Practice,* 34(4), 230–235.

Giroux, H. A. (1988). *Teachers as intellectuals: Toward a critical pedagogy of learning.* Granby, MA: Bergin & Garvey.

Gruber. (2000). Erfahrung erwerben. In C. Harteis, H. Heid and S. Kraft (Eds.), *Kompendium weiterbildung* (S. 121–130). Opladen: Leske + Budrich.

Henninger, M. & Mandl, H. (2000). Vom Wissen zum handeln – ein ansatz zur förderung kommunikativen handelns. In H. Mandl & J. Gerstenmaier (Eds.), *Die kluft zwischen wissen und handeln empirische und theoretische lösungsansätze* (S. 198–219). Göttingen: Hogrefe.

Huber, S. G. (2002). *Qualifizierung von Schulleiterinnen und Schulleitern im internationalen Vergleich.* Innsbruck: StudienVerlag.

Huber, S. G. (2003). *Preparing school leaders for the 21st century: An international comparison of development programs in 15 countries.* Lisse: Swets & Zeitlinger.

Huber, S. G. & West, M. (2002). Developing school leaders: A critical review of current practices, approaches and issues, and some directions for the future. In K. Leithwood and P. Hallinger (Eds.), *International handbook of educational leadership and administration.* Dordrecht: Kluwer Academic Press.

Imants, J. & de Jong, L. (1999). *Master your school: the development of integral leadership.* Paper presented at the International Congress for School Effectiveness and Improvement, San Antonio, Texas.

Jackson, D. & West, M. (1999). *Leadership for sustained school improvement.* Paper presented at ICSEI1999, San Antonio, Texas, USA.

Joyce, B. & Showers, B. (1988/1995). *Student achievement through staff development.* New York: Longman.

Leithwood, K. A. (1992). The move toward transformational leadership. *Educational Leadership*, 49(5), 8–12.

Murphy, J. (1992). *The landscape of leadership preparation: Reframing the education of school administrators.* Newbury Park, CA: Corwin Press.

Rosenbusch, H. S. (1997). Organisationspädagogische Perspektiven einer Reform der Schulorganisation. *SchulVerwaltung*, 10, 329–334.

Senge, P. (1990). *The fifth discipline.* New York: Doubleday.

Siebert, H. (1996). *Didaktisches Handeln in der Erwachsenenbildung: Didaktik aus konstruktivistischer Sicht.* Neuwied: Luchterhand.

Storath, R. (1995). *'Praxisschock' bei Schulleitern. Eine Untersuchung zur Rollenfindung neu ernannter Schulleiter.* Neuwied: Luchterhand.

# 18

# School Leadership Preparation and Development in Global Perspective: Future Challenges and Opportunities

Dr. Philip Hallinger

*Professor and Executive Director, College of Management, Mahidol University, Bangkok, Thailand*

As noted in the *Introduction*, the impetus for this volume derived from the rapid pace of developments in the field of school leaders preparation and development. In the space of just a few years, educational policymakers throughout the world seemed to discover the need for more skillful school leadership. This led quite naturally to a search for *systemic* solutions to increase the leadership capacity of educational institutions to implement urgent reforms throughout the world.

In the search for solutions, educational policymakers and practitioners (e.g., principals, teachers and system administrators) began to ask a variety of questions:

- What is the state-of-the-art with respect to school leadership?
- Can we develop skillful leaders with the capacities needed to improve schools?
- What are best practices in selection, preparation and development of school leaders?
- How can we best organize development experiences for prospective and practicing school leaders?
- What should the curriculum consist of and what instructional methods will have the greatest impact on learners' knowledge, skills, and attitudes?
- How can we increase the likelihood that training experiences will carry over to the workplace?
- Should we set standards of performance and how can we evaluate the impact of our programs on individuals and on their schools?

These are but a few of the questions that policymakers have asked in recent years. Consequently, these are central questions around which this book was organized. The reader should not, however, make the mistake of thinking that the answers to these questions come easily. This book contains the most recent thinking of some of the best thinkers in educational leadership internationally of the past two decades. Even so, their knowledge and experience at best provides a framework and direction for action. As Hargreaves and Fullan (1998) conclude with respect to the implementation of change more generally:

> There is no easy answer to the 'how' question. Singular recipes ... oversimplify what it will take to bring about change in your own situation. Even when you know what research and published advice tells you, no one can prescribe exactly how to apply [it] to your particular school and all the unique problems, opportunities, and peculiarities that it contains. (p. 106)

The same holds true for the design and implementation of programs of school leader preparation and development. Solutions must be crafted to the local context. What works in North America could be a design for failure in Thailand or England. This is *not* to say that global commonalities do not exist; there are many, but that each nation must proceed carefully in learning from and applying knowledge gained from each other's experience.

With that in mind, this final chapter is structured as a conversation with someone setting out to design new programs of school leader preparation and development. What would be the critical themes, ideas and guidelines they might carry away from this volume at the turn of the millennium? How would the thinking of these scholars shape the thinking of local policymakers and practitioners?

## New Millennium Themes in School Leader Preparation and Development

The authors of this volume directly represent perspectives from the United States, Canada, England, Wales, Germany, Singapore, Thailand, Hong Kong, Taiwan, Australia, and China. Indirectly, the book speaks to current practices in school leader preparation in at least 10 other nations as well. The global themes that emerged from this volume include the following:

1. Evolving from passive to active learning;
2. Creating systemic solutions that connect training to practice;
3. Crafting an appropriate role and tools for using performance standards;
4. Creating effective transitions into the leadership role;
5. Evaluating leadership preparation and development;
6. Developing and validating an *indigenous* knowledge base across cultures;
7. Creating a research and development role for universities.

### Evolving from passive to active learning

As noted in a number of chapters, a persisting critique of educational leadership preparation and development as practiced in the past has focused on methods of teaching and learning. Critics have called attention to several shortcomings in this regard:

- inadequate attention to the *affective* domains of leadership.

- over-reliance on lecture and discussion and a lack of learning that involves students actively in their learning;
- insufficient opportunities for students to practice what they learn and weak linkages between the classroom and workplace application.

I will address each of these shortcomings in light of what this volume suggests about state-of-the-art instruction in school leader preparation and development.

*Addressing the affective dimensions of leadership*

Bridges' (1977) noted 25 years ago that leadership development inside and outside of education has traditionally focused on the cognitive side of the leader's role. He noted some of the potentially dysfunctional consequences such as 'analysis paralysis'. The chaotic changes that characterize life in schools of this era further handicap our capacities for rational planning (Hargreaves & Fullan, 1998). Moreover, the shift from a unitary, order-taking administrator to multiple school-based leaders places even greater importance on developing the affective capacities of future leaders (Leithwood, 1996).

Yet, to examine a curriculum in educational administration in most universities around the world, one would believe that emotions continue to play little or no part in leadership. Despite Bridges' irrefutable analysis, the emotional side of leadership remains outside the realm of most leadership preparation programs. This must change and should be reflected in the classroom practices and processes as well as the content of leadership development programs.

The instructional approaches noted above hold greater promise for addressing this omission than do traditional approaches to 'learning to lead.' For example, Littky and Schen assert that a stated goal of their program is to help develop moral courage among participants. Would most programs be willing to make such a bold statement? I suspect not. Yet it is only through engagement with real problems under working pressures that the affective, moral and ethical dimensions of leadership can be addressed in a meaningful fashion.

*Active learning*

This volume identifies a variety of approaches that more actively involve learners in making sense of what they learn and how to apply it in schools. Problem-based learning (cf Copland; cf Hallinger & Kantamara), simulation (cf Hallinger & Kantamara), learning technologies (cf Chong et al.; cf Hallinger & Kantamara; cf Tomlinson), apprenticeships and other forms of workplace learning (cf Chong et al.; cf Littky & Schen; cf Tomlinson) all are designed to facilitate the transfer and application of knowledge. The authors describe these instructional approaches in some depth and share a combination of empirical findings and anecdotal observations about their use in school leader preparation and development.

For example, empirical research provides initial confirmation that problem-based learning can achieve positive results for adult learners when implemented systematically (cf Copland; cf Hallinger & Kantamara). PBL appears to be useful for developing thinking and problem-solving skills, fundamental knowledge, teamwork skills, and important affective orientations to the school leader's role. Even the most vocal spokespersons for PBL do not, however, propose this method as the *single* preferred approach for use in developing school leaders. At Stanford University, for example, the PBL portion of the curriculum has never exceeded 40% of the total coursework (Bridges & Hallinger, 1995, 1997). To achieve the best results PBL should be used as part of a repertoire with other methods depending upon a program's learning goals.

Simulation is another active learning approach that warrants further use in school leader preparation and development. Simulation places the learner in a situation that requires active solution of a problem or completion of a task. Simulation can be combined with PBL and also with learning technology, or it can be designed as a stand-alone method (Hallinger, Crandall, & Ng Foo Seong, 2000; cf Hallinger & Kantamara; Kantamara, 2000; Wolfe, 1993).

Not surprisingly, the use of learning technologies has also expanded greatly in the past decade. Learning technologies include multi-media based instruction, computer simulation, and the use of the Internet for instruction, information sharing, and the development of collegial on-line networks (Bransford, 1993; cf Caldwell; cf Chong et al.; cf Hallinger & Kantamara; cf Huber; cf Ming-dih; cf Tomlinson). There is little question that learning technologies will continue to play an expanding role in program delivery, even if we cannot predict all of the directions this will take.

Two chapters in the volume focused specifically on attempts to strengthen the linkage between the classroom and the workplace (cf Chong et al.; Littky & Schen). Problem-based learning seeks to create classroom contexts that mirror the realities of the work-place and to place knowledge in context (cf Copland; Hallinger & Kantamara). In contrast, apprenticeship and workplace learning approaches seek to contextualize learning by placing the learning in the actual work context; a seemingly sensible idea, but one that requires considerable resources and effort to do it well. Littky and Schen (cf) articulate this position well:

> We formulated a mission, philosophy and design that flow from our experience and from research on adult development. The Principal Residency Network's mission is to 'develop a cadre of principals who champion educational change through leadership of innovative, personalized schools.' Our philosophy is grounded in teaching 'one student at a time' at all educational levels. We believe that the best learning takes place in small communities that integrate academic and applied learning, promote collaborative work, and encourage a culture of lifelong learning.

This approach is sensible and well suited to the needs of school leadership in this era. Like most things that are worthwhile, this approach takes commitment, sustained attention, and time. As Littky and Schen report, this apprenticeship approach is being tried more widely in the United States. It will be interesting to see if its promise is fulfilled.

In conclusion, a reading of this volume would lead to the following three conclusions concerning the organization and delivery of instruction in school leader preparation and development programs:

- Neither PBL, simulation, learning technology, apprenticeship nor any other single approach is the *silver bullet* that will create excellence in the professional learning of school leaders.
- However, empirical research and craft knowledge both suggest that each of these aforementioned learning methods deserves systematic consideration and

application in comprehensive programs of school leadership preparation and development.
- Moreover, programs that fail to incorporate active learning methods into their curricula are doomed to irrelevance in the coming era.

**Creating systemic solutions that connect training to practice**

Possibly the most salient concept for thinking about professional development in the past decade has been the *learning organization*. A learning organization is one that organizes in ways that enhance the capacity of staff to learn individually and collectively (Hallinger, 1998; Leithwood, 1994). This concept is especially salient as we think about the professional development of school leaders.

Chong and his colleagues (cf) ask what the concept of the learning organization means to individual principals. They suggest that school leaders have much to gain in a lifelong learning mode.

- Learning to use the tools of information technology;
- Learning to think creatively and critically; and
- Learning to live and work globally in a networked world over their life span.

The ugly reality of professional development is that it is very difficult to organize learning opportunities for busy school staff at convenient times and locations, never mind with the desired support. Singapore's approach seeks to:

> produce innovative principals who can take their schools to new heights of performance, learning experiences need to be intensified by locating them in the real setting of schools. Thus, much of the learning is in the school workplace, and is supported by learning in the university class- and tutorial-room, in business environments, and in educational institutions both in Singapore and overseas. (cf Chong et al.)

To the extent that schools can develop structures and norms that foster the norms and practices of learning organizations, they can mitigate obstacles to professional learning noted by several authors in this volume (e.g., cf Bolam; cf Caldwell; cf Chin; cf Hallinger; cf Ming-dih). This portends as one of the key areas of exploration among practitioners and staff developers in school leader preparation in the coming decade.

**Crafting an appropriate role and tools for performance standards**

Two of the most influential scholars in school leadership of the past two decades – Ken Leithwood (Canada) and Joseph Murphy (USA) – each authored chapters concerning a relatively new phenomenon in school leader preparation: performance standards. As with many of the trends noted in this volume, the push towards use of performance standards was not even on the map 15 years ago, but has emerged as a driving force in program designed today throughout the world. As Leithwood (cf) has noted in his chapter, performance standards for school leaders are not only being used to shape program designs in the United States (the subject of the Murphy chapter), but also in New Zealand, Australia, and the United Kingdom.

I would make several observations here. First, the degree of similarity among the standards across four different countries discussed by Leithwood is notable. This suggests the emergence of a global consensus on the role of an effective principal! This is both surprising and a phenomenon that we would not have expected even a decade ago. The development of a common notion of what makes for a productive school and the roles of principals and teachers (and even parents and the community) in bringing that about reflects an unanticipated outcome of the process of globalization.

Second, these standards in essence define what these educational agencies seek in their principals. Many facets of this 'definition' are tied to the local context of education. That is, the context shapes perception of the needed capacities among leaders. Since globalization is increasingly leading to convergence on a variety of system features among educational systems across the world, it is not surprising to find these similarities. Yet, to what extent the educational systems in individual nations – particularly outside of the Western industrialized nations – will find this 'definition' appropriate could vary more widely.

Third, the fact that noted scholar such as Leithwood and Murphy share such different degrees of confidence in performance standards should give pause to the rush towards wholesale adoption. Leithwood's critique provides a useful direction for improving the design and implementation of standards in school leadership development. Moreover, I would add two more cautionary points as well.

The design of standards reflects a *system* view of school leadership development. For example, contrast this approach with the perspective advocated by Littky and Schen (cf) of developing school leaders *One Principal at a Time*. The standard-driven model clearly reflects a top-down and perhaps even simplistic perspective on improving the practice of school leaders. As suggested above, it neglects the local context and makes assumptions about our ability to define *effective leadership* that go beyond the firm boundaries of our knowledge base (Hallinger & Heck, 1996).

Moreover, it remains unclear whether performance measures designed to accompany the standards reach standards of validity that make them feasible for making important personnel decisions. Proponents of the use of standards in school leader preparation and development have clearly surged to the forefront of the field on the crest of the accountability-based reforms of the past decade. However, it remains to be seen whether this infatuation is a short-term romance or a last marriage with the field.

**Creating effective transitions into the leadership role**

Induction into the principalship or other school leadership positions has emerged as a key issue in school leader preparation and development. Heck (cf), Tomlinson (cf), Littky & Schen (cf), as well as Chong and colleagues (cf) all provide examples of how current programs are attempting to address the linkage between preparation and induction. Heck (cf), in particular, provides a useful framework for viewing the induction of principals:

> As these glimpses suggest, the socialization framework is a useful lens for understanding new administrators' evolving professional practice and, to some extent, their subsequent career paths. Almost all of the new principals appeared to focus on the overlapping commonalities among their growing understanding of the principal's role from their planned induction and organizational experience, their use of organizational support networks, and their developing awareness of

> their own evolving professional practice. The new principals expressed this as being able to see the 'big picture' and being able to view situations comprehensively, as opposed to focusing on individual factors.

This is equally applicable to the in-service arena. Policymakers generally believe that updating the knowledge base of practicing school leaders will lead to more successful program implementation. This belief has only been partially borne out in practice (Hallinger, 1992). Evaluations of leadership development efforts find that exposure to new knowledge via training bears only a small relationship to change in practice at the school (Marsh, 1992). As research on teacher change has concluded, pre-service and in-service programs of training and development that fail to incorporate coaching and support following the introduction of new ideas and skills yields few lasting effects (Fullan, 1991).

Leadership development intended for behavioral change must include a support component that all too often is absent. Sometimes team participation will afford a means of support. In other cases, a school leader may find support from a colleague who acts as a coach. The operative principle is that school leaders need the same support components for behavioral change as teachers: motivation to learn, time to learn, resources for learning, a model, a coach, and opportunities for practice.

**Developing a life-long learning perspective towards school leadership development**

The changing knowledge base and context for school leadership makes lifelong learning a fundamental facet of the professional role. 'In complex, rapidly changing times if you don't get better as a teacher [or principal] over time, you don't merely stay the same. You get worse' (Stoll & Fink, 1996, quoted in Hargreaves & Fullan, 1998, p. 49). Annual attendance at a convention no longer suffices as a leader's efforts at professional development. Life-long learning has become a *necessary* and *fundamental* facet of the school leader's role. The increasing complexity of the school leader's role demands ongoing efforts to maintain currency. It is no different than for a doctor who would find it difficult to think about continuing to practice without ongoing engagement with the profession's knowledge base.

Fortunately, encouragement of a norm of lifelong learning stands as one of the hallmark achievements of the principal's center movement started at Harvard University in the early 1980's. Widespread adoption of local principal's centers as well as statewide leadership academies has fostered the expectation that school leaders must be learners in order to lead schools (Hallinger, 1992). This is one of the least unanticipated but most important outcomes of school leadership development efforts to date (Hallinger, 1992).

The past two decades have seen a demonstrable progress in the attitude of school leaders towards the notion of lifelong learning. This must, however, be strengthened further through government policy as well as through the active engagement of the profession in charting the course of professional learning (cf, Lam). Local school authorities need to examine the implicit expectations as well as their policies with respect to professional development. Do they expect school leaders to engage in ongoing development? Do governments and local education agencies provide resources to support both learning and implementation? Do policies provide a framework of support for prospective and current leaders (cf Caldwell; cf Chong et al.)?

The profession must also take a hard look at its responsibilities. A recurrent theme in this book has been the necessity of school leaders to take responsibility for their professional learning. Amazingly, this has *not* been a traditional norm among educators. The time has come for professional educators to engage with parties inside of the educational profession (e.g., universities, research institutions) as well as outside the profession (e.g., governments, corporations, community institutions) to define the agenda for professional learning and development in the coming years (see Hargreaves & Fullan, 1998).

**Evaluating leadership preparation and development**

Ten years ago I edited a volume of the journal Education and Urban Society devoted to the theme, 'Evaluating School Leadership Development' (Hallinger, 1992). The volume included articles describing empirical studies of school leadership development in the United States (Hallinger & Anast, 1992; Marsh, 1992), Sweden (Ekholm, 1992), Canada (Leithwood & Steinbach, 1992), and Great Britain (Wallace, 1992). In that volume I drew the following conclusion concerning research in this field.

> Policymakers will be particularly keen to know if these training interventions *made a difference* in the practice of school leadership and school performance. Unfortunately, we cannot be sure since none of the studies were designed to address these questions. Indeed, a frustration that results from reading this set of evaluations is the inability to speak with confidence about the *impact* of the interventions on administrative practice in schools. (Hallinger, 1992, p. 308)

Unfortunately, since 1992 changes in the landscape of international research on school leader preparation and development have lagged well behind the pace of changes in policy and practice. There remain relatively few examples of high quality research on training and development interventions and no evidence of programmatic research anywhere in the world! As I shall note in the succeeding paragraphs, adding to our knowledge in this domain represents a potentially important contribution for universities and scholars in the coming decade.

While descriptive research focusing on program description is useful, the more important contributions will be studies of impact (e.g., Copland, cf; Leithwood & Steinbach, 1992; Ng Foo Seong, 2001). Policymakers will be keen to find out whether the increased investment in school leader preparation and development is making a difference. Moreover, program designers should be equally eager to find out which interventions yield greatest results and in what domains.

**Developing and validating an *indigenous* knowledge base across cultures**

The contents of this book confirm that school leader preparation and development has emerged as a global enterprise. This reflects an optimistic belief in the capacity to develop more effective school leaders as well as in the impact of leadership on school improvement. Despite this optimism, the knowledge base on which to build leadership for school change remains uncertain, unevenly distributed, and poorly integrated into training programs. Thus, Evans concludes:

> Over the past few decades the knowledge base about … change has grown appreciably. Some scholars feel that we know more about innovation than we

ever have. ... But although we have surely learned much, there remain two large gaps in our knowledge: training and implementation. (Evans, 1996, p. 4)

Evan's observation is especially salient for non-Western, developing nations where the need for educational change is acute, but the knowledge base is less mature than in the industrialized West (e.g., see Bajunid, 1996; Cheng, 1995; Hallinger, 1995). For example, when Asian school leaders receive formal administrative training, they generally learn Western-derived frameworks. This knowledge base, which is not without critics in the West, usually lacks even the mildest forms of cultural validation (Cheng, 1995; Hallinger & Kantamara, cf; Swierczek, 1988).

This has led scholars in the Asia Pacific region to advocate steps to develop an 'indigenous knowledge base' on school leadership (Bajunid, 1996; Cheng, 1995; Dimmock and Walker, 1998; Hallinger & Leithwood, 1996, 1998; McDonald & Pratt, 1997). The chapters contributed by Hallinger and Kantamara (cf) and by Lam (cf) both add ammunition to this argument. As Hallinger and Kantamara suggest:

The findings from this project highlight the inherent limitations of applying knowledge gained in one cultural context to another. While we have only begun to understand elements of successful school improvement in Thailand, there is no question that substantial culturally derived differences exist when compared with Western nations. We believe that many of these differences are shared by other Asian nations, though this awaits empirical verification.

I, personally, view this as one of the exciting challenges in our field today. There is as much to be learned from the differences in our cultural approaches to leadership and learning as there is from the similarities. Cooperative approaches to curriculum sharing are called for in this global era. No doubt much of what is taught will transfer across cultures with minor to moderate adaptations. However, in general, knowledge sharing across nations should be accompanied by systematic research and development into the theoretical models that underlie those approaches in our local cultures.

Unfortunately, we have not seen enough of this type of systematic work being done to date. It is easy to sell knowledge-based products to other countries, especially developing countries that may assume that Western products are sophisticated and validated. In the long run, however, the real benefits of partnership will accrue from cooperative exploration of knowledge creation and application (e.g., see Bajunid, 1996). This leads to the next thematic area: the role of universities in school leader preparation and development.

### Creating a research and development role for universities in school leader preparation and development

This global era affords a more open sharing of information be it about research findings, national policies, training curriculum, or professional development practices than in the past. This book affirms the importance of proceeding carefully and slowly when transferring training curriculum from one context to another. Indeed, this frames an important role for universities in the coming era.

While traditionally universities have been the primary providers of school leader preparation and development services, they have not generally distinguished themselves in this role (Murphy, 1992, 1993). Exceptions to the rule certainly exist (cf Chong et al.; cf Copland). In general, however, I do not view universities as currently organized as well equipped to provide high quality learning experiences for the development of school leaders. For reasons well articulated by Murphy (1992) and others (Cooper & Boyd, 1987, 1988; Crowson & MacPherson, 1987; Hallinger & Murphy, 1991; Hart & Weindling, 1996; cf Littky & Schen), universities find it exceedingly difficult to create lasting, well-designed bridges to the world of practice.

At the same time, I do believe that universities have an important role to play in the enterprise of school leader preparation and development. This role ought to focus upon knowledge creation, knowledge adaptation, development of training models and materials, validation of curriculum materials, assessment of individuals, and evaluation of programs. In developing nations this seems even more important as universities represent the only institutions well equipped for these tasks. Unfortunately, too often universities frantically compete for training resources and ignore the more fundamental and lasting role they should serve of generating and validating knowledge resources in their societies.

## Conclusion

As Roland Barth has eloquently observed: 'One definition of the at-risk learner is any student who leaves school before or after graduation with little possibility of continuing learning' (1997, p. 12). This observation is as salient to the learning of school leaders of today and tomorrow as it is to students at-risk in our secondary schools. Taken together, the perspectives and recommendations offered in this volume point towards the need to root the development of school leaders in the fostering of schools as learning communities and educators as members of a learning profession. These are ambitious tasks, but worthy ones that will strengthen the profession of teaching.

## References

Bajunid, I. A. (1996). Preliminary explorations of indigenous perspectives of educational management: the evolving Malaysian experience. *Journal of Educational Administration*, 34(5), 50–73.

Barber, M. (1999). *A world class school system for the 21st Century: The Blair Government's education reform strategy*. No. 90 in a Seminar Series of the Incorporated Association of Registered Teachers of Victoria (IARTV), December [ISBN 1 876323 31 0] [reprint of a paper presented at the Skol Tema Conference in Stockholm in September 1999].

Barth, R. (1997). *The principal learner: A work in progress*. The International Network of Principals' Centers, Harvard Graduate School of Education, Cambridge, MA.

Bridges, E. (1977). The nature of leadership. In L. Cunningham, W. Hack & R. Nystrand (Eds.), *Educational administration: The developing decades* (pp. 202–230). Berkeley, CA: McCutchan.

Bridges, E. & Hallinger, P. (1995). *Implementing problem-based leadership development*. Eugene, OR: ERIC Clearinghouse for Educational Management.

Bridges, E. & Hallinger, P. (1997). Using problem-based learning to prepare educational leaders. *Peabody Journal of Education*, 72(2), 131–146.

Caldwell, B. (2002). A blueprint for successful leadership in an era of globalization in learning. In P. Hallinger (Ed.), *Reshaping the landscape of school leadership development: A global perspective*. Lisse, Netherlands: Swets & Zeitlinger.

Cheng, K. M. (1995). The neglected dimension: Cultural comparison in educational administration. In K. C. Wong & K. M. Cheng (Eds.), *Educational leadership and change: An international perspective* (87–104). Hong Kong University Press, Hong Kong.

Chong, K. C., Stott, K. & Low, G. T. (2002). Developing Singapore school leaders for a learning nation. In P. Hallinger (Ed.), *Reshaping the landscape of school leadership development: A global perspective*. Lisse, Netherlands: Swets & Zeitlinger.

Cooper, B. & Boyd, W. (1987). The evolution of training for school administrators. In J. Murphy & P. Hallinger (Eds.), *Approaches to administrative training in education* (pp. 3–27). Albany, NY: SUNY Press.

Cooper, B. & Boyd, W. (1988). The evolution of training for school administrators. In D. Griffiths, R. Stout & P. Forsyth (Eds.), *Leaders for America's schools* (251–272). Berkeley, McCutchan.

Copland, M. A. (2000). Problem-based learning and prospective principals' problem-framing ability. *Educational Administration Quarterly*, 36(4), 584–606.

Crowson, R. & MacPherson, B. (1987). The legacy of the theory movement: Learning from the new tradition. In J. Murphy & P. Hallinger (Eds.), *Approaches to administrative training in education* (pp. 45–66). Albany: State University of New York Press.

Ekholm, M. (1992). Evaluating the impact of comprehensive school leader development in Sweden. *Education and Urban Society*, 24(3), 365–385.

Fullan, M. (1991). *The new meaning of educational change*, Teachers College Press, New York.

Hallinger, P. (1998). Educational change in the Asia-Pacific region: The challenge of creating learning systems. *Journal of Educational Administration*, 36(5), 492–509.

Hallinger, P. (1992). School leadership development: Evaluating a decade of reform. *Education and Urban Society*, 24(3), 300–316.

Hallinger, P. & Anast, L. (1992). The Indiana Principals' Academy: Assessing school reform for principals. *Education and Urban Society*, 23(3), 410–430.

Hallinger, P., Crandall, D., Ng Foo Seong, D. (2000). Systems thinking/Systems changing: A computer simulation for learning how to maker schools smarter (141–162). In K. Leithwood & K. S. Louis (Eds.), *Intelligent learning systems*. New York: JAI Press.

Hallinger, P. & Heck, R. (1996). Reassessing the principal's role in school effectiveness: A review of empirical research, 1980–1995. *Educational Administration Quarterly*, 32(1), 5–44.

Hallinger, P. & Murphy, J. (1991). Developing leaders for future schools. *Phi Delta Kappan*, 72(7), 514–520.

Hargreaves, A. & Fullan, M. (1998). *What's worth fighting for out there*. New York: Teachers College Press.

Hart, A. & Wending, D. (1996). Developing successful school leaders. In K. Leithwood, J. Chapman, D. Corson, P. Hallinger & A. Hart (Eds.), *International Handbook of Educational Leadership and Administration* (pp. 309–336). Dordrecht, Netherlands: Kluwer Academic Publishers.

Heck, R. (2002). Examining the impact of professional preparation on beginning school administrators. In P. Hallinger (Ed.), *Reshaping the landscape of school leadership development: A global perspective*. Lisse, Netherlands: Swets & Zeitlinger.

Huber, S. (2002). School leader development: Current trends from a global perspective. In P. Hallinger (Ed.), *Reshaping the landscape of school leadership development: A global perspective*. Lisse, Netherlands: Swets & Zeitlinger.

Kantamara, P. (2000). *Learning to lead change in Thai schools*. Unpublished doctoral dissertation, Nashville, TN: Vanderbilt University.

Lam, J. (2002). Balancing stability and change: Implications for professional preparation and development of principals in Hong Kong. In P. Hallinger (Ed.), *Reshaping the landscape of school leadership development: A global perspective*. Lisse, Netherlands: Swets & Zeitlinger.
Leithwood, K. & Steinbach, R. (1992). Improving the problem-solving expertise of school administrators. *Education and Urban Society*, 23(3), 317–345.
Littky, D. & Schen, M. (2002). Developing school leaders: One principal at a time. In P. Hallinger (Ed.), *Reshaping the landscape of school leadership development: A global perspective*. Lisse, Netherlands: Swets & Zeitlinger.
Marsh, D. (1992). School principals as instructional leaders: The impact of the California School Leadership Academy. *Education and Urban Society*, 24(3), 386–410.
Murphy, J. (1993). *Preparing tomorrow's leaders: Alternative designs*. University Park, PA: University Council for Educational Administration.
Murphy, J. (1992). *The landscape of leadership preparation: Reframing the education of school administrators*. Thousand Oaks, CA: Corwin Press.
Murphy, J. & Shipman, N. (2002). Developing standards for school leadership development: A process and rationale. In P. Hallinger (Ed.), *Reshaping the landscape of school leadership development: A global perspective*. Lisse, Netherlands: Swets & Zeitlinger.
Ng Foo Seong, D. (2001). *Impact of an intelligent simulation system on knowledge acquisition among school leaders*. Unpublished doctoral dissertation, Nashville, TN: Vanderbilt University.
Stoll, L. & Fink, D. (1996). *Changing our schools*. Buckingham, England: Open University Press.
Tomlinson, H. (2002). Supporting school leaders in an era of accountability: The National College for School Leadership in England. In P. Hallinger (Ed.), *Reshaping the landscape of school leadership development: A global perspective*. Lisse, Netherlands: Swets & Zeitlinger.
Walker, A., Bridges, E. & Chan, B. (1996). Wisdom gained, wisdom given: Instituting PBL in a Chinese culture. *Journal of Educational Administration*, 34(5), 98–119.
Wallace, M. (1992). Developing a training initiative for British school managers. *Education and Urban Society*, 24(3), 346–364.
Wolfe, J. (1993). A history of business teaching games in English-speaking and post-socialist countries: The origination and diffusion of a management education and development technology. *Simulation & Gaming*, 24, 446–463.

# Author Biographical Information

## About the Editor

***Dr. Philip Hallinger***
Professor and Executive Director of the Graduate College of Management, Mahidol University (Thailand). Professor Hallinger was formerly Professor in the Dept. of Leadership and Organizations at Peabody College of Vanderbilt University (USA) where he directed the *Vanderbilt International Institute for Principals* for 12 years. He has been an active researcher in the fields of school leadership, leadership development, school effectiveness and school improvement. His prior work has been noted as instrumental in introducing or building on several new perspectives in school leadership including instructional leadership, the role of social culture, cognition and school leadership, and problem-based learning.

Professor Hallinger's has been a senior co-editor (with Ken Leithwood) of the *International Handbook of Research on School Leadership and Administration* (First Edition, 1996; Second Edition, In-press), and co-author (with Edwin Bridges) of *Implementing Problem-Based Leadership Development* (1995). His most recent research and practice has focused on the principal's contribution to school effectiveness and improvement (with Ronald Heck), the use of problem-based learning in leadership development (with Edwin Bridges), and the development and use of computer simulations in leadership development (with David Crandall, Pornkasem Kantamara and David Ng Foo Seong). His work over the past 10 years has involved the development of principals' centers in Southeast Asia.

## About the Authors

***Dr. Ray Bolam***
Semi-retired and continues to work as Professor of Education in the School of Social Science at Cardiff University and as Visiting Professor at the Universities of Leicester and Bath. At Cardiff University, he was Head of the School of Education (1996–99) and Director of the National Professional Qualification for Headteachers (NPQH) Centre for Wales (1996–2001).

***Dr. Brian J. Caldwell***
Professor and Dean of Education at the University of Melbourne (Australia). Professor Caldwell is internationally recognized for his efforts in the conceptualization and

implementation of self-managing schools in New Zealand and Australia. He has written extensively on educational reform, self-managing schools, and strategic school leadership.

***Dr. Joseph Meng-chun Chin***
Professor of Educational Administration at National Chengchi University (Taiwan). Professor Chin's research interests include philosophy of education and principal leadership.

***Dr. Chong Keng Choy***
Associate Professor of Educational Administration at the National Institute of Education (Singapore). He was programme consultant to the *Leaders in Education Programme* launched in March 2001 in Singapore. Dr. Chong's research interests are in the areas of Futures, Mentoring, Executive Development, School Management, and Organisation Design.

***Dr. Michael Copland***
Assistant Professor of Education at the University of Washington (USA). A former school principal, Dr. Copland's was formerly Director of the *Prospective Principals Program* at Stanford University. His research has focused primarily on cognition and problem-solving among principals, and the use of problem-based learning in school leader preparation.

***Feng Daming***
Associate Professor of Educational Administration at East China Normal University (ECNU) and Deputy Director of the Training Department at the National Training Center for High School Principals, Ministry of Education (People's Republic of China). He is also the member of Shanghai Core Group of Professional Development for School Leaders, Shanghai Municipal Education Commission, where he has been instrumental in bringing international perspectives into the training and development of school leaders.

***Bruce Davis***
From 1989 to 1996 Bruce Davis was Secretary for the Department of Education and the Arts in Tasmania, Australia, which included responsibility for the state's education system. From 1996 to 1999 he was chief executive of the Australian Principals Centre. He presently advises education systems on continuing professional education for principals.

***Dr. Ronald H. Heck***
Chair and Professor of Educational Administration and Policy at the University of Hawaii at Manoa, (USA). Professor Heck has been one of the international leaders in research on school leadership and school effects on learning.

***Dr. Stephan Huber***
Researcher at the Research Centre for School Development and Management, University of Bamberg (Germany). Dr. Huber has been conducting research in the fields of school development, school leadership, and staff and leadership development nationally and internationally. At the moment he is in charge of preparing a concept to establish an Academy for Educational Leadership in Germany.

***Dr. Pornkasem Kantamara***
Received her doctorate from Peabody College, Vanderbilt University. She is currently a Lecturer at Asian University of Science and Technology (Thailand). Her teaching and research interests include human resource development and change.

***Dr. Y.L. Jack Lam***
Currently Chair Professor and Head of the Department of Educational Administration and Policy, Faculty of Education, The Chinese University of Hong Kong. Between 1973 and 1998, he had been Graduate Chair of Brandon University (Canada), as well as a visiting professor in 17 universities in Canada, U.S., China, Taiwan, Australia and Macau.

***Dr. Kenneth Leithwood***
Professor of Educational Administration and Associate Dean for Graduate Research at the University of Toronto/Ontario Institute for Studies in Education (Canada). Professor Leithwood has been internationally recognized for his extensive program of research on school leadership including on issues of transformational leadership, problem-solving, school improvement, and leadership development.

***Dr. Ming-Dih Lin***
Associate Professor in the Department of Educational Administration in the Faculty of Education Chung Cheng National University (Taiwan). His research interests include the principalship, school improvement, and educational reform.

***Dr. Dennis Littky***
Formerly a nationally recognized school principal. He currently serves as Principal for the Metropolitan Regional Career and Technical Center in Providence, RI (USA) and Co-Director of the Big Picture Company, a non-profit organization dedicated to catalyzing and supporting national school reform. Dr. Littky is Founding Director of the Principal Residency Network, a principal certification program that prepares aspiring educational leaders through school-based residencies with outstanding mentor principals.

***Dr. Joseph F. Murphy***
Chair of the Interstate School Leaders Licensure Consortium of the Council of Chief State School Officers (USA). Recently President of the Ohio Leadership Academy and a Professor at Ohio State University, Professor Murphy is Professor of Leadership and Organizations at Vanderbilt University. Professor Murphy has written extensively on a wide variety of issues on school leadership policy and practice.

***Dr. Molly Schen***
Director of Program Development for the Principal Residency Network and managing director of the Big Picture Company in Rhode Island (USA). She is currently working on materials and strategies to train principals in the six core leadership areas of the Principal Residency Network: moral courage, moving the vision, relationships & communication, instructional leadership, management through flexibility and efficiency, and public support.

***Neil J. Shipman***
Was a school leader for 25 years prior to serving as the Director of the Interstate School Leaders Licensure Consortium project from 1995–2001. He has also served as

a professor of educational leadership, and is currently an education consultant, residing in Chapel Hill, North Carolina (USA).

***Dr. Kenneth Stott***
Associate Professor and Associate Dean of Leadership Programmes in the Graduate Programmes and Research Office of the National Institute of Education (Singapore). Dr. Stott has co-authored a number of books, including *Teams, Teamwork and Teambuilding* and *Marketing Your School.*

***Rosanne Steinbach***
Research Associate in the Centre for Leadership Development, at the University of Toronto/Ontario Institute for Studies in Education (Canada). As a long-time member of OISE's research team in school leadership, Ms. Steinbach has co-authored numerous articles and books on principal leadership, problem-solving, and school improvement.

***Dr. Low Guat Tin***
Associate Professor in Policy and Management Studies at the National Institute of Education of the Nanyang Technological University (Singapore). Dr Low has been an instrumental contributor to NIE's programmatic efforts in school leadership development over the past two decades. She has co-authored a number of books on school leadership, including *Successful Women in Singapore, Developing Executive Skills, Management Tools.*

***Dr. Harry Tomlinson***
Formerly a nationally recognized headmaster in the United Kingdom, Dr. Tomlinson is now Professor of Educational Management at Leeds Metropolitan University (England). Professor Tomlinson has been an active contributor to school leadership development as a practitioner, researcher, and trainer for three decades. Most recently, he has been actively involved in the development of the United Kingdom's new school management initiatives.

CONTEXT OF LEARNING
Classrooms, Schools and Society
ISSN 1384-1181

1. Education for All.
   Robert E. Slavin
   1996. ISBN 90 265 1472 7 (hardback)
   ISBN 90 265 1473 5 (paperback)

2. The Road to Improvement: Reflections on School Effectiveness.
   Peter Mortimore
   1998. ISBN 90 265 1525 1 (hardback)
   ISBN 90 265 1526 X (paperback)

3. Organizational Learning in Schools.
   Edited by Kenneth Leithwood and Karen Seashore Louis
   1999. ISBN 90 265 1539 1 (hardback)
   ISBN 90 265 1540 5 (paperback)

4. Teaching and Learning Thinking Skills.
   Edited by J.H.M. Hamers, J.E.H. van Luit and B. Csapó
   1999. ISBN 90 265 1545 6 (hardback)

5. Managing Schools towards High Performance: Linking School Management Theory to the School Effectiveness Knowledge Base.
   Edited by Adrie J. Visscher
   1999. ISBN 90 265 1546 4 (hardback)

6. School Effectiveness: Coming of Age in the Twenty-First Century.
   Pam Sammons
   1999. ISBN 90 265 1549 9 (hardback)
   ISBN 90 265 1550 2 (paperback)

7. Educational Change and Development in the Asia-Pacific Region: Challenges for the Future.
   Edited by Tony Townsend and Yin Cheong Cheng
   2000. ISBN 90 265 1558 8 (hardback)
   ISBN 90 265 1627 4 (paperback)

8. Making Sense of Word Problems.
   Lieven Verschaffel, Brain Greer and Erik de Corte
   2000. ISBN 90 265 1628 2 (hardback)

9. Profound Improvement: Building Capacity for a Learning Community.
   Coral Mitchell and Larry Sackney
   2000. ISBN 90 265 1634 7 (hardback)

10. School Improvement Through Performance Feedback.
    Edited by A.J. Visscher and R. Coe
    2002. ISBN 90 265 1933 8 (hardback)

11. Improving Schools Through Teacher Development: Case Studies of the Aga Khan Foundation Projects in East Africa.
    Edited by Stephen E. Anderson
    2002. ISBN 90 265 1936 2 (hardback)

12. Reshaping the Landscape of Shool Leadership Development
    A Global Perspective
    Edited by Philip Hallinger
    2003. ISBN 90 265 1937 0 (hardback)